INTRODUCTION TO FIRE
PUMP OPERATIONS

INTRODUCTION TO FIRE PUMP OPERATIONS

SECOND EDITION

Dr. Thomas B. Sturtevant

DELMAR
CENGAGE Learning™

Australia • Brazil • Japan • Korea • Mexico • Singapore • Spain • United Kingdom • United States

Introduction to Fire Pump Operations, Second Edition
Dr. Thomas B. Sturtevant

Vice President, Technology and Trades SBU: Alar Elken

Editorial Director: Sandy Clark

Acquisitions Editor: Alison Weintraub

Development Editor: Jennifer A. Thompson

Marketing Director: Dave Garza

Channel Manager: William Lawrensen

Marketing Coordinator: Mark Pierro

Production Director: Mary Ellen Black

Production Manager: Larry Main

Production Editor: Ruth Fisher

For product information and technology assistance, contact us at
Cengage Learning Customer & Sales Support, 1-800-354-9706

For permission to use material from this text or product, submit all requests online at **www.cengage.com/permissions**
Further permissions questions can be emailed to
permissionrequest@cengage.com

ISBN-13: 978-0-7668-5452-9

ISBN-10: 0-7668-5452-3

Delmar
Executive Woods
5 Maxwell Drive
Clifton Park, NY 12065
USA

Cengage Learning is a leading provider of customized learning solutions with office locations around the globe, including Singapore, the United Kingdom, Australia, Mexico, Brazil, and Japan. Locate your local office at **international.cengage.com/region**

Cengage Learning products are represented in Canada by Nelson Education, Ltd.

For your lifelong learning solutions, visit **www.cengage.com/delmar**

Visit our corporate website at **www.cengage.com**

Notice to the Reader

Publisher does not warrant or guarantee any of the products described herein or perform any independent analysis in connection with any of the product information contained herein. Publisher does not assume, and expressly disclaims, any obligation to obtain and include information other than that provided to it by the manufacturer. The reader is expressly warned to consider and adopt all safety precautions that might be indicated by the activities described herein and to avoid all potential hazards. By following the instructions contained herein, the reader willingly assumes all risks in connection with such instructions. The publisher makes no representations or warranties of any kind, including but not limited to, the warranties of fitness for particular purpose or merchantability, nor are any such representations implied with respect to the material set forth herein, and the publisher takes no responsibility with respect to such material. The publisher shall not be liable for any special, consequential, or exemplary damages resulting, in whole or part, from the readers' use of, or reliance upon, this material.

Printed in the United States of America
6 7 11 10 09

*This book is dedicated to Karen, my loving wife,
for her careful review and honest evaluation of the
manuscript and more importantly for her love,
encouragement, and commitment to me and our
three wonderful children Rachel, Hannah, and James.*

CONTENTS

APPENDIXES 359

PREFACE

This book was written for several reasons. First, and perhaps most important, it was written because I enjoy the subject matter. Of all the positions I have held in the fire service, the most memorable and enjoyable were those as a pump operator. Second, it was written because the majority of books on the subject are outdated. I have been teaching from the same textbooks that I learned from when I took classes at a community college some 15 years ago. I'm not implying that these older texts are of poor quality, rather that they are simply outdated. In fact, their extended existence pays tribute to those who wrote them. Granted, some concepts of pump operations have remained relatively unchanged over the years, but pump operations, related standards, and terminology have not been stagnant. Finally, it was written because I wanted a single resource for relevant information on pump operations. When teaching pump operations, I found the "perfect" textbook to be several existing texts combined with information I picked up during my career in the fire service. So, this book attempts to place all the information needed to operate a pump efficiently and effectively within the same cover.

INTENDED AUDIENCE

Introduction to Fire Pump Operations, second edition, is a straightforward, reader-friendly text designed for firefighters who are aspiring to meet the professional qualifications for vehicle driver/pump operator. It is intended to be used for training in fire departments, academies, and college fire programs. Although the basic flow of the text is the same as the first edition, the material in the second edition has been updated and expanded. The most important aspect of this edition is that it addresses all the requirements identified in NFPA 1002, Chapters 4 and 5, 2003 edition, for driver pump operator. In addition, the text addresses the objectives listed in the Fire Protection Hydraulics and Water Supply model curriculum course established at the National Fire Academy's Fire and Emergency Services Higher Education (FESHE) conferences. A Correlation Guide to the Standard 1002, and FESHE course outcomes can be found on pages xv and xviii, respectively.

HOW TO USE THIS BOOK

The basic goal of this text is to provide one location for the knowledge required to efficiently and effectively carry out the duties of a pump operator. Therefore, the textbook includes all the requirements of NFPA 1002, *Fire Apparatus Driver/Operator Professional Qualifications*,

2003 edition, for driver pump operators. Other relevant NFPA standards are also introduced and discussed. In addition, the textbook covers the objectives contained in FESHE's Fire Protection Hydraulics and Water Supply model core curriculum course. Because of this, the textbook can be used for state or local pump operator training and certification as well as in college-level courses.

It is my view that the position of pump operator is vital to the mission of the fire service. It is a position that requires knowledge and skills gained from classroom lecture, practical hands-on training, and experience. To that end, the location of the training/education, whether it be a fire department, state training agency, college, or vocational school, doesn't matter as much as the quality of information provided. Apparatus equipped with pumps are expensive pieces of equipment requiring extensive knowledge and skills to safely operate them.

The text is divided into four sections:

- The first section serves as an introduction to the duties and responsibilities of the pump operator and includes preventive maintenance and driving emergency vehicles.

- The second section focuses on the operating principles, theories, and construction of pumps as well as the systems and components typically used in conjunction with fire pumps.

- The third section presents the three interrelated fire pump operation tasks or activities of securing a water supply, operating the pump, and maintaining discharge pressures. This section builds on the foundation laid in the first two sections of the text and provides basic step-by-step procedures for operating the pump and related components.

- The fourth section focuses on water flow calculations to include hydraulic theory, friction loss principles, and fireground pump discharge pressure calculations. This section brings to light the importance of understanding the relationship between the amount of water needed and the amount that is actually flowing.

The purpose for dividing the text into the four sections is to present like subjects together. For example, the first section focuses on the driver and emergency apparatus. The theory, construction, and operating principles of pumps and related components, for the most part, are all contained within the second part. The operation of pumps and related components are primarily contained within the third part of the text. Finally, the majority of calculations are contained within the fourth section. By structuring the book in this way, the text can easily be rearranged depending on the focus and preference of the student and instructor.

FEATURES OF THIS BOOK

- Meets the 2003 edition of NFPA Standard 1002, Chapters 4 and 5, job performance requirements. An *NFPA Correlation Grid*, specifying page numbers in which content meets objectives, is found on page xv.

- Meets the Fire and Emergency Services Higher Education (FESHE) course curriculum for Fire Protection Hydraulics and Water Supply. A *FESHE Correlation Grid* specifying where content meets course outcomes can be found on page xviii.

- *Review Questions*, *Practice Problems* and *Activities* are included at the end of each chapter, allowing students to apply the lessons they have learned in the chapter.
- *Practice Problems* are integrated throughout the hydraulics sections, carefully building on each concept presented. Additional practice problems can be found at the end of each chapter, as well as a *list of formulas* at the end of Water Flow Calculation chapters.

NEW TO THIS EDITION

There are many updates to the book for the second edition to ensure that the content remains up-to-date, accurate, and user-friendly:

- Logically reorganized to follow the order of NFPA Standard 1002, the book focuses on four sections: Section 1—*Pump Operator and Emergency Vehicles*, Section 2—*Pump Construction and Peripherals*, Section 3—*Pump Procedures*, Section 4—*Water Flow Calculations*.
- References to all NFPA Standards, including 1500, 1901, 1911, in addition to 1002 are up-to-date.
- Correlation to NFPA Standard 1002 and the FESHE course outcomes appear at the beginning of the book for a quick reference.
- Enhanced instructor supplement package, including an Instructor's Guide with lesson plans, answers to review questions, and skill sheets, as well as an Instructor's CD-ROM containing PowerPoint and electronic version of the Instructor's Guide in Microsoft Word.

SUPPLEMENT PACKAGE

In order to further meet the needs of our instructors, the second edition offers an enhanced supplement package:

- The *Instructor's Guide* includes detailed lesson plans featuring Training Tips and correlation to the PowerPoint presentations, answers to review questions, practice problems, and activities in the book, as well as skill sheets correlating to the job performance requirements outlined by the 2003 NFPA Standard 1002. *(Order #: 0-7668-5453-1)*
- An *Instructor's CD-ROM* is also available, including the electronic version of the Instructor's Guide in Microsoft Word, PowerPoint presentations that outline the key points in each chapter, a correlation grid to the 2003 edition of Standard 1002, and a correlation to the Fire and Emergency Services Higher Education (FESHE) course outcomes for Fire Protection Hydraulics and Water Supply. *(Order #: 0-7668-5454-X)*

ACKNOWLEDGMENTS

The author and Delmar, Cengage Learning would like to thank the following reviewers for the comments and suggestions they offered during the development of the first and second editions of this project. Our gratitude is extended to:

Steve Aranbasich
Henderson Fire Department
Henderson, NV

Richard Arwood
Memphis Fire Department
Memphis, TN

Dennis Childress
Orange County Fire Authority
Rancho Santiago College
Orange, CA

Gary Courtney
New Hampshire Technical College
Laconia, NH

Victor Curtis
Mesa Fire Department
Mesa, AZ

Jim Duffy
Henderson Fire Department
Henderson, NV

Doug Hall
Red Rocks Community College
Westminster, CO

Keith Heckler
Rothfuss Engineering Company
Jessup, MD

Attila Hertelendy
University of Nevada
Fire Science Academy
Carlin, NV

Robert Kinniburgh
Charlotte Fire Department
Charlotte, NC

Ric Koonce
J Sargeant Reynolds Community College
Richmond, VA

Chief Dave Leonardo
Loudonville Fire Department
Loudonville, NY

Chief Dave Leonardo
Verdoy Fire Department
Latham, NY

Captain Bob Sanborn
Bowling Green Fire Department
Bowling Green, KY

Clarence E. White
Maryland Fire Rescue Institute
University of Maryland
College Park, MD

Douglas Whittaker
Onondaga Community College
Syracuse, NY

We would also like to thank the many manufacturers who willingly provided information, artwork, and photographs.

Our special thanks are also extended to Dr. Charles Waggoner for his support and encouragement of this project, as well as for his review of several manuscript sections.

Finally, we wish to thank the Coalmont Fire Department and the Red Bank Fire Department for their assistance and for the photographs of their personnel and equipment.

SUGGESTIONS ENCOURAGED

Those who reviewed and commented on this text during its preparation provided valuable insight and suggestions, which in turn produced a better text. Those who read and use the text will also, undoubtedly, have valuable insight and suggestions. I strongly encourage any and all comments and feel that to have someone comment on this text is truly an honor, which will, in turn, produce a better text.

ABOUT THE AUTHOR

Dr. Thomas B. Sturtevant is a program manager for the Emergency Services Training Institute (ESTI) within the Texas Engineering Extension Service, itself a member of the Texas A&M University System. He currently manages the Emergency Management Administration online bachelor degree program with West Texas A&M University and the Department of Defense Emergency Services Training and Education program. He also manages ESTI's curriculum development and accreditation/certification with National Professional Qualification System. He was a tenured assistant professor at Chattanooga State Technical Community College, Tennessee, where he held the positions of dean of Distance Education and coordinator of Fire Science Technology. Dr. Sturtevant was a fire protection specialist with the Tennessee Valley Authority and held various firefighting positions with the San Onofre Nuclear Generating Station, California, and the United States Air Force. He is a Certified Fire Protection Specialist and has an education doctorate in Leadership for Teaching and Learning and a masters in Public Administration from the University of Tennessee. His research and consulting efforts focus on program evaluation and emergency service professional development.

NFPA 1002
CORRELATION GUIDE

The following table correlates the NFPA 1002, 2003 edition requirements to this textbook's chapters.

NFPA 1002 Requirements	Text Chapter
Prerequisites for certification to NFPA 1002 include:	
1.4.1 Appropriate driver's license(s)	1
1.4.2 NFPA 1500 Medical exam	1
5.1 NFPA 1001—Firefighter I	1
General Requirements	
4.2 Preventive Maintenance	2
4.3 Driving/Operating	3
Apparatus Equipped with Fire Pump	
5.1.1 Preventive Maintenance	2
5.2 Operations	
Note: *Each requirement listed in this section is covered within one or more chapters of the text. Each requirement, then, is listed on the left with the specific chapter(s) identified on the right.*	
5.2.1 Produce effective hand or master streams, given the sources specified in the following list, so that the pump is engaged,	8
all pressure control and vehicle safety devices are set,	5, 8, 9
the rated flow of the nozzle is achieved and maintained,	6, 8, 9, 11, 12
and the apparatus is continuously monitored for potential problems:	9
Internal tank, pressurized sources to include hydrant and supply line from another hydrant, static source, transfer from internal tank to external source	7, 8

NFPA 1002 Requirements	Text Chapter
(A) *Requisite Knowledge*	
Hydraulic calculations for friction loss and flow using both written formulas and estimation methods,	11
safe operation of the pump,	8, 9
problems related to small-diameter or dead-end mains, low pressure and private water supply systems, hydrant coding systems, and reliability of static sources.	7
(B) *Requisite Skill*	
The ability to position a fire department pumper to operate at a fire hydrant and at a static water source	7, 8
power transfer from vehicle engine to pump	4, 8
draft	8
operate pumper pressure control systems	5, 8
operate the volume/pressure transfer valve	4, 5, 8
operate auxiliary cooling systems	5
make the transition between internal and external water sources	8
and assemble hose lines, nozzles, valves, and appliances	6
5.2.2 Pump a supply line of 2½-inch or larger, given a relay pumping evolution the length and size of the line and the desired flow and intake pressure, so that the correct pressure and flow are provided to the next pumper in the relay.	7, 8, 9
Requisite knowledge and skills are the same as listed for 5.2.1	
5.2.3 Produce a foam fire stream, given foam-producing equipment, so that properly proportioned foam is provided.	5, 6
(A) *Requisite Knowledge*	
Proportioning rates and concentrations, equipment assembly procedures, foam system limitations, and manufacturer's specification	5, 6, 8
(B) *Requisite Skills*	
The ability to operate foam proportioning equipment and connect foam stream equipment.	5, 6, 8
5.2.4 Supply water to fire sprinkler and standpipe systems, given specific system information and a fire department pumper, so that water is supplied to the system at the correct volume and pressure.	9

NFPA 1002 Requirements	Text Chapter
(A) *Requisite Knowledge*	
Calculation of pump discharge pressure	10, 11, 12
hose layouts	6, 7, 8
location of fire department connection; alternative supply procedures if fire department connections are not usable; operating principles of sprinkler systems as defined in NFPA 13, NFPA 13D, and NFPA 13R; fire department operations in sprinklered properties as defined in NFPA 13E; and operating principles of standpipe systems as defined in NFPA 14.	9
(B) *Requisite Skills* are the same as listed for 5.2.1 and 5.2.2.	

FIRE AND EMERGENCY SERVICES HIGHER EDUCATION (FESHE)

In June 2001, The U.S. Fire Administration hosted the third annual *Fire and Emergency Services Higher Education Conference*, at the National Fire Academy campus, in Emmitsburg, Maryland. Attendees from state and local fire service training agencies, as well as colleges and universities with fire-related degree programs attended the conference and participated in work groups. Among the significant outcomes of the working groups was the development of standard titles, outcomes, and descriptions for six core associate-level courses for the *model fire science* curriculum that had been developed by the group the previous year. The six core courses are *Fundamentals of Fire Protection, Fire Protection Systems, Fire Behavior and Combustion, Fire Protection Hydraulics and Water Supply, Building Construction for Fire Protection,* and *Fire Prevention.*[1]

FIRE AND EMERGENCY SERVICES HIGHER EDUCATION (FESHE) CORRELATION GUIDE

The following table correlates the Model Curriculum Course Fire Protection Hydraulics and Water Supply requirements to this textbook's chapters.

Name:	Fire Protection Hydraulics and Water Supply	
Course Description:	This course provides a foundation of theoretical knowledge in order to understand the principles of the use of water in fire protection and to apply hydraulic principles to analyze and to solve water supply problems.	
Course Requirements		**Text Chapter**

I. Water as an extinguishing agent
  A. Physical properties
  B. Terms and definitions
		4, 10
		4, 10, 11

II. Math review
  A. Fractions
  B. Ratios, proportions, and percentage
  C. Powers and roots
		Practical Problems in
		Mathematics for
		Emergency Services
		ISBN 0-7668-0420-8

(continued)

Course Requirements	Text Chapter
III. Water at rest	
A. Basic principles of hydrostatics	
1. Pressure and force	4, 10
2. Six principles of fluid pressure	4, 10
3. Pressure as a function of height and density	4, 10
4. Atmospheric pressure	4, 5
B. Measuring devices for static pressure	5, 11
IV. Water in motion	
A. Basic principles of hydrokinetics	10, 11
B. Measuring devices for measuring flow	5, 11
C. Relationship of discharge velocity, orifice size, and flow	4
V. Water distribution systems	
A. Water sources	7
B. Public water distribution systems	7
C. Private water distribution systems	7
D. Friction loss in piping systems	11
E. Fire hydrants and flow testing	7
VI. Fire pumps	Section II
A. Pump theory	4
B. Pump classifications	4
C. Priming systems	4, 5
D. Pump capacity	4
E. Pump gauges and control devices	5
F. Testing fire pumps	8
VII. Fire streams	
A. Calculating fire flow requirements	11 & 12
B. Effective horizontal and vertical reach	6
C. Appliances for nozzles	6
D. Performance of smooth-bore and combination nozzles	6
E. Hand-held lines	6
VII. Fire streams (cont.)	
F. Master streams	6
G. Nozzle pressures and reaction	6, 10
H. Water hammer and cavitations	9
VIII. Friction loss	
A. Factors affecting friction loss	
B. Maximum efficient flow in fire hose	
C. Calculating friction loss in fire hose	11
D. Friction loss in appliances	
E. Reducing friction loss	
IX. Engine pressures	
A. Factors affecting engine pressure	11, 12

Course Requirements	Text Chapter

X. Standpipe and sprinkler systems
  A. Standpipe systems
    1. Classifications
    2. Components
    3. Supplying standpipe systems
  B. Sprinkler systems
    1. Classifications
    2. Components
    3. Supplying sprinkler systems
	
	9
	
	9

Section

1

PUMP OPERATOR AND

EMERGENCY VEHICLES

This section of the book covers the following sections of the National Fire Protection Association (NFPA) 1002 *Standard for Fire Apparatus Driver/Operator Professional Qualifications*, 2003 edition (herein referred to as *NFPA 1002*): 1.4, General Requirements; 4.2, Preventive Maintenance; 4.3, Driving/Operating; and 5.1 General. Chapter 1 presents the qualifications, duties, and responsibilities of the driver/operator; preventive maintenance of emergency vehicles is discussed in Chapter 2; and driving emergency vehicles is presented in Chapter 3.

PUMP OPERATOR DUTIES AND RESPONSIBILITIES

Learning Objectives

Upon completion of this chapter, you should be able to:

- List four names used to identify the individual responsible for pump operations.
- List and discuss the duties of a fire pump operator.
- Define fire pump operations and discuss its three interdependent activities.
- Discuss the basic requirements every pump operator should possess to include requirements contained in NFPA 1002.
- Explain the importance of studying fire pump operations.
- Explain the role of laws and standards.
- Explain the scope and purpose of the NFPA 1002.
- List and explain four additional NFPA standards that pertain to pump operations.

NFPA 1002
Standard for Fire Apparatus Driver/Operator
Professional Qualifications
(2003 Edition)

This chapter addresses parts of the following
knowledge elements within sections

1.4:

"shall be licensed to drive all vehicles they are expected to operate."

"shall be subject to periodic medical evaluation . . ."

"shall meet the objectives of Chapter 4 for each type of apparatus they will be expected to operate."

5.1:

prior to certification, the driver pump operator must meet the requirements of NFPA 1001, Fire Fighter I and the requirements of NFPA 1002 in Sections 5.1 (General) and 5.2 (Operations).

INTRODUCTION

My first real experience as a driver/pump operator came one day when both assigned drivers on my shift called in sick. My rank at the time was firefighter/ relief driver. Still wet behind the ears, I eagerly accepted the challenge of driving Engine 3, an old 750-gallons-per-minute pumper. After conducting the daily inspection, I thought it wise to quickly review pump procedures. I pulled out the old black notebook and began looking through the steps for basic operations. Before I finished, the house bell sounded for a reported smell of smoke at Mack's restaurant. Jumping into the driver's seat, I experienced the excitement and thrill of my first chance to prove myself as a driver/pump operator. Although excited, I handled myself in a calm manner. My captain, on the other hand, had a stressed look on his face knowing he had to deal with a rookie driver. Enroute I began to think about what I would do when we arrived on scene.

That's when I began to lose my confidence. I couldn't remember the steps I had been taught. The more I tried to think the steps through, the more confused I became. Then I began thinking about "what if" questions. What if I had to draft, what if I had to relay pump or, worse yet, what if I had to do hydraulic calculations? I kept thinking, "When we get on scene, please let me get water to the nozzle." By that time I didn't care about pressures or flows as long as I was able to get water out the nozzle. Fortunately for me, the call ended up being a false alarm and we returned without laying a line or having to engage the pump.

Back in the station, my captain praised me for my professionalism and my ability to remain calm while driving. I knew, however, how ill-prepared I was for the duty of driver/pump operator. It was then that I decided to learn more about pump operations. At first I thought I could simply read a couple of books or take some classes on pump operations; however, as I began studying, I quickly became amazed at how much I didn't know and how much there was to learn. It seemed to me back then that I would never learn it all. Through the years, as both a pump operator and an educator, I have continued to learn about pump operations and can safely say that I still don't know it all. I have come to the conclusion that becoming the very best pump operator requires a constant and healthy vigilance for opportunities to learn and practice.

This chapter focuses on the individual assigned to the duties of operating the fire pump and serves as a general overview of fire pump operations. In addition, the importance of studying fire pump operations and the impact of standards and laws is discussed. The remaining chapters in Section 1 focus on preventive maintenance and driving emergency vehicles.

MISPERCEPTIONS ABOUT FIRE PUMP OPERATIONS

Several misperceptions about fire pump operations appear to be pervasive in the fire service. For example, some believe that most of the activities related to fire pump operations occur in the first few minutes after arrival on the scene. Perhaps this perception exists because during this time visual work by the pump operator such as connecting supply lines, opening discharges, and setting pump pressures can be seen. It is true that initial pump operation activities are numerous and challenging in the first minutes of an operation. However, there is no autopilot on fire pumps that allows hands-free operation after the initial setup is complete. Fire pump instrumentation must be continually observed and adjustments must be made to maintain appropriate pressures and flows. Additionally, because conditions can change rapidly on the fireground, water flow requirements can change rapidly as well. Fire pump operators must be constantly ready to adapt pumping operations to meet changing fireground needs.

Another misperception is the belief that fire pump operations is more complex and unpredictable than other operations or activities on the fireground. Perhaps this perception stems from the fear and misunderstanding of scientific hydraulic theory and principles. In reality, scientific theory and principles prevail in both fire suppression and pump operations. Fire pump operations can actually be more predictable and controllable than a structural fire.

Finally, fire pump operation is often thought of as a nonglamorous and unimportant job or as a necessary evil for future advancement within the fire service. Certainly pictures seen on TV and in newspapers rarely depict the lonely fire pump operator hard at work. Rather, the focus is on dramatic rescues and on firefighters directing elevated streams of water at the fire. Keep in mind that without pump operations, suppression efforts would be literally nonexistent, and dramatic rescues would either turn into risky rescues or would not be possible at all.

FIRE PUMP OPERATORS

Today, the position of fire pump operator is a noble one in the fire service. The titles given to the position range from traditional names to ones that reflect the changes in the position. New technology and automation coupled with the ever-increasing responsibility of pump operators require new levels of maturity, education, and skill. Because of this, selection of future pump operators is vital to the overall mission of the fire service: to save lives and property.

What's in a Name?

The first fire pumps used in America were manual piston pumps. These pumps were simple devices that used manpower to operate them (Figure 1–1). Because of their simplicity, operating these pumps required minimal training, making unnecessary the need for a designated operator.

Figure 1–1 *A typical early American piston fire pump. Extensive manpower was needed to operate these pumps. Source: Ferdinand Ellsworth Cary, "The Complete Library of Universal Knowledge" (1904).*

Figure 1–2 *The introduction of steam to power fire pumps required trained engineers to operate them. Source: Ferdinand Ellsworth Cary, "The Complete Library of Universal Knowledge" (1904).*

The steam engine was introduced in the late 1800s as a power source for fire pumps and it changed the requirements for fire pump operators (Figure 1–2). Steam as a power source was complicated and dangerous, and the need for a designated person to operate these sophisticated pumps was realized. Engineers familiar with the concepts of steam and pressure were assigned to each apparatus. One of the most important duties of an engineer was to develop enough pressure to power the pump when it arrived at the fire scene (Figure 1–3). The engineer's duties began to change as steam and then the combustion engine powered both the pump and the apparatus. This position became an integral part of the fire service and the term *engineer* evolved into one of the more popular titles assigned to the person who operated pumps.

Today, the term engineer is still used to identify pump operators; however, other terms have also come into use, including wagon driver or tender, chauffeur, apparatus operator, motorized pump operator, driver/pump operator, and apparatus engineer or driver. For the purpose of this text, the term **pump operator** is used to describe the individual responsible for operating the fire pump, driving the apparatus, and conducting preventive maintenance.

pump operator
the individual responsible for operating the fire pump, driving the apparatus, and conducting preventive maintenance

What Do They Do?

Pump operators have a tremendous amount of responsibility. The consequences of improper action or lack of action can be devastating. For example:

Figure 1–3 *English-style self-propelled steam fire engine. Note the engineer riding on the back, working to build up steam. Source: Ferdinand Ellsworth Cary, "The Complete Library of Universal Knowledge" (1904).*

⚠Safety
Because the consequences are so great, one rule that applies to all duties of a pump operator is that of constant attention to safety.

- Improper inspection and testing may result in mechanical damage and failure or reduced life span of the apparatus, pump, and equipment.
- Inadequate water supplies may increase the duration of suppression activities resulting in increased risk to personnel and property.
- Careless driving techniques increase the risk of accidents resulting in expensive damage to equipment as well as injury and loss of life to both personnel and civilians.

Because the consequences are so great, one rule that applies to all duties of a pump operator is that of constant attention to safety. With safety in mind, the duties of pump operators can be grouped in three areas: preventive maintenance, driving the apparatus, and pump operations.

Preventive Maintenance An often overlooked and underemphasized duty is that of maintaining the apparatus in a ready state at all times (i.e., in a safe and efficient working condition). To do so, pump operators must inspect their apparatus and equipment regularly (Figure 1–4). In addition, pump operators must conduct periodic tests to ensure that the equipment is in peak working condition. Only through diligent inspection and testing can the pump operator be confident that the apparatus and fire pump will perform in an efficient and safe manner when most needed.

Figure 1–4 *The duty of preventive maintenance helps ensure the apparatus is in a ready state at all times.*

fire pump operations
the systematic movement of water from a supply source through a pump to a discharge point

Driving the Apparatus Being the "Best Pump Operator in the World" will be of little value if the pump never makes it to the scene. Therefore, an important duty of the pump operator is ensuring the safe transportation of personnel and equipment to the emergency scene (Figure 1–5). Although this sounds easy, statistics indicate that driving emergency vehicles is a dangerous activity. In fact, the *NFPA Journal* (July/August 2001) reports that from 1990 to 2000 approximately 25% of all firefighter fatalities each year occurred while responding to or returning from an emergency. In addition, during this same time period, the number of vehicle accidents steadily increased (see Figure 1–6).

Pump Operations Perhaps the most obvious duty of a pump operator is that of operating the pump (Figure 1–7). **Fire pump operations** can be defined as the systematic movement of water from a supply source through a pump to a discharge point. This definition identifies the three interdependent activities of pump operations: water supply, pump procedures, and discharge maintenance (Figure 1–8). Each of these activities must be carried out, and each affects the outcome of the operation.

- *Water Supply.* The first activity encompasses securing a water supply. The focus of this activity is the movement of water from a source to the intake side of the pump. Water supplies can be secured from three sources. First, water can be secured from the apparatus itself. Pumping apparatus tend to have at least a minimal water supply on board. According to NFPA

Figure 1–5 *Ensuring the safe arrival of personnel and equipment is another important duty of the pump operator.*

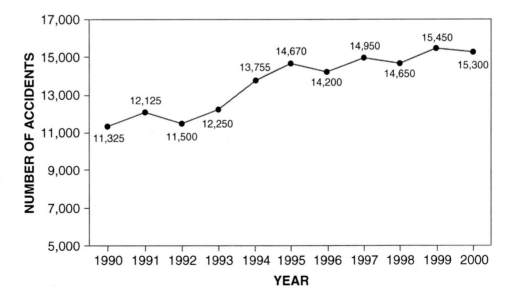

Figure 1–6 *Annual number of emergency vehicle accidents from 1990 to 2000. Source: NFPA Journal, July/August 2001.*

Figure 1–7 *One duty of the pump operator is that of operating the pump.*

Figure 1–8 *The systematic movement of water from a supply source through a pump to a discharge point defines fire pump operations. Courtesy of Greg Burrows.*

1901, *Automotive Fire Apparatus*, an onboard supply of at least 300 gallons must be provided. This water source is typically the quickest and easiest to secure, yet it is often the most limited. Second, water can be secured from pressurized sources, including hydrants, relay operations, and elevated water supplies. Third, water can be secured from static sources, including ponds, lakes, rivers, and tanker shuttle supplies. Each water supply source has unique characteristics and challenges for securing and maintaining adequate flows. More information on water supply is provided in Section 3.

■ **Note**
According to NFPA 1901, an onboard water supply must be at least 300 gallons.

- *Pump Procedures.* The second activity relates to the procedures used to operate the pump. This activity moves water from the intake side of the pump and delivers it to the pump's discharge gates. Although a variety of pump sizes and configurations exist, the general operation of most pumps is basically similar. Activities include opening and closing valves and increasing and decreasing engine pressure. Also included is the operation of pump peripherals such as pressure relief devices and priming devices.

- *Discharge Maintenance.* The third activity centers around the maintenance of discharge lines. This activity moves water from the discharge gate to the nozzle. Hose size, nozzles, and appliances, as well as hydraulic calculations, are used to determine appropriate flow rates and pressures. To maintain flow rates and pressures, discharge gates are opened and closed and engine speed is increased or decreased.

These three activities—water supply, pump procedures, and discharge maintenance—are interdependent in that each activity is dependent on the others as well as affected by the others. They are dependent because if one activity is missing, pump operations cannot exist. Obviously, without a water supply the pump will not work. Likewise, without discharge lines the movement of water to a discharge point cannot occur. These activities are affected by each other because the ability to accomplish each activity influences the other. For example, with a limited water supply the pump may not be able to provide appropriate discharge flows and pressures. Proper pump procedures cannot overcome excessively long discharge lines. Pump operations can therefore be viewed as the balancing of these three activities. Section 3 of this text is devoted to these three activities.

WHAT DOES IT TAKE TO BE A PUMP OPERATOR?

Basic Qualifications

Prior to the initiation of pump operator training, every individual should possess certain basic qualifications. For example, according to NFPA 1002, a *medical evaluation* should be conducted to ensure that the individual is medically fit to perform the duties of a pump operator. The medical evaluation should include a comprehensive physical as well a physical performance evaluation. In addition, NFPA 1002 requires pump operators to possess a valid driver's license and be licensed for all vehicles they are expected to drive. Prior to starting pump operation, individuals should possess at least *a basic level of education*. Math skills and communications skills such as writing and reading are essential to both learning pump operations as well as to conducting pump operator duties. Finally, *maturity and responsibility* should be exhibited by pump operator candidates because the potential damage to expensive equipment and the risk to lives demand that pump operators be mature and responsible.

Knowledge

As stated earlier, pump operator duties include conducting preventive maintenance inspections, tests, driving emergency vehicles, and manipulating the pump and its peripherals. Obviously, pump operators should possess the knowledge required to perform the activities associated with any of these duties. Simply memorizing the steps to perform the activities does not provide the type of knowledge pump operators should possess. A pump operator's knowledge should include the *understanding of the process* (the how, the why, and the variables), not just the steps. With this type of knowledge, pump operators can quickly troubleshoot problems, anticipate and correct potential problems, and utilize equipment in a safe manner.

Skills

The old saying that you cannot put a fire out with books has some validity when discussing pump operations. Although learning pump operations in the classroom is the vital first step, you simply cannot effectively and safely operate a pump from just reading a book. Operating a pump requires skills achieved through hands-on training. Acquiring these skills is not just a one-time event. Pump operators must continually hone their skills through practice, practice, practice.

Learning Process

The old method of taking the pumper out and learning how to pump by following specific steps until they are memorized provides an inefficient pump operator. The need to quickly adapt to changing fireground flow demands, to troubleshoot pump problems, and to safely operate the pump requires pump operators to be intimately familiar with pump operating principles, pump construction, pump peripherals, nozzle theory, hose, and so on. Acquiring the knowledge and skills of a pump operator requires both classroom lecture and hands-on training.

The first step, however, is to ensure that the basic qualifications are met. The second step is to provide the knowledge required to safely and efficiently perform the duties of a pump operator. The third step is to provide hands-on training with a pumper to develop and sharpen manipulative skills. In reality, presenting knowledge and skills can be a mixed process but should be progressive and logical in presentation. Computer-controlled pump simulators (Figure 1–9) and pump software provide an intermediate step between lecture and hands-on training. Pump operator skills can be practiced safely and efficiently without the threat of damaging expensive equipment or injury to personnel. After demonstrating competency on the simulators, training is moved outside with actual pumping apparatus. The concept is the same as a flight simulator for training and recertifying pilots.

Selection Process

The position of pump operator is generally considered a promotion in most departments; therefore, a selection process is typically used to determine the most qualified candidate for the position. Although the process varies from one department to another, the selection process ought to include the evaluation of the candidate's basic qualifications. In addition, the process should include, as a minimum, evaluation of the requirements found in NFPA 1002.

The first step in the selection process should be to ensure that candidates have the basic qualifications. Next, a written or oral evaluation should be administered to determine knowledge of pump operations. Finally, a practical evaluation should be administered to determine the candidate's skill level. The length of

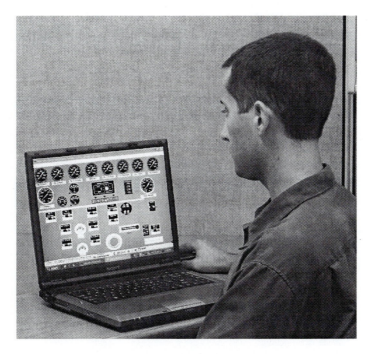

Figure 1–9

Computer-controlled pump simulators provide realistic hands-on training. Courtesy CS Inc.

service of a candidate provides little information on the ability to function in the capacity of pump operator.

WHY STUDY FIRE PUMP OPERATIONS?

One reason for studying fire pump operations is *economics*. Fire apparatus and their pumps are major expenditures for any fire organization. The lack of preventive inspections and maintenance as well as improper use can quickly cause unneeded expensive repairs and replacements. In reality, the apparatus and equipment belong to a community or business. Citizens of the community and shareholders of the business expect and deserve the most efficient use of their money and resources.

One of the more important reasons for studying fire pump operations is that of *safety* of firefighters. Improper operation of fire pumps can cause serious injury to personnel. A simple interior attack can turn into disaster if suppression streams are inadvertently stopped, leaving attack teams with no protection from the heat and progression of fire. Improperly managed pressures and water-pressure surges can cause hose lines to overpower firefighters, tossing them around with ease, or to break loose from firefighter's hands and turn into unpredictable killers. To ensure the safety of the firefighters, pump operators must be knowledgeable about proper inspection, testing, and operating procedures for their equipment.

Another reason for studying pump operations is the desire to be promoted. As previously stated, the position of pump operator is typically considered a *promotion*. In most cases, this means a pay raise. In addition, it often means more responsibility with greater chances for future advancement within the department.

Finally, *professionalism* is another important reason for studying fire pump operations. Simply stated, it is doing the best job that you can. Doing the best job that you can means, in part, operating the pump in an efficient and effective manner to assist in the organization's overall goal of saving lives and property. The more efficiently and effectively pump operations are carried out, the greater the chances of realizing that goal.

❗Safety
● To ensure the safety of the firefighters, pump operators must be knowledgeable about proper inspection, testing, and operating procedures for their equipment.

ROLE OF LAWS AND STANDARDS

The most significant role of laws and standards to fire pump operations is that of safety. Although adherence to laws and standards will not guarantee safety, it will help reduce the risk and seriousness of accidents and injuries. Lack of adherence will increase the risk of accidents and injuries as well as increase the chance that individuals, departments, and communities will be held liable. Increasingly, the fire service and its members are being challenged to defend themselves in the legal system for negligence, omissions, or wrongful acts. Knowledge of and adherence to laws and standards will assist pump operators in performing their duties in a safe and legally blameless manner.

Laws

laws
rules that are legally binding and enforceable

Simply stated, **laws** are rules that are legally binding and enforceable. Of particular interest to pump operators are laws enacted at the state and local level. These laws can affect pump operators on an almost daily basis. A major focus of these laws tends to be on emergency vehicle driving regulations. For example, nearly all state and local laws require that emergency vehicles, regardless of whether they are en route to an emergency, come to a complete stop when crossing an intersection against a red light. State and local laws also tend to identify the requirements of emergency vehicle warning devices, which include location and visibility requirements, as well as specifying when warning devices must be used. Finally, state and local laws typically identify exemptions, limitations, and immunity, if any, extended to the operation of emergency vehicles.

Standards

standards
guidelines that are not legally binding or enforceable by law unless they are adopted as such by a governing body

Standards are guidelines that are not legally binding or enforceable by law unless they are adopted as such by a governing body. The standards of most concern to fire pump operators are those published by the National Fire Protection Association (NFPA, see Appendix A). The NFPA publishes consensus standards related to equipment specifications and requirements, procedural guides, and professional qualifications. The following is an overview of NFPA standards related to pump operations.

NFPA 1002

Fire Apparatus Driver/Operator Professional Qualifications

NFPA 1002, *Fire Apparatus Driver/Operator Professional Qualifications* NFPA 1002 identifies the minimum job performance requirements (JPRs) for individuals responsible for driving and operating fire department vehicles, as well as specific JPRs for pump operators, aerial operators, tiller operators, wildland apparatus operators, and aircraft rescue and firefighting apparatus operators. General requirements include those things that everyone who operates emergency vehicles are subject to. For example, individuals are subject to medical evaluation requirements as required by NFPA 1500, *Fire Department Occupational Safety and Health Program.* JPRs are written as outcomes that describe what an individual is expected to be able to do. The prerequisite knowledge and skills required to complete the JPRs are also included when appropriate.

Chapter 4 of this standard deals specifically with individuals who drive and operate apparatus equipped with fire pumps. One important requirement identified in this chapter is that, prior to being certified as a fire department driver/pump operator, individuals must meet the requirements of Fire Fighter I per NFPA 1001, *Fire Fighter Professional Qualifications.* The JPRs listed in this chapter relate to the duties of a pump operator as discussed earlier in this text: preventive maintenance, driving, and pump operations.

NFPA 1500, *Fire Department Occupational Safety and Health Program* NFPA 1500 identifies the requirements for an occupational safety and health program for the fire service. This standard specifies several areas related to pump operations that must be included in an occupational safety and health plan for the fire service.

Chapter 3 of this standard requires the establishment of a training and education program. For example, this section requires that adequate initial training and retraining be provided by qualified instructors. In addition, this section requires that individuals who drive or operate emergency vehicles meet the requirements of NFPA 1002.

Chapter 4 identifies specific requirements for the safe use of emergency vehicles. One such requirement is that all new apparatus purchased by a fire department be in compliance with NFPA 1901 (discussed shortly). In addition, the standard requires that safety issues be addressed through the development of standard operating procedures (SOPs). These SOPs should include requirements for safely driving emergency vehicles such as obeying traffic laws during emergency and nonemergency response, vehicle speed, crossing intersections, and the seating of personnel on the apparatus (see Appendix B).

■ **Note**

NFPA 1500 requires that safety issues be addressed through the development of standard operating procedures (SOPs).

A medical evaluation program is also required to be included in an occupational safety and health plan. This program must include a medical evaluation to ensure that members meet the requirements of NFPA 1582, *Medical Requirements for Fire Fighters*. In addition, members are required to demonstrate physical performance as well as participate in a physical fitness program.

Finally, a hearing protection program must be included in a department's safety and health program. Pump operators can be exposed to excessive levels of noise for extended periods of time. This standard requires that when noise levels exceed 90 decibels, hearing protection must be provided.

NFPA 1582, *Comprehensive Occupational Medical Program for Fire Departments* NFPA 1582 addresses the minimum medical requirements for firefighting personnel including full-time or part-time employees and paid or unpaid volunteers. Medical exams specified by this standard are preplacement, periodic, and return to duty. Examples of items covered are vision, hearing, heart, extremities, and specific medical conditions such as asthma and diabetes.

NFPA 1001, *Fire Fighter Professional Qualifications* NFPA 1001 identifies the performance requirements for individuals who perform interior structural firefighting operations. Two levels of certification are identified in this standard, Fire Fighter I and Fire Fighter II. Keep in mind that according to NFPA 1002, prior to starting pump operations training, individuals must be certified to NFPA 1001, Fire Fighter I.

NFPA 1901

Automotive Fire
Apparatus

NFPA 1901, *Automotive Fire Apparatus* NFPA 1901 identifies requirements for new fire apparatus. Specific requirements are included for pumper fire apparatus (Chapter 3), initial attack fire apparatus (Chapter 4), mobile water supply fire apparatus (Chapter 5), aerial fire apparatus (Chapter 6), quint fire apparatus (Chapter 7), special service fire apparatus (Chapter 8), and mobile foam fire apparatus (Chapter 9). Other related fire apparatus standards include:

NFPA 1906 *Wildland Fire Apparatus*

NFPA 1911 *Service Tests of Fire Pump Systems on Fire Apparatus*

NFPA 1912 *Fire Apparatus Refurbishing*

NFPA 1914 *Testing Fire Department Aerial Devices*

NFPA 1915 *Fire Apparatus Preventive Maintenance Program*

NFPA 1925 *Marine Fire-Fighting Vessels*

Miscellaneous NFPA Standards The related NFPA standards of interest to pump operators are as follows:

NFPA 13E, *Recommended Practice for Fire Department Operations in Properties Protected by Sprinkler and Standpipe Systems*

NFPA 291, *Recommended Practice for Fire Flow Testing and Marking of Hydrants*

NFPA 1142, *Standard on Water Supplies for Suburban and Rural Fire Fighting*

NFPA 1145, *Guide for the Use of Class A Foams in Manual Structural Fire Fighting*

NFPA 1410, *Standard on Training for Initial Emergency Scene Operations*

NFPA 1451, *Standard for a Fire Service Vehicle Operations Training Program*

NFPA 1961, *Standard on Fire Hose*

NFPA 1962, *Standard for the Inspection, Care, and Use of Fire Hose, Couplings, and Nozzles and the Service Testing of Fire Hose*

NFPA 1964, *Standard for Spray Nozzles*

SUMMARY

A variety of titles is used to identify the individual responsible for pump operations. For the purpose of this text, pump operator is used. The three major duties of a pump operator include (1) preventive maintenance, (2) driving, and (3) pump operations. Pump operation is defined as the movement of water from a source through a pump to a discharge point. To ensure that the best qualified individuals fill pump operator positions, the selection should be based on knowledge and skills rather than seniority. Safety is one of the most important reasons for studying pump operations. Safety is also the primary focus of laws and standards. Knowledge of and adherence to laws and standards will assist pump operators in the safe and legally blameless performance of their duties.

REVIEW QUESTIONS

Key Terms and Concepts

On a separate sheet of paper, identify and/or define each of the following.

1. Fire pump operations
2. Pump operator duties
3. Laws
4. Standards

Multiple Choice and True/False

Select the most appropriate answer.

1. Which of the following is the first activity of fire pump operations?
   a. driving the apparatus
   b. securing a water supply
   c. gaining relevant knowledge and skills
   d. preventive maintenance

2. Although a variety of pump sizes and configurations exists, the general operation of most pumps is basically similar.
   True or False?

3. All of the activities related to fire pump operations occur in the first few minutes after arrival on scene.
   True or False?

4. It is well understood that fire pump operations are more complex and unpredictable than other fireground operations.
   True or False?

5. Which of the following is the correct title for NFPA 1901?
   a. *Standard for Automotive Fire Apparatus*
   b. *Fire Department Vehicle Driver/Operator Professional Qualifications*
   c. *Initial Attack Fire Apparatus*
   d. *Service Tests of Pumps on Fire Department Apparatus*

6. Which of the following is the correct title for NFPA 1002?
   a. *Pumper Fire Apparatus*
   b. *Fire Apparatus Driver/Operator Professional Qualifications*
   c. *Initial Attack Fire Apparatus*
   d. *Service Tests of Pumps on Fire Department Apparatus*

7. According to NFPA 1002, prior to starting pump operations training, individuals must be certified to NFPA 1001, Fire Fighter II.
   True or False?

Short Answer

On a separate sheet of paper, answer/explain the following questions.

1. List at least four names used to identify the individual responsible for pump operations.

2. List the three major duties of pump operators and discuss several activities for each.

3. Define fire pump operations.

4. List and define the three activities of pump operations. Explain why they are interdependent.

5. Which of the three duties of pump operators do you consider to be the most important? Why?

6. Explain the role of laws and standards.

7. List and discuss four NFPA standards related to pump operations.

8. List the basic qualifications, knowledge, and skills every pump operator should possess. Identify the specific requirements contained in the 2003 edition of NFPA 1002.

9. Explain why it is important to study fire pump operations.

ACTIVITIES

1. Identify pumping apparatus in your organization. Include as much detail as you can such as manufacturers, types and size of pumps, on-board water capacity, and so on.

2. List and explain the requirements and selection process of fire pump operators for your organization.

3. Research specific state and local laws related to the operation of emergency vehicles your organization must comply with. Be sure to identify standards referenced or adopted as law.

4. Identify water supply sources available for your response district or area.

PRACTICE PROBLEM

1. Your supervisor has asked you to assist in the development of a new engineer promotional process. You have been asked to provide a list of the broad categories in which each candidate will be evaluated. You have also been asked to include a justification for each category you identify. Provide your response.

BIBLIOGRAPHY

Buff, Sheila. *Fire Engines in North America*. Secaucus, NJ: The Wellfleet Press, 1991.
Excellent historical review of fire apparatus in America, from hand pumps to modern apparatus, with many photos.

Callahan, Timothy, and Bahme, Charles W. *Fire Service and the Law*. Quincy, MA: National Fire Protection Association, 1987.
A practical overview of fire-service law, providing insight on specific fire-service issues.

Lyons, Paul R. *Fire in America*. Boston, MA: National Fire Protection Association, 1976.

A good historical review of fire protection development in America.

Rosenbauer, D. L. *Introduction to Fire Protection Law*. Boston, MA: National Fire Protection Association, 1978.
Provides comprehensive presentation of fire protection law with case studies and court case decisions.

Teele, Bruce W. *NFPA 1500 Handbook*. Quincy, MA: National Fire Protection Association, 1993.
This text provides the foundation for a basic occupational safety program guide for the fire service.

Chapter 2

EMERGENCY VEHICLE PREVENTIVE MAINTENANCE

Learning Objectives

Upon completion of this chapter, you should be able to:

- Explain the importance of preventive maintenance.
- List and discuss four NFPA standards that contain apparatus-related preventive maintenance requirements.
- Explain the role of documentation in a preventive maintenance program.
- Identify and discuss four types of pump tests.
- Explain the need for diligence to safety while conducting preventive maintenance activities.
- Explain the importance of cleaning and servicing vehicles.
- List and discuss three levels of responsibility with a preventive maintenance program.
- List and explain the three basic steps for conducting a preventive maintenance vehicle inspection.
- Conduct a preventive maintenance vehicle inspection, explaining what to look for with each component.

NFPA 1002
Standard for Fire Apparatus Driver/Operator Professional Qualifications
(2003 Edition)

This chapter addresses parts of the following knowledge elements within sections

4.2.1 and 5.2.1:

Manufacturer specifications and requirements, policies, and procedures of the jurisdiction.

The ability to use hand tools, recognize system problems, and correct any deficiency noted according to policies and procedures.

4.2.2:

Departmental requirements for documenting maintenance performed, understanding the importance of accurate record keeping.

The ability to use tools and equipment and complete all related departmental forms.

INTRODUCTION

As stated in Chapter 1, the three duties of a pump operator include conducting preventive maintenance, driving the apparatus, and operating the pump. Driving the apparatus is certainly a critical duty of the pump operator. The pump operator and the apparatus will be of little value if they never make it to the scene. Operating the pump is also an important duty of the pump operator. Certainly, suppression activities will be limited, and perhaps even more dangerous, if the pump is not operated properly. Both of these duties rely, in part, on the proper operation of both the apparatus and pump. The duty of preventive maintenance, then, is to ensure that the apparatus, pump, and related components are in a ready state and in peak operating efficiency. Critical to carrying out the duty of preventive maintenance is a preventive maintenance program.

A good preventive maintenance program is important because:

- Mechanical failures can jeopardize the success of an operation as well as cause injury to both emergency personnel and civilians.
- Improper maintenance has been cited as a key factor in a number of emergency response accidents and pump failures.
- The time and money spent on preventive maintenance is significantly less than the potential damage likely to occur when preventive maintenance is not conducted. Poor **preventive maintenance** can increase the frequency and cost of repairs and reduce vehicle reliability.

preventive maintenance
proactive steps taken to ensure the operating status of the apparatus, pump, and related components

- Criminal and civil liability may occur when emergency apparatus are not properly maintained.
- Manufacturers and insurance companies as well as current standards require it.

There are three main levels of responsibility within a preventive maintenance program. The first level rests with the *fire department*. In general, the fire department is ultimately responsible for the establishment, implementation, and monitoring of a preventive maintenance program. The next two levels of responsibility are typically divided between those activities conducted by *pump operators* and those conducted by *certified mechanics*. The specific activities of each depends on the level of training and the type of preventive maintenance activity being conducted. In general, certified mechanics conduct those activities that require the apparatus to be taken out of service, activities that require several hours to conduct, and activities that focus on detailed service or repair. Pump operators typically conduct noninvasive activities that do not require the apparatus to be taken out of service. For example, checking and adding oil to the engine may be accomplished by the pump operator, whereas changing the oil and oil filter would most likely be conducted by a mechanic.

This chapter focuses on those preventive maintenance activities typically conducted by pump operators (Figure 2–1). First, general aspects of preventive

Figure 2–1 *Pump operators are responsible for ensuring the apparatus is in a ready state at all times.*

maintenance programs are discussed. Next, specifics about conducting the vehicle inspection are presented followed by a discussion of cleaning and servicing the apparatus. A brief discussion of pump-related inspections and testing is presented. A more detailed explanation of pump testing is provided in Section 3 of this textbook. Finally, the importance of safety while conducting preventive maintenance activities is stressed.

PREVENTIVE MAINTENANCE PROGRAMS

Preventive maintenance (PM) can be defined as proactive activities taken to ensure that apparatus, pump, and related components remain in a ready state and in peak operating performance. Typically, these activities can be grouped into inspecting, servicing, and testing. Inspections are conducted to verify the *status* of components. For example, during a daily inspection, pump operators verify oil, water, and fuel levels. Servicing activities such as cleaning, lubricating, and topping off fluids help maintain vehicles in peak operating *condition*. Tests are conducted to determine the *performance* of components, typically the fire pump and related equipment. For example, during annual pump service tests, pump operators test the ability of the pump to flow its rated capacity. Each of these major activities is discussed in more detail later in this chapter.

So, what should pump operators inspect, service, and test? Basic requirements are established in national standards and specific requirements are listed in manufacturers' recommendations. In addition, both standards and manufacturer recommendations provide guidance on when preventive maintenance activities should occur and the type of documentation that should be maintained.

■ Note
Minor changes in the performance of an apparatus are simply warning signs for major problems looming ahead.

It should be noted that preventive maintenance does not occur solely during inspecting, servicing, and testing of apparatus. Rather, it is an ongoing process by the pump operator to ensure that the apparatus, pump, and related equipment are operating properly. This constant monitoring should occur during emergencies, while driving the apparatus, and while training. The individual most in tune with the apparatus is the pump operator. There is no better person to detect minor changes in the performance of the apparatus. Often, minor changes are simply warning signs for major problems looming ahead. The immediate detection of small changes can increase safety as well as the operating life of the apparatus.

Standards

The establishment of responsibility as well as specific requirements of a preventive maintenance program can be found in several NFPA standards.

NFPA 1002: *Fire Apparatus Driver/Operator Professional Qualifications* NFPA 1002 identifies the minimum job performance requirements for driver/operators. This standard requires that pump operators be able conduct and document routine tests, inspections, and servicing functions to ensure that the apparatus is in a ready state. Specific driver operator certifications provided for in this standard include (1) pumper, (2) aerial, (3) tiller, (4) wildland, (5) aircraft rescue fire-fighting apparatus.

NFPA 1500: *Fire Department Occupational Safety and Health Program* Chapter 4, section 4–4 of NFPA 1500 presents requirements for inspections, maintenance, and repair of apparatus. One of the more important requirements of this section relates to the specific requirement that a fire department establish a preventive maintenance program. As part of this program, fire department vehicles must be routinely inspected. Specifically, vehicles must be inspected at least weekly, within 24 hours after any use, when repairs or major modifications have taken place, and prior to being placed in service. If the vehicles are used on a daily basis, then the inspection should also be on a daily basis. Section 4–4 of NFPA 1500 also requires that preventive maintenance be conducted by qualified personnel. Therefore, personnel must be trained to the level of preventive maintenance they are expected to complete. Several other important requirements of this section that relate to preventive maintenance include the following:

- The use of manufacturer's instructions as minimum criteria for inspecting, testing, and repairing of apparatus.
- The establishment of a list of major defects that automatically place a vehicle out of service.
- That fire pumps must be tested in accordance with NFPA 1911, *Service Tests of Fire Pump Systems on Fire Apparatus.*

NFPA 1901: *Automotive Fire Apparatus* NFPA 1901 identifies specific minimum requirements that apply to all new automotive fire apparatus. The standard first presents general requirements required by all new apparatus for items such as the chassis, electrical system, cab, and body. Next the standard provides specific requirements by apparatus type. One important requirement of this standard is the responsibility of the manufacturer to conduct and document certification inspections and tests of vital components such as the pump, water tank capacity, apparatus weight, and load analysis. The requirement ensures that the apparatus will provide the appropriate level of performance when delivered to the fire department.

NFPA 1911: *Service Tests of Fire Pump Systems on Fire Apparatus* NFPA 1911 requires that pumps with a rated capacity of 250 gpm or greater be service tested at least annually and after any major repair or modification. This requirement ensures that the pump continues to provide the proper level of performance. The standard provides detailed information on the conditions, testing equipment, and procedures for conducting the service test. In some departments, it is the responsibility of the pump operator to conduct this annual service test.

NFPA 1915: *Fire Apparatus Preventive Maintenance Program* The purpose of this standard is to help ensure fire apparatus are maintained in a ready state and in safe operating condition. The standard requires that preventive maintenance inspections be conducted as required by the manufacturer and when any defects or deficiencies are reported or suspected. This standard also requires that written criteria be established that require an apparatus to be taken out of service. Finally, specific requirements for the inspection and maintenance of apparatus components and systems are provided within the standard.

Preventive Maintenance Schedules

The scheduling of preventive maintenance activities depends on several factors. Perhaps most important, the schedule should be based on manufacturers' recommendations. Manufacturers specify how often certain components should be inspected and tested. The schedule depends on how often the apparatus and pump are used. Apparatus used frequently requires a more frequent schedule of preventive maintenance activities. Finally, the schedule should be based on NFPA standards. Several examples include the following:

NFPA 1500 suggests daily inspections for vehicles used on daily bases and weekly inspections when vehicles are not used for extended periods.

NFPA 1911 requires an annual pump service test and when the pump has undergone major repair or modification.

NFPA 1915 requires that, at a minimum, vehicle inspections be performed according to manufacturer's recommended intervals.

NFPA 1901 focuses on requirements for predelivery and acceptance inspections and test.

Often, preventive maintenance schedules include inspection and testing at the following frequencies: daily, weekly, monthly, and annually.

Documentation

Documenting preventive maintenance activities is important for several reasons. First, documentation assists with keeping track of needed maintenance and repairs, which might otherwise easily be forgotten or endlessly postponed. Second, documentation provides the ability to determine maintenance trends. For example, documentation may show that, over time, engine oil is increasingly being added to a vehicle. Third, chapter 4 of NFPA 1500 requires that inspections, maintenance, repair, and service records be maintained for all vehicles. Fourth, preventive maintenance documentation may be required for a warranty claim. Finally, documentation can be used to establish proper preventive maintenance in a legal dispute. When an accident occurs, it is a good bet that preventive maintenance documentation will be closely scrutinized.

The pump operator may be required to maintain several preventive maintenance documents, including:

- Daily, weekly, and periodic inspections forms
- Weekly, monthly, and annual pump test result forms
- Fuel, oil, and mileage forms
- Maintenance and repair request forms
- Equipment inventory forms

Today, many departments use computers to assist with the documentation of preventive maintenance activities (Figure 2–2). Preventive maintenance software allows information to be transferred to the computer from hard copy forms (Figure 2–3). Preventive maintenance information can also be transferred to the computer through the use of handheld scanners/computers. When this technology is used, pump operators can enter data directly into the handheld device while conducting inspections or pump tests. When the inspection or pump test is completed, the data is then downloaded to a computer. After preventive maintenance information is transferred to a computer, it can then be processed and printed out in a variety of formats, sorts, and lists. Preventive maintenance software can be purchased commercially or specifically designed for a department. In addition, public domain and shareware software are available and can easily be obtained through a number of online systems.

Figure 2–2

Computers can be used to maintain, track, schedule, sort, and print preventive maintenance information.

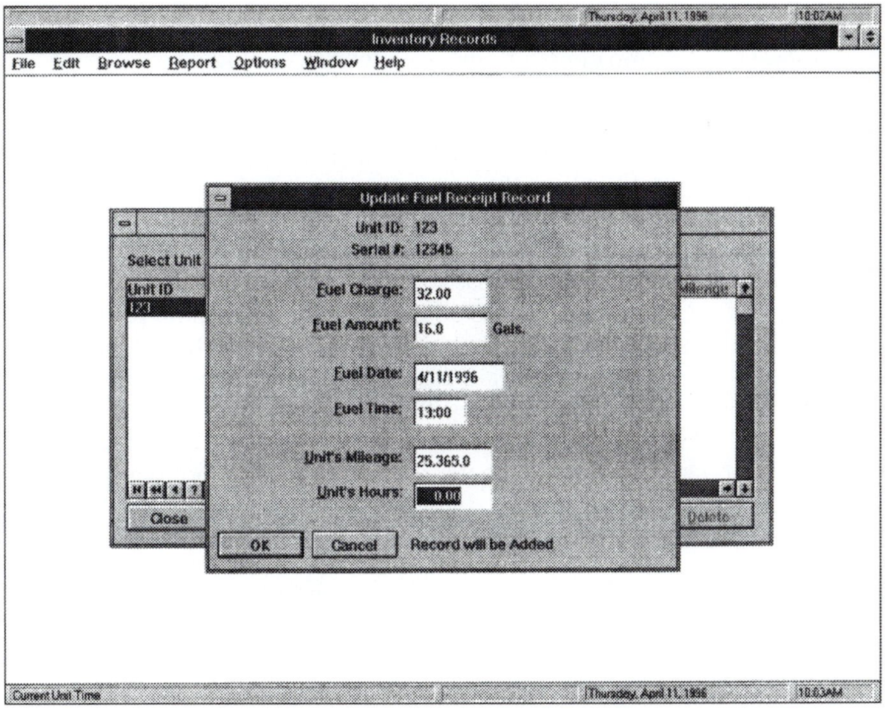

Figure 2–3
Preventive maintenance software can be used to transfer information onto a computer. Courtesy FirePrograms for Windows.

INSPECTIONS

Inspections are the most frequently conducted preventive maintenance activity. In general, inspections include checking components on a daily, weekly, or monthly basis. For example, a daily inspection might include a visual check of a tire's general condition, while a weekly inspection might include actually taking the tire's air pressure. The goal of an inspection is to ensure that the apparatus, pump, and related components are in a safe operating condition. Inspections typically include checking components for:

- Operability, position, or status
- Fluid level, leaks
- Condition, damage, wear, and corrosion

Inspection Process

The inspection process includes three basic steps. Step one is preinspection. Prior to conducting the inspection, the pump operator should review the previous

■ **Note**

Special attention should be given to recent repairs, modifications, or changes to the apparatus.

inspection report and, if possible, receive a debriefing from the previous shift's pump operator. Special attention should be given to recent repairs, modifications, or changes to the apparatus. In addition, the pump operator should review the manufacturer's preventive maintenance documentation when inspecting new or unfamiliar vehicles. In doing so, the pump operator will gain insight into the current status of the apparatus.

Step two is the actual inspection itself. The preventive maintenance information contained in national standards and provided by manufacturers do not specify the process to be used for inspecting a vehicle. However, the inspection should be conducted in a systematic, routine process to help ensure that all components are inspected. During the inspection, the status of components is recorded on the vehicle inspection form. A common sequence used to conduct the inspection includes the following:

1. Outside the vehicle (all sides, top, bottom, and compartments)
2. Engine compartment
3. Inside cab
4. Pump and related components

Additional information about what to inspect within each of these areas is provided later in this chapter.

Step three is the postinspection activity of documenting and reporting the inspection results. In some cases, this means signing the inspection form by the pump operator and the captain. In other cases, it means transferring the results to an electronic preventive maintenance system as discussed previously in this chapter. Regardless, all abnormal findings should be reported immediately.

Inspection Forms

Inspection forms are important for more than simply documenting results of an inspection. If properly designed, the form acts as a guide to prompt the pump operator on what components to inspect. To be of greatest value, the form should be laid out in a manner that corresponds to the general steps in which the inspection will routinely be conducted (Figure 2–4). Finally, the form should include instructions for conducting the inspections, how to document satisfactory and unsatisfactory conditions, and the steps to take when reporting abnormal findings. See Appendix E for examples of vehicle inspection forms.

safety-related components
those items that affect the safe operation of the apparatus and pump, and that should be included in apparatus inspections

What to Inspect

There are two criteria for determining what components to include in a preventive maintenance inspection. The first criterion concerns safety-related components. **Safety-related components** are those items that affect the safe operation of the

Figure 2–4
*Inspection forms
can be used to
guide the pump
operator through
the inspection.*

■ Note

**Safety-related compo-
nents and manufacturers'
recommendations make
up the majority of items
that should be inspected.**

**manufacturers'
inspection
recommendations**
those items
recommended by the
manufacturer to be
included in apparatus
inspections

apparatus and pump. In essence, all safety-related components should be inspected. For example, NFPA 1500 suggests the following be inspected routinely:

tires

brakes

warning systems

windshield wipers

head lights and clearance lights

mirrors

The second criterion for determining what to inspect concerns **manufacturers'
inspection recommendations**. For example, manufacturers typically recommend
that the following be inspected:

engine oil

coolant level

transmission oil

brake system

belts

Together, safety-related components and manufacturers' recommendations
make up the majority of items that should be inspected. The following components
are typically included in daily, weekly, or monthly inspections depending on frequency of use and manufacturer's recommendation and are listed and discussed
in the sequence presented earlier in this chapter.

Outside the vehicle The inspection should begin as you approach the vehicle. As
you walk around the vehicle once, notice its general condition and look for the
following:

Obvious damage

Determine if the vehicle is leaning to one side (not caused by surface grade)

Liquid under the vehicle (oil, coolant, grease, water, foam, fuel, or transmission fluid)

Missing parts or equipment

Hazards around the area to vehicle movement

Next, begin a systematic inspection of the vehicle. For example, start on the driver's side toward the front of the vehicle and move in a clockwise direction until you arrive at the front of the vehicle. Examples of common components to inspect in each of the areas include the following:

Tires, Wheels, and Rims

Proper inflation. Take a pressure reading of each tire to ensure it is properly inflated. The proper tire pressure is listed on the side of the tire. Improper inflation can cause tire damage and difficulty with vehicle handling (steering, stopping, and traction).

Tread wear. Ensure that tire tread wear is not excessive, and look for cuts, nicks, or other damage such as impaled objects. No fabric should be visible through the tire tread or sidewall. NFPA 1915 requires tire replacement when tire wear exceeds state or federal standards. The Department of Transportation (DOT) requires a tread depth of $\frac{4}{32}$ inches for major grooves in the front tire and $\frac{2}{23}$ inches tread depth for all other tires. Some tires have raised patches, called wear bars, in the grooves of tire treads to help determine wear. Look in the grooves; if the wear bars are noticeable, the tire treads are within the standard. Some tires do not have wear bars. In such cases, a penny can be used to assess tread wear. Place a U.S. penny into the tire tread groove with Lincoln's head down. If the tread is at or beyond the top of Lincoln's head, the tread is at or above $\frac{3}{32}$ of an inch of tread (Figure 2–5). Excessive tire wear can also affect vehicle handling.

Rim condition. Look for damage such as dents and nicks. Also check the wheel lug nuts for tightness and for damage such as rust. Damage to a rim can cause the tire to lose pressure and even come off.

Valve stem. Check the general condition of the valve stem to ensure it is not loose, cracked, or otherwise damaged.

Tires. Ensure that tires do not come in contact with any part of the vehicle's body. Dual tires should not be in contact with each other. Inspect the splash guards to ensure they are properly attached to the vehicle and are in good condition.

Cab Doors and Vehicle Compartments

While moving around the vehicle, check the cab and all compartments to ensure that doors open, close, and latch properly and all equipment is present and properly stowed.

Figure 2–5
Conducting a Penny Test can help determine tread wear.

Pump and Related Components

Check control valves to ensure smooth unrestricted operation. Water tank levels should be visually inspected, generally by using the tank vent located on top of the vehicle. Foam tanks levels should also be inspected.

Other Outside Inspection Components

Glass. Ensure all glass is clean, secure, and void of large cracks and nicks. All stepping surfaces, platforms, and handrails should be clean, secure, and in good working order.

Saddle tanks. Check that saddle tank caps are secured, the tank is securely mounted, and check for leaks.

Suspension. Check suspension for damage (cracks, rust, or separation) to springs, spring hangers, U-bolts, and shock absorbers.

Air tanks. Air tanks should be drained to remove moisture. When the vehicle is started, the time it takes to refill the tank should be noted.

Windshield wiper blade. Check for any obvious defects, for proper tension, and that they are free from debris.

Equipment. Equipment such as extinguishers, ladders, and hose should be free from damage and properly secured.

Pump intake and discharge. Caps and connected hose should be checked for proper tightness. Connects should be tight, but not so tight that they become difficult to remove. Valves should be operated to ensure free movement and that they are in the proper position, usually closed.

Reflective striping. Inspect for obvious damage. At some point, it will be important to note its reflective capability at night.

Hose bed. Inspect for proper hose loading. Hose bed covers should be checked for damage and should be securely fastened.

Engine compartment When the outside inspection is complete, the next area to inspect is the engine compartment. Gaining access to the engine compartment usually means raising the hood or the cab. In either case, make sure to properly secure the hood or the cab prior to placing any part of your body into the engine compartment. Also, be sure to properly latch the hood and cab after the inspection is finished. Most, if not all, of the engine compartment inspection items can and should be completed when the engine is not running. No specific sequence is suggested other than to attempt to establish a routine to ensure all items are inspected. Examples of common components to inspect within the engine compartment include the following:

Belts and Wires

Visually inspect all belts and wires for obvious damage. Both should be free from dirt and debris. Belts should be checked for any nicks, cuts, or excessive wear. Also check to ensure belts are aligned with their pulley and the belt is properly tensioned. Wires should be inspected for loose connections, worn or frayed insulation, and for exposed metal.

Engine Oil

Engine oil works as a lubricant for the engine and as a means of cooling and cleaning internal engine parts. Check the engine oil level with the dipstick. Follow these simple steps to check the oil level:

1. Ensure the vehicle is not running and is on level ground. The engine oil can be checked prior to or after running the engine. The benefit of checking the oil prior to running the engine is that cool oil stays on the dipstick better, making it easier to measure the oil level. In addition, the cooler the engine, the less risk of incurring an accidental burn. If the oil is checked after the operation of the engine, wait several minutes to allow the oil to drain from the engine into the pan to ensure an accurate reading.

2. Remove the engine oil dipstick and wipe off the end with a rag.

3. Slowly insert the dipstick fully and then remove it again.

4. Read the level of the oil. The oil level should be within the range marked on the dipstick usually indicated with "Full" and "Add" markings.

5. If the level is below the "Add" marking, oil should be added. Slowly add oil through the fill opening as needed, being careful not to overfill. Wait a few minutes to let the oil settle and then recheck the level. Only trained individuals should add engine oil.

6. Note when and how much oil was added.

In some locations, driver operators are trained to add engine oil as needed. In other locations, only certified mechanics can add engine oil. The correct engine oil type should be used. Mixing different oil types should be avoided. Consult the manufacturer's manual for the recommended oil type. Engine oils are classified using a rating system developed by the

American Petroleum Institute (API) and the Society of Automotive Engineers (SAE).

The *API classification* uses a two-letter system to identify the type of engine and the service class. The first letter identifies the type of engine and is either an S (Service) indicating the oil is for use with gasoline engines or C (Commercial) indicating use with diesel engines. The second letter indicates the service class. The letters have been sequentially assigned over the years and range from A to L. The L indicates the most recent and advanced performance properties. The SL rating is the most current and includes performance properties of each earlier category. The current diesel engine oil categories include CF, CG, and CH.

The *SAE classification* uses a numbering system to grade or rate engine oil viscosity. Engine oil viscosity ratings can range from SAE 5 (low) to SAE 50 (high). Sometimes a rating includes a W after the number, as in SAE 10W and SAE 20W, which means the oil is rated for flow at 0 degrees Fahrenheit. Obviously this oil is rated for use in colder climates. Without the W, as in SAE 30 and SAE 50, the ratings are measured at 210 degrees Fahrenheit. Multi-viscosity oils, as in SAE 20W50, include performance characteristics of a 20W oil at 0 degrees Fahrenheit and a 50 weight oil at 210 degrees Fahrenheit.

Vehicle Batteries

Vehicle batteries are located in the engine compartment or in a separate compartment. Ensure the battery is secure and check for obvious damage and signs of corrosion such as dirty battery top, corroded or swollen cables, corroded terminal clamps, loose hold-down clamps, loose cable terminals, or a leaking or damaged battery case. For older batteries, carefully remove the caps to check the electrolyte (water) level and refill with distilled water as needed. Newer batteries are sealed and do not require internal inspection. Battery cables should be checked for corrosion and to ensure they are securely attached to the battery post. Corrosion can be removed with a rag or wire brush and cleaned with a mild base solution such as baking soda and water. Appropriate personal protective equipment, including eye and hand protection, should be worn when inspecting and cleaning batteries. Care should be exercised when recharging batteries. While charging, batteries produce hydrogen gas. Adequate ventilation and control of ignition sources should be provided while recharging batteries. Manufacturer's recommendations should be consulted before a battery is recharged with a charger or jumped from another vehicle.

Radiator Coolant

The radiator system should be checked for obvious signs of damage such as dents, leaks, and debris. The coolant level should also be checked. Some systems allow the coolant level to be checked without opening the radiator fill cap. Use extreme care if the radiator fill cap must be removed to check the coolant level. Remove the cap only when the engine is cool. Add approved coolant when needed.

Windshield Washer Fluid

Check the level of the windshield washer reservoir and fill with an approved fluid as needed.

Cooling Fan

Check the cooling fan for obvious damage such as cracks and missing blades. Ensure the fan can operate free from obstructions.

Air Intake

Inspect the air intake system for damage. Some systems have an automatic indicator for when to change the air filter.

Power Steering

The power steering system should be checked for damage and leaks. The fluid level should be inspected and topped off when needed.

Brake Fluid

Inspect the brake lines and master cylinder for damage and leaks. The brake fluid in the master cylinder should be checked and brake fluid added as necessary.

Automatic Transmission Fluid

The automatic transmission fluid should be checked and filled in the same manner as the engine oil. Some manufacturers suggest checking the fluid level while the engine is running.

Inside cab With the outside of the vehicle and engine compartment inspected, it is time to inspect the inside of the cab. First, take a moment to adjust the seats and all mirrors and inspect all glass such as the windshield and side windows. Mirrors and glass should be clean and free of damage such as large cracks. The seat belt restraining system should be inspected. Check for proper and unobstructed operation and any obvious sign of damage such as worn or torn webbing. Seat belt buckles should open, close, and latch properly. Check for operation and positioning of adjustable steering wheels. Next, make sure the emergency brake is on and that the transmission is in park (automatic transmission) or neutral (manual transmission). Also, all electrical switches should be in the off position so the battery is not excessively loaded. Finally, start the vehicle, being sure to listen for any unusual noises or vibrations. The manufacturer's instructions for starting the vehicle should be followed. If the vehicle is inside, the exhaust should be ventilated to reduce carbon monoxide and other toxic fumes. After starting the vehicle, the following items should be checked:

Gauges

Gauges should be checked to ensure that the devices or components they are measuring are operating within designed limits as specified by the manufacturer. Most gauges provide visual references for normal operation. Examples of gauges to check include the following:

Air pressure gauge. Note the time it takes for the air pressure to build to its minimum operating pressure. NFPA 1901 requires new apparatus to have quick build-up times that can reach operating pressure within 60 seconds.

Oil pressure gauge

Vacuum pressure gauge

Hydraulic pressure gauge

Fuel level gauge. Ensure adequate fuel is available and refill the tanks as directed by department policy or, if no policy exists, refill until the fuel level reaches the three-quarter mark. Refueling should occur as soon after the vehicle is inspected as possible. According to NFPA 1901, the fuel tank must be sufficient in size to drive the pump for at least $2\frac{1}{2}$ hours at its rated capacity when pumping at draft. The tank fill opening should also have a label indicating the appropriate fuel type.

Oil temperature gauge

Engine coolant temperature gauge

Ammeter and/or voltmeter

Speedometer (should read zero while parked) and odometer

Tachometer measures engine speed in revolutions per minute (rpm).

Lights and Warning Devices

Check all lights for proper operation. This may require an individual on the outside to help. Both running lights (headlights high/low beam, brake lights, turns signals, and backup lights) and emergency warning lights should be checked. In the cab, lights and illumination devices should also be checked.

Audible warning systems should be checked for proper operations. Check to ensure that no one is in close proximity of audible warning devices prior to conducting the operations check.

Check the backup warning device, being sure to depress the brakes before placing the transmission into reverse.

Brake and Clutch Pedals

The brake pedal should be depressed and should stop prior to reaching the floorboard.

The clutch pedal, in manual shift transmissions, should also be depressed for proper operation. While depressing the pedal, note the distance (free play) of the pedal before resistance is felt (throw-out bearing contacting the clutch plate). Insufficient or excessive free play should be reported immediately. Consult the manufacturer's recommendations for specific guidance on testing the clutch pedal free play.

Steering Wheel

The steering wheel should be checked for proper operation. In most cases, it is not necessary, nor is it advisable, to turn the steering wheel so that the wheels turn. Doing so can cause excessive strain and damage to the steering system. Rather, turn the steering wheel until just before the wheels turn. This distance should not exceed 10 degrees in either direction. For a 20-inch steering wheel that would be approximately 2 inches of movement. Excessive play may cause difficulty in steering and may indicate a problem with the steering system.

Fire Pump and Related Components

The pump should be engaged following the manufacturer's procedures to ensure proper operation. Pumps can be powered in several different ways and are discussed in detail in Chapter 4. After engaging the pump, ensure pump panel gauges are operating properly.

Other Cab Components

Radios for proper operation

Heating, air-conditioning, and defroster for correct operation.

Computer equipment for operation and communication with base stations if applicable.

Windshield wipers and washer controls for proper operation

Automatic snow chains when vehicles are so equipped and during appropriate seasons. The vehicle may need to be in motion for this check to be completed.

Pump Inspection and Test

Daily, weekly, monthly, and periodic pump inspections and tests are conducted for two primary reasons: (1) to ensure that components are in proper working order, and (2) to keep the components in working order, for example, testing the priming pump to ensure that it is operating properly while also lubricating its close-fitting parts. Examples of inspections and tests include the following:

Inspect: intake strainers, primer reservoir, discharge gauges, pressure relief strainer

Test: priming system, transfer valve, dry vacuum, pressure-regulating device

Detailed information on pump testing is provided in Chapter 9 of this textbook.

CLEANING AND SERVICING

Cleaning and servicing a vehicle are two important tasks of a vehicle operator's responsibility. Cleaning refers to the washing, vacuuming, and waxing of a vehicle to remove dirt, grime, and debris and to protect the finish from the elements. Servicing refers to minor maintenance activities necessary to keep the vehicle in good working order such as topping off fluids, lubricating parts, replacing parts, and tightening connections. Sometimes these tasks are completed during preventive maintenance inspections. For example, minor corrosion around battery terminals can be quickly wiped away with a rag during the inspection. In addition, engine oil can easily be added during the inspection. These tasks can also be done when needed or at specific intervals. For example, it may be department policy to clean and wash vehicles on a specific day. In addition, a vehicle may need to be washed after an emergency run or training exercise. Regardless, all vehicle cleaning and servicing should be accomplished according to the manufacturer's recommendations.

Keeping vehicles clean at all times is important for several reasons. First, it helps maintain good public relations. After all, municipal apparatus usually belongs to the tax-paying public. Second, it helps ensure the pump, systems, and equipment operate as intended. Finally, maintaining clean apparatus is important to help ensure the vehicles can be inspected properly; dirt and grime could cover defects or potential problems.

SAFETY

As with any activity carried out by pump operators, safety should be considered when conducting preventive maintenance inspections and tests. One way to help ensure safety is to not rush through inspections and tests. Hurrying to finish can increase the risk of an accident and increase the chance that a safety problem is overlooked. Another important safety consideration is to ensure that the work area is free from hazards. This can be accomplished by walking around the apparatus, looking under and above, for slippery surfaces and loose equipment. Finally, increased safety can be achieved by always keeping it in mind. Several common safety considerations for preventive maintenance inspections and tests include:

- Checking for loose equipment before raising a tilt cab.
- Not smoking around the engine compartment and fuels.
- Wearing appropriate clothing (no loose jewelry, wearing safety glasses, gloves, etc.).
- Considering vapor and electrical hazards.
- Always being careful when opening the radiator cap.
- Using the proper tool for the work being conducted.
- Being sure to secure equipment and close all doors prior to moving the apparatus.

SUMMARY

Preventive maintenance is an important pump operator duty to ensure that the apparatus, pump, and related components are in a safe and ready operating status. A good preventive maintenance program helps guard against mechanical failure and costly repairs. Typically the fire department is responsible for the overall administration of the preventive maintenance program. Pump operators and certified mechanics share the responsibility of ensuring that the apparatus is in a safe operating condition.

Preventive maintenance activities typically include inspections and pump tests. Inspections, consisting of both safety related components and manufacturers' recommendations, verify the condition, status, and operability of the apparatus,

pump, and related components. Pump tests determine the operating performance of the pump and related components. NFPA standards and manufacturers' recommendations provide a basis for what and how often components should be inspected and tested. Documentation of inspection and tests results assist with tracking needed maintenance and repairs. In addition, the forms can be used to guide the pump operator through the inspection and test.

Preventive maintenance is an ongoing process that should occur any time the pump operator, or others, are operating, cleaning, or training with the apparatus. As with other duties and activities conducted by the pump operator, diligence to safety should be maintained at all times.

REVIEW QUESTIONS

Key Terms and Concepts

On a separate sheet of paper, identify and/or define each of the following.

1. Preventive maintenance
2. Safety-related components
3. Manufacturers' inspection recommendations
4. Pump tests

Multiple Choice and True/False

Select the most appropriate answer.

1. Which of the following NFPA standards establishes responsibility and specific requirements of a preventive maintenance program?

    a. 1500   **c.** 1911

    b. 1901   **d.** All of the above are correct.

2. Which of the following NFPA standards requires that pump operators be able to conduct and document routine tests, inspections, and servicing functions to ensure that the apparatus is in a ready state?

    a. 1500   **c.** 1911

    b. 1901   **d.** 1002

3. NFPA _____ requires that pumps with a rated capacity of 250 gpm or greater be service tested at least annually.

    a. 1500   **c.** 1911

    b. 1901   **d.** 1002

4. A preventive maintenance inspection should begin

    a. with the engine compartment.

    b. with the pump.

    c. with an inventory of equipment.

    d. with checking previous inspection reports.

5. NFPA _____ requires that manufacturer's, certification, and acceptance tests be conducted on new apparatus.

    a. 1500   **c.** 1911

    b. 1901   **d.** 1002

Short Answer

On a separate sheet of paper, answer/explain the following questions.

1. Explain why a preventive maintenance program is so important.

2. What are the three levels of responsibility within a preventive maintenance program?

3. List and briefly discuss four NFPA standards that contain preventive maintenance requirements related to pump operations.

4. List and explain the components to inspect during a preventive maintenance inspection.

5. List the two criteria for determining what components to include in a preventive maintenance inspection.

ACTIVITIES

1. Conduct a preventive maintenance inspection of an apparatus.

2. Develop a daily preventive maintenance inspection form.

3. Develop a list of common hazards that may be encountered while conducting preventive maintenance inspections and tests.

PRACTICE PROBLEM

1. Your department has recently experienced significant maintenance problems with several apparatus. Your chief has asked you to review the preventive maintenance program and identify any weaknesses. What will you use as the basis for evaluating the program?

Chapter 3

DRIVING EMERGENCY VEHICLES

Learning Objectives

Upon completion of this chapter, you should be able to:

- Explain the basic concepts contained in most state emergency driving laws.
- Discuss the role of NFPA 1500 and NFPA 1002 in emergency vehicle safety.
- Discuss the basic attributes of a good driver.
- Explain how speed, braking, road conditions, and environmental conditions affect safe driving.
- Discuss how emergency vehicle characteristics (weight, width, length, height, and loads) affect safe driving.
- Explain the importance of safe driving to emergency response.

NFPA 1002
Standard for Fire Apparatus Driver/Operator Professional Qualifications
(2003 Edition)

This chapter addresses parts of the following knowledge elements within section 4.3 as follows:

4.3.1, 4.3.6

Effects on vehicle control of liquid surge, braking reaction time, and load factors;

Effects of high center of gravity on roll-over potential, general steering, reactions, speed, and centrifugal force;

Applicable laws and regulations;

Principles of skid avoidance, night driving, shifting, and gear patterns;

Negotiating intersections, railroad crossings, and bridges; weight and height limitations for both roads and bridges;

Identification and operation of automotive gauges;

Operation limits;

Effects on vehicle control of liquid surge, braking reaction time, general steering reactions, speed, and centrifugal force;

Applicable laws and regulations;

Principles of skid avoidance, night driving, shifting, and gear patterns;

Negotiating intersections, railroad crossings, and bridges; weight and height limitations for both roads and bridges;

Identification and operation of automotive gauges;

Proper operation limits;

Ability to operate passenger restraint devices;

Maintain safe following distances;

Maintain control of the vehicle while accelerating, decelerating, and turning;

Operate under adverse environmental or driving surface conditions, and use automotive gauges and controls.

4.3.2, 4.3.3, 4.3.4, 4.3.5

Vehicle dimensions, turning characteristics, the effects of liquid surge, spotter signaling, and principles of safe vehicle operation;

Ability to use mirrors, judge vehicle clearance;

INTRODUCTION

Each of the three pump operator duties—preventive maintenance, driving the apparatus, and operating the pump—is important. If one of the duties is not properly carried out, the ability to deliver water to the discharge point would, at best, be inefficient. Although each of the duties is important, a strong argument can be made that driving the apparatus is perhaps the most critical duty of a pump operator.

There are several reasons why driving the apparatus can be considered the most critical pump operator duty. First, a pump operator who inefficiently operates or improperly maintains a pump may still be able to provide at least some water to a discharge location. However, the best pump operator in the world and the best maintained pump will be of little value if the apparatus does not make it to the scene. Second, driving the apparatus is by far the most common duty carried out by the pump operator. More time is spent behind the wheel than behind the pump panel.

Finally, and perhaps most important, the consequences of improperly carrying out the duty of driving the apparatus can be significant. Each year from 1990 to 2000, fire department emergency vehicles were involved in more than 11,000 accidents while responding to or returning from incidents (Figure 3–1). During that same period, an average of 21 firefighter fatalities per year occurred while responding to or returning from incidents. This figure represents almost a quarter of all firefighter fatalities for each of the years. In addition, some 5,661 injuries occurred while responding to or returning from incidents. These statistics do not include all accidents involving apparatus nor do they include civilian death and injury

Figure 3–1

Estimated number of emergency vehicle accidents while responding to or returning from incidents. Source of data: NFPA Journal, *November/December, 2002.*

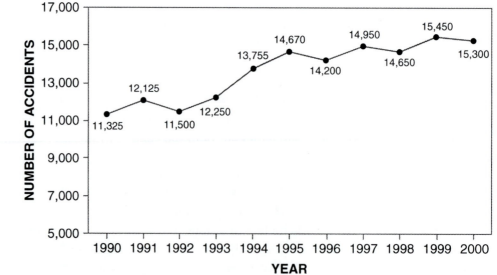

information. Rather, they include only those fire department accidents, injuries, and fatalities that occurred while responding to or returning from an incident. In addition, they do not take into account minor occurrences and near misses.

It is a fact that the earlier the fire department arrives on an emergency scene, the better the chances of saving lives and property. It makes sense then, that fire departments endeavor to keep response times to a minimum. A few examples of attempting to keep response times to a minimum include:

- Using 911 to quickly report and mobilize emergency forces
- Strategically locating stations for shorter responses
- Continuously training to quickly deploy suppression and rescue activities
- Equipping emergency apparatus with warning devices to move through traffic
- Providing certain privileges to responding apparatus

The need for a quick response is an integral part of almost every emergency response activity. Unfortunately, the urge to get to the scene as quickly as possible often overshadows the safe operation of emergency vehicles. When this occurs, the chance for an accident occurring increases significantly.

!**Safety**
Apparatus drivers should be thinking of a quick yet safe response.

When thinking in terms of getting to the emergency scene, apparatus drivers should be thinking of a quick yet safe response. Another way of looking at it is to think of a response in terms of effectiveness and efficiency. Driving the apparatus so fast that an accident occurs is not an effective response. Conversely, driving the apparatus too slowly is not an efficient response. The term *effective and efficient response* connotes the meaning of both a safe and timely response.

Accidents involving responding apparatus can produce several significant outcomes. First, assistance to those who summoned help will be delayed because additional units must be dispatched. Not only must additional units be dispatched for the original incident, additional units must be dispatched for the second incident that involves the responding apparatus. Second, fire department members and civilians could be seriously or fatally injured. In addition, department and civilian vehicles and property could sustain extensive damage. Third, the city, the department, and the driver could face civil and/or criminal proceedings. Finally, the image presented by the accident will last a long time in the minds of the public.

!**Safety**
Laws and standards assist the driver by providing a set of rules and guidelines that focus on the safe operation of emergency vehicles.

Fire department apparatus accidents occur for a variety of reasons. However, several common factors appear in a vast majority of the accidents. One such factor is associated with apparatus operators not following laws and standards related to emergency response. Emergency response laws and standards are established, in part, as a result of analyzing factors related to accidents involving emergency vehicles. The basic goal or intent is to help protect emergency responders as well as the general public. It makes sense then, that when emergency response laws and standards are not followed, the potential for an accident increases and that emergency responders and the general public incur unnecessary risk. A second factor relates to operators not being fully aware of both driver and apparatus limitations.

Driver operators must fully understand their own driving limitation as it relates to reaction time, vision, and attention as well as their vehicle's characteristics as it relates to weight, dimensions, and load. A third factor is the lack of appreciation for driving considerations such as weather and traffic. That operators sometimes do not fully appreciate, understand and/or compensate for emergency response driving demands and conditions is a fourth factor. Knowledge of and skillful safe driving techniques and practices can dramatically reduce the chance of an accident and may also help reduce wear and tear on emergency vehicles.

This chapter discusses the pump operator's duty of driving fire apparatus, focusing on laws and standards, driver and apparatus limitations, and driving considerations related to the safe operation of fire apparatus.

LAWS AND STANDARDS

Apparatus drivers assume a tremendous amount of responsibility. They are responsible for the safety of the apparatus, equipment, and those riding on the apparatus. Finally, they are held responsible for the safety of the public at large by properly operating emergency vehicles.

Laws and standards assist the driver by providing a set of rules and guidelines that focus on the safe operation of emergency vehicles. Laws are rules that drivers must obey. Standards provide suggested guidelines for drivers. Laws and standards assist the driver in ensuring the safety of the apparatus and of personnel riding on the apparatus, as well as the public at large. Therefore, drivers must be familiar with the laws and standards that govern emergency response for their department, city, and state.

Emergency Vehicle Laws

Laws enacted by state and local governments affect apparatus drivers on a daily basis. State laws, referred to as *statutes*, set the overall rules and standards for emergency driving within the state. All government and nongovernment entities involved with emergency response must obey these laws. Local laws, referred to as *city code* or *ordinances*, set specific requirements for emergency response within the city, county, or other form of local government. Local laws may be more restrictive or provide more detail than state laws. For example, a state may allow emergency vehicles to travel through a stop light or sign after slowing down, whereas a city may require emergency vehicles to come to a complete stop prior to traveling through a stop light. Local laws are based in part on state law and, in turn, tend to use the same language and terminology as their state's law.

Although laws governing emergency response vary somewhat across the states, most state laws tend to address the same basic components. Most state laws, for example, identify or define what is considered to be an **authorized emergency vehicle**. Typically this means fire department vehicles, ambulances, rescue

■ Note
Drivers must be familiar with the laws and standards that govern emergency response for their department, city, and state.

authorized emergency vehicles
legal terminology for vehicles used for emergency response, such as fire department apparatus, ambulances, rescue vehicles, and police vehicles equipped with appropriate identification and warning devices

vehicles, and police vehicles with appropriate identification and warning devices. In addition, state laws typically identify the individual empowered to designate an "authorized emergency vehicle." In most cases, a commissioner, fire chief, police chief, or county sheriff is the individual so named.

Most state laws require that emergency vehicle drivers obey the same laws as other vehicle operators unless specifically exempt from doing so. It is important to keep in mind that exemptions extended to emergency vehicle drivers are not automatic. State laws typically define several conditions that must exist for exemptions to be extended:

- Only authorized emergency vehicles are covered.
- The exemptions are only provided when responding to an emergency.
- Audible and visual warning devices must be operating when taking advantage of the exemption.

Finally, state laws typically stress the duty to drive with due regard for the safety of all persons. Therefore, the exemptions previously stated are not available when their use may bring harm to the public.

In order to assist the driver of an emergency vehicle with an efficient response, state laws typically exempt emergency vehicle drivers from several normal driving regulations. One exemption typically found in most state laws is the ability to park emergency apparatus regardless of normal requirements (Figure 3–2). Other exemptions include the ability to pass through a red light or stop signs

Figure 3–2 *State laws typically exempt authorized emergency vehicles from several normal driving regulations.*

and exceeding the maximum speed limit. Finally, emergency vehicle drivers are often exempt from normal direction or movement of travel. Again, these exemptions are typically limited to response to emergencies and are only extended when safely executed.

Section 55-8-108 of the Tennessee Code Annotated is provided as an example of a state law related to emergency vehicle drivers:

(a) The driver of an authorized emergency vehicle, when responding to an emergency call . . . but not upon returning from a fire alarm, may exercise the privileges set forth in this section, but subject to the conditions herein stated.

(b) The driver of an authorized emergency vehicle may:

(1) Park or stand, irrespective of the provisions of this chapter;

(2) Proceed past a red or stop signal or stop sign, but only after slowing down as may be necessary for safe operation;

(3) Exceed the speed limits so long as life or property are not thereby endangered; and

(4) Disregard regulations governing direction of movement or turning in specified directions.

(c) The exemptions herein granted to an authorized emergency vehicle shall apply only when such vehicle is making use of audible and visual signals meeting the requirements of this state. . . .

(d) The foregoing provisions shall not relieve the driver of an authorized emergency vehicle from the duty to drive with due regard for the safety of all persons, nor shall such provisions protect the driver from the consequences of the driver's own reckless disregard for the safety of others.

See Appendix C for several other states' laws related to emergency vehicle drivers.

Standards

Two standards of most interest to apparatus drivers are established by the National Fire Protection Association (NFPA) (Figure 3–3). NFPA 1500, *Fire Department Occupational Safety and Health Program*, identifies minimum requirements for an occupational safety and health program for the fire service. NFPA 1002, *Fire Apparatus Driver/Operator Professional Qualifications*, identifies the minimum job performance requirements for individuals responsible for driving and operating fire department vehicles. Copies of both standards should be available at fire departments.

NFPA 1500 (2002 Edition) NFPA 1500, Chapter 6, "Fire Apparatus, Equipment, and Driver/Operator," discusses minimum safety-related requirements associated with emergency vehicles, drivers of emergency vehicles, and operation of emergency vehicles. Because minimum safety requirements are identified, emergency vehicle drivers should be familiar with this standard. In addition, the standard *may*

Figure 3–3 *NFPA standards play an important role in emergency vehicle safety. Reprinted with permission from National Fire Protection Association, Quincy, MA 02269.*

carry the weight of law in that the standard can be viewed as the legal minimum standard in court. Therefore, the standard may be used to judge a driver's action or inaction in judicial proceedings. Finally, the standard *will* carry the weight of law if adopted as such by a local government.

Emergency Vehicles: According to NFPA 1500, safety and health should be considered when purchasing, maintaining, and operating emergency response vehicles. In addition, new fire apparatus, wildland fire apparatus, and marine firefighting vessels shall be specified and ordered to meet their respective NFPA standard (1901, 1906, and 1925). Refurbishing fire apparatus shall meet the applicable requirements of NFPA 1912.

Emergency Vehicle Drivers: NFPA 1500 requires that emergency vehicles be operated by qualified individuals. First, emergency vehicle drivers must possess a valid driver's license. Because emergency vehicle driving regulations vary, the type of license required will depend on state and local requirements. In addition, emergency vehicle drivers must meet the appropriate requirements established in NFPA 1002. Finally, emergency vehicle drivers must complete a driver training program prior to operating emergency vehicles (Figure 3–4).

Operating Emergency Vehicles: NFPA 1500 assigns responsibility for the safe operation of the vehicle under all conditions to the driver. The standard requires that emergency vehicle drivers obey all normal driving laws when not responding. Although this may sound too basic to be mentioned, drivers

Figure 3–4
According to NFPA 1500, emergency vehicle drivers must complete a driver training program prior to operating emergency vehicles.

may neglect or forget this rule during nonemergency operations. To reinforce this basic rule and to further assist in the safe operation of emergency vehicles, standard operating procedures (SOPs) should be developed for both emergency and nonemergency operations. The first priority of emergency response-related SOPs should be the safe arrival of the fire apparatus to the incident. These SOPs should include, as a minimum, specific criteria or requirements for speed, crossing intersections, traversing railroad crossings, and use of emergency warning devices.

The safety hazards encountered at intersections during emergency response are significant. Therefore, NFPA 1500 requires that emergency vehicle drivers come to a complete stop when any intersection hazard is present. Specifically, the standard requires that drivers bring the apparatus to a complete stop when any of the following exists:

❗Safety
● Safety hazards encountered during emergency response are significant.

- As directed by a law enforcement officer
- At red traffic lights or stop signs
- At negative right-of-way and blind intersections
- When all lanes of traffic in an intersection cannot be accounted for
- When stopped school bus with flashing warning lights is encountered
- At unguarded railroad grade crossing (also for nonemergency)
- When other intersection hazards are present

Figure 3–5
Firefighters must be seated and properly secured whenever the apparatus is in motion.

Riding on Apparatus: NFPA 1500 also addresses the safety of those riding on the apparatus. The standard requires that all personnel riding on the apparatus be seated and properly secured with seat belts prior to moving the apparatus and any time the vehicle is in motion (Figure 3–5). The only exceptions to this requirement are for patient treatment, the loading of hose, and tiller driver training. While the vehicle is in motion, seat belts are not to be released or loosened at any time including while donning self-contained breathing apparatus (SCBA) or personal protective equipment (PPE). Finally, personnel riding in open cab apparatus not enclosed on at least three sides and the top should wear helmets and eye protection. In addition, helmets are recommended to be worn if enclosed cab seats do not provide head and neck protection.

NFPA 1002 NFPA 1002 identifies the minimum requirements for fire department vehicle drivers. The standard is divided into general requirements and apparatus-specific requirements. General requirements include those items that all drivers must meet. Apparatus-specific requirements focus on those items associated with the main function of the apparatus.

General requirements center around the safe driving of apparatus. Specifically, drivers must be licensed for the apparatus they are expected to operate. In addition, all drivers are subject to medical evaluation requirements as identified in NFPA 1500. Further, drivers must demonstrate the ability to conduct and document general vehicle inspection, testing, and servicing functions. Finally, drivers must demonstrate the ability to safely maneuver the apparatus over a predetermined route in compliance with all applicable state, local, and department rules and regulations.

THE DRIVER AND APPARATUS

The basic tools used to carry out the duty of driving apparatus are, of course, the driver and the apparatus (Figure 3–6). Although all drivers are not the same, there are several characteristics common among the best drivers. Fire department apparatus also vary, yet have several common characteristics. Understanding these

Figure 3–6 *Safe drivers understand their own limits as well as the limits of their apparatus.*

characteristics and the influences that affect them is vital to the safe operation of fire department apparatus.

Driver

What makes the best drivers react in a safe and appropriate manner? The answer to this question could be a rather long list of specific items; however, four broad characteristics are discussed to present the important common characteristics:

- Knowledge and skill
- Attitude
- Physical and psychological fitness
- Maturity and responsibility

First, apparatus drivers must possess appropriate knowledge and skill for the position they hold. *Knowledge* provides the ability to think through situations and identify courses of action grounded in safety. *Skill* provides the ability to execute the course of action in a safe manner. For example, all apparatus drivers must understand and be knowledgeable of intersection crossing considerations to evaluate potential hazards and identify a safe course of action, as well as be able to skillfully execute the plan by maneuvering the apparatus safely through the intersection. When drivers are knowledgeable and skillful in their jobs, they are better able to cope with situations in a safe manner.

Second, the best drivers have a *good attitude* toward safety. When this characteristic is present, safety becomes an integral part in every action of the driver. Safety becomes a habit with a good attitude toward safety. Drivers with a good attitude are not overconfident of their ability to drive emergency vehicles. Rather competent drivers show respect for, and knowledge of, the vehicle's limitations.

Third, drivers must be *physically and psychologically fit.* The stress and rigors associated with driving fire apparatus can take their toll on even the best drivers. Being physically and psychologically fit will help reduce the negative effects of stress. Being physically fit allows the driver to control the vehicle and to react quickly and appropriately. Being psychologically fit ensures that the driver is alert and can concentrate and think clearly.

Finally, emergency vehicle operators should be mature and responsible. *Mature* individuals have a balance of knowledge, skill, and experience and are able to utilize them in decision making and actions they take. *Responsible* individuals internalize their accountability for safety to themselves as well as to others. The potential risk to emergency personnel and to the public dictate that this characteristic be present in all emergency vehicle drivers.

Apparatus

In much the same way as drivers will react differently in a given situation, different apparatus will react differently in a given situation. Therefore, drivers should

Figure 3–7 *The physical characteristics of individual apparatus must be considered when driving the apparatus. Courtesy KME Fire Apparatus.*

■ **Note**

Vehicles should be routinely inspected to detect and correct any deficiency that may create unsafe driving conditions.

also be familiar with their apparatus. Specifically, drivers should be familiar with the *physical characteristics* (Figure 3–7) of the apparatus such as length, height, width, and weight. Driving over bridges (Figure 3–8), through tunnels, and around traffic will be affected by these physical characteristics. Drivers should also be familiar with the condition of the apparatus. Vehicles should be routinely inspected to detect and correct any deficiency that may create unsafe driving conditions. Daily inspections provide the opportunity to assist in the safe operation of the vehicle. Drivers should also be familiar with the *handling characteristics* of the vehicle such as turning, braking, and shifting. For example, large capacity tankers will handle differently than a minipumper when it comes to stopping and turning. Finally, drivers should also be familiar with *vehicle control systems* and *emergency warning systems*.

Figure 3–8 *Drivers should know the weight of their vehicle and the weight limits of bridges within their jurisdiction.*

THE ROLE OF OFFICERS

Although drivers of emergency apparatus are responsible for the safety of the apparatus, those riding on the apparatus, and the public at large, assigned officers are responsible for the actions of the driver. The relationship between the driver and officer has both a legal and practical aspect. Legally, the officer shares in the safe operation of the apparatus with the driver. Both can be held accountable for unsafe actions. In effect, the officer and driver act as a team to ensure that safety is considered in every aspect of driving. As a team, safety can be enhanced by:

- Confirming that the crew is seated and properly secured
- Identifying and communicating hazards to each other

- Allowing the officer to operate warning devices while the driver concentrates on driving
- Evaluating high hazard areas together, such as intersections and railroad crossings

SAFE-DRIVING CONSIDERATIONS

An important requirement of all drivers is to keep the apparatus under control at all times. Even small mistakes may cause drivers to quickly lose control of their apparatus. One common mistake occurs when an apparatus is operated as if it were a personal vehicle. This happens, in part, because of habits formed while driving personal vehicles. Drivers subjected to stress or who are preoccupied will compensate by relying on habits. Another common mistake is the failure to maintain an appreciation for the unique driving requirements and characteristics of large vehicles. Fire apparatus tend to be large heavy vehicles requiring constant vigilance toward safety to ensure that control is maintained. Finally, the failure to adequately change driving techniques to compensate for changing driving conditions is another common error.

Maintaining control of fire apparatus requires an understanding of basic safe-driving concepts and techniques.

Defensive Driving

Drivers must understand the need to operate their vehicles in a defensive posture. In essence, defensive driving includes a constant awareness of operating vehicles safely to avoid accidents. Key elements for defensive driving include anticipating and planning for:

- The limitations of themselves and their apparatus
- Both appropriate and inappropriate actions of other drivers on the road
- The effects of speed, braking, and weather conditions
- Possible hazards while driving
- Adequate distances from other vehicles with regard to braking and turning
- The need to yield the right of way

Driving Preparations Before driving the emergency vehicle, several important preparations must be accomplished. At beginning of each shift, the driver should conduct a preventive maintenance inspection as discussed in Chapter 2. This is also a good time for the driver to adjust mirrors, seats, and restraints and look over all in-cab instrumentation and devices. In short, the driver should prepare the vehicle for immediate use. When an alarm is received, the driver should walk around the vehicle checking for obstructions and disconnecting all ground lines such as

total stopping distance
the distance of travel measured from the time a hazard is detected until the vehicle comes to a complete stop

perception distance
the distance the apparatus travels from the time a hazard is seen until the brain recognizes it as a hazard

reaction distance
the distance of travel from the time the brain sends the message to depress the brakes until the brakes are actually depressed

braking distance
the distance of travel from the time the brake is depressed until the vehicle comes to a complete stop

traction
friction between the tires and road surface

battery chargers, exhaust systems, and heaters. The vehicle should be started as soon as possible to allow the engine to warm and in some cases build air pressure. The vehicle's operating guide and standard operating procedures should be followed when starting the engine. After the engine starts, all dashboard instrumentation, such as oil and air pressure, engine rpm, and ammeter, should be checked for normal operation indication. The driver is responsible for ensuring everyone is seated with fastened seat belts and that bay doors are fully open. Finally, the parking brake is disengaged, the transmission placed in road gear, and the accelerator slowly depressed. Attention should be given to ensure the vehicle has cleared the bay doors prior to initiating a turn. Failure to do so may result in damage to the side of the apparatus or bay opening.

Speed

Speed affects safe driving in two important ways: It affects the ability to stop the apparatus and it affects the ability to steer the apparatus.

Speed and Braking   Simply stated, the faster the apparatus travels, the greater the distance it takes to stop. Doubling apparatus speed increases the stopping distance an estimated four times. Whenever possible, reduce speed to decrease stopping distance.

   Total stopping distance is measured from the time a hazard is detected until the vehicle comes to a complete stop. Total stopping distance consists of perception distance, reaction distance, and braking distance (Figure 3–9). **Perception distance** is the distance the apparatus travels from the time the hazard is seen until the brain recognizes it as a hazard. For an alert driver traveling at 55 mph, the time it takes to perceive the need to stop is about 3/4 of a second. During this time, the vehicle will have traveled approximately 60 feet. **Reaction distance** is the distance of travel from the time the brain sends the message to depress the brakes until the brakes are depressed. During this time, the apparatus will have traveled another 60 feet in 3/4 of a second. **Braking distance** is the distance of travel from the time the brake is depressed until the vehicle comes to a complete stop. During this time, the vehicle will have traveled approximately 170 feet in about $4\frac{1}{2}$ seconds. Thus, the total stopping distance for a heavy vehicle with good brakes traveling 55 mph on flat, dry, and hard pavement is about 290 feet, approximately the length of a football field, (60 + 60 + 170 = 290) in 6 seconds ($\frac{3}{4} + \frac{3}{4} + 4\frac{1}{2} = 6$).

   The stopping distance noted occurs when the brakes are applied properly. If the brakes are improperly applied, **traction** (friction between the tires and the road) is lost. The result is both an increased stopping distance as well as loss of steering control. Road conditions that affect traction and, consequently, stopping distance and steering include:

   • Just after it starts raining (water/oil mixtures)
   • Standing water (potential for hydroplaning)

■ Note
The faster the apparatus travels, the greater the distance it takes to stop.

Figure 3–9 *Total stopping distance for an apparatus traveling at 55 mph on dry pavement is approximately 290 feet.*

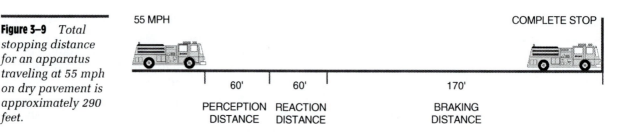

55 MPH   COMPLETE STOP

60'   60'   170'

PERCEPTION   REACTION   BRAKING
DISTANCE   DISTANCE   DISTANCE

- Snow (packed snow and ice formed from melting snow)
- Bridges (ice will form sooner on bridges than on other road surfaces)
- Black ice (a thin clear layer of ice that makes the road appear wet)

Speed and Steering Steering is a function of traction. When a vehicle's speed exceeds the ability of its tires to hold traction on the road, steering is virtually impossible. The road conditions previously mentioned can reduce traction, necessitating slower speeds to maintain steering control. Speed can also affect steering when entering a curve. Excessive speed in curves can cause the tires to lose traction, especially if the brakes are applied. When tires maintain traction and the speed is excessive in a curve, the vehicle may overturn.

Controlling Skids Skidding occurs when tires lose traction against the road surface. Road conditions do not cause skidding. Rather, excessive speed, improper braking, improper tire or tread wear, or improper steering cause skids. It should be understood that skidding can occur on *any* road surface including dry surfaces. When an apparatus begins to skid, it is out of control. Immediate and proper action is required to regain traction and control of the apparatus. The following can be used to help regain control:

- React quickly but do not panic or overreact.
- Take your foot off the accelerator and let the engine slow the apparatus.
- Do not disengage the clutch or slam on the brakes.
- Apply brakes slightly but at least to the extent that wheels no longer turn.
- Turn the front wheels in the direction of the skid (Figure 3–10).

Centrifugal Force and Weight Shifts

■ Note
The key to safely navigating a curve is to maintain traction.

The effects of both centrifugal force and weight shifts (weight transfer) can adversely affect the ability to maintain safe control of apparatus. *Centrifugal force*, the tendency to move outward from the center, occurs when apparatus navigate a curve. The key to safely navigating a curve is to maintain traction. When traction is lost, the apparatus will be out of control and will move in an outward direction

DIRECTION
OF TRAVEL

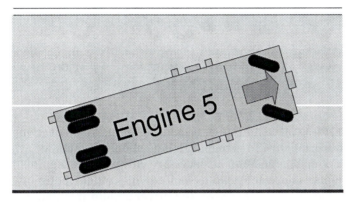

Figure 3–10 *Turning the wheels into a skid will help regain control of the apparatus.*

(Figure 3–11). When apparatus change direction of travel (such as navigating curves) or when they change velocity (sudden stopping), a *shift in weight* occurs on the apparatus. These shifts can be severe enough to cause loss of control or even overturn the apparatus. Recall that a body in motion will tend to stay in motion. Consider a 1,000-gallon water tanker traveling at 50 mph. If the brakes are suddenly applied to stop the vehicle, the onboard water will attempt to stay in motion (Figure 3–12). In essence, the weight of water shifts in the direction of travel as it attempts to stay in motion. The result is that it may take a substantially longer distance to stop the vehicle. In addition, the weight shift may cause the vehicle to turn over when taking a curve at too great a speed.

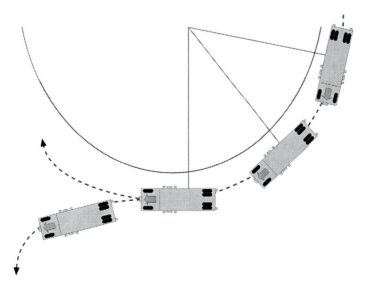

Figure 3–11 *When traction is lost in a curve, centrifugal force will move the apparatus in an outward direction.*

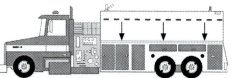

WHEN THE TANKER STOPS ABRUPTLY, THE WATER ATTEMPTS TO STAY IN MOTION.

TRAVELING AT A STEADY SPEED, THE WEIGHT OF WATER IS EXERTED EVENLY ON THE APPARATUS.

Figure 3–12 *When the tanker stops, the on-board water attempts to stay in motion.*

spotter
an individual used to assist in backing up an apparatus

wheel chock
device placed next to wheels to guard against inadvertent movement of the apparatus

Backing Apparatus Improper backing accounts for an overwhelming number of avoidable accidents. In general, backing should be avoided whenever possible. When backing an apparatus is necessary, always have a guide or **spotter**. Some departments require at least one spotter while others require two, one in front and one in the rear. In either case, the driver should maintain visual contact at all times with the spotter. The use of guides, however, does not relieve the driver of the responsibility to safely back the apparatus.

Parking Apparatus Drivers must ensure the control of their apparatus when in a stationary position through proper parking techniques. Proper parking of apparatus is essential to guard against inadvertent rolling on a grade caused by loss of the parking brake or incorrect engagement of the pump. One way to maintain control of a stationary apparatus is to chock (or block) the wheels. NFPA apparatus standards (NFPA 1901 to 1904) require that two **wheel chocks** be mounted and readily accessible on apparatus. Apparatus should be chocked (Figure 3–13) when the apparatus remains in a stationary (parked) position for any length of time. Further, chocking the wheels should be the first thing taken care of when the driver, or other crew member, exits the cab. If parked on an incline, the two chocks should be placed on the downhill side of both rear wheels. When on a level grade, the two chocks can be placed on both sides of a rear tire. Chocks should be placed snug against the tire. The driver is responsible for making sure that wheel chocks are removed and properly stowed prior to moving the apparatus.

A second method used to ensure the safe control of a stationary apparatus is to properly align the front wheels. When parked next to a curb, the rotation of the front wheels should point toward the curb. When no curb is present, the front wheels should be positioned to roll the apparatus away from the road (Figure 3–14). When moving the apparatus from a stationary position, the driver must remember the alignment of the front tires to avoid hitting the curb.

Environmental Conditions

Laws governing the rules of the road are established for ideal conditions, that is, dry roads and good visibility. Drivers must make appropriate adjustments when adverse environmental conditions are encountered. Environmental conditions

Figure 3–13 *Drivers should chock the wheels when the apparatus will be stationary for any length of time. Courtesy Ziamatic Corporation.*

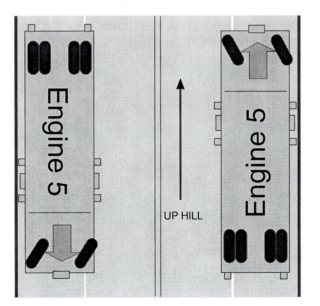

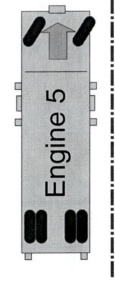

UP HILL

WHEN A CURB IS PRESENT, THE ROTATION OF TIRES SHOULD BE TOWARD THE CURB.

WHEN NO CURB IS PRESENT, THE TIRES SHOULD BE POSITIONED TO ROLL THE APPARATUS AWAY FROM THE ROAD

Figure 3–14 *To further assist with safely controlling an apparatus, drivers should properly align wheels when parked near curbs.*

affect driving by reducing visibility and traction. *Limited visibility* reduces the chances of early hazard recognition, while *reduced traction* affects the ability to steer and stop the vehicle. Examples of environmental conditions that affect vision and traction are:

- Night driving
- Fog or rain
- Snow and ice
- Glare
- Wind

Mechanical devices found on apparatus, such as engine, transmission, and driveline retarders used to assist brakes with slowing and stopping, can both assist and hinder traction under adverse environmental conditions. Drivers should be familiar with manufacturers' recommendations for their use and operation. Automatic chain devices provide the increased traction of snow chains on an as-needed basis. These devices are especially helpful when snow and ice conditions are encountered while driving the apparatus (Figure 3–15).

Road Considerations

Although a driver can do little to improve the surface of a road, maintaining constant awareness of road surface conditions and reacting appropriately will help ensure the safe operation of emergency vehicles. Road conditions that require attention include:

- Sharp curves
- Severe grades

Figure 3–15 *An automatic chain device allows the driver to engage snow chains from the cab while the apparatus is in motion. Courtesy Onspot of North America, Inc.*

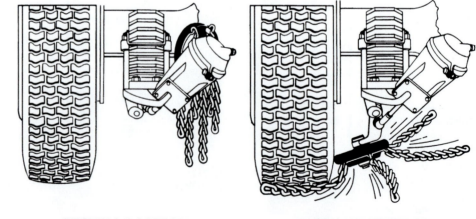

RESTING POSITION   WORKING POSITION

- Soft or limited road shoulders
- Unguarded railroad crossings
- Poor pavement (Figure 3–16)
- Center crowns and curves that are not banked

Each of these conditions can cause the driver to lose control if appropriate action is not taken. In most cases, this means recognizing that the condition exists and reducing the speed of the vehicle.

Drivers must also become familiar with any unimproved or unpaved roads within their jurisdiction. Specifically, drivers must know which roads are safe to use given the weight, width, and length of emergency vehicles. Care must be taken while driving on these surfaces in that loose dirt, gravel, ruts, potholes, gulleys, uneven grade, and other deformities will decrease the ability to maintain safe control of emergency vehicles. The safest thing to do is to not drive on unimproved road surfaces. When it is necessary to drive on these surfaces, the driver should reduce considerably the operating speed of the vehicle.

Figure 3–16 *Poor road conditions may impact the safe operation of emergency vehicles.*

EMERGENCY RESPONSE

Emergency response begins well before an alarm is ever received because drivers must ensure that apparatus are ready for immediate response. Drivers must be familiar with their response districts. Hazards such as congested areas, road construction, limited height and weight factors of bridges and tunnels, and school zones must be identified and avoided if possible. Knowledge and skill in safe-driving techniques must be mastered. During emergency responses, all safe-driving techniques must be followed. Exemptions from driving regulations do not relieve the driver of the responsibility to maintain control of the apparatus at all times. Although safe-driving techniques must be followed for both nonemergency and emergency driving, several additional factors must be considered when responding to an emergency.

When the alarm comes in, a physiological response is initiated that increases the pulse rate, causes the pupils to dilate, and pumps adrenaline into the bloodstream. This type of stress is often called the "fight or flight" syndrome. The purpose is to rapidly prepare the body for a burst of strength and quickness with a heightened alertness. This condition can be both beneficial for and detrimental to the emergency driver. On the detrimental side is the tendency of the stress to encourage speed over safety. Perhaps this is caused by the "flight" aspect. On the beneficial side is the driver's heightened alertness and reaction. Drivers must learn to control the negative side of stress while capitalizing on the positive side of stress.

Prior to leaving the station, several activities must be completed. First, primary and alternate response routes must be selected. Knowledge of response routes allows the driver to concentrate on safety rather than on street signs. Second, the driver must ensure that the apparatus is ready for the response; everyone must be seated and secured, exhaust ducts and chargers must be disconnected, bay doors must be fully open, and emergency warning devices must be activated.

During the response, the driver must aggressively avoid being distracted from safely operating the apparatus. Distractions can be especially dangerous when first leaving the station, when confronted with intersections or other hazardous locations, and when approaching the emergency scene. Distractions are often insidious in that drivers do not realize that they are being distracted. These distracters can easily accumulate to the point at which the driver experiences sensory overload. When this occurs, the driver may not process and react to information correctly. Several potential distractors include:

- The initial excitement of the response
- Thinking about response routes
- Thinking about the nature of the emergency
- Focusing on the emergency on approach
- Emergency lights and sirens

As previously mentioned, the driver and officer work as a team to ensure the safe arrival of the apparatus and crew. Both the driver and officer must attempt to stay focused on safety during the response. However, the officer may be required to gather information on the particular incident. For example, the officer may need to review prefire plans, material safety data sheets (MSDS), and the *emergency response guidebook*. Although this activity is necessary, looking up information can take time and shifts the officer's attention away from response hazard recognition. The use of onboard computers and software allows officers to quickly access required information (Figure 3–17). The less time it takes to gather information, the more time the officer has to assist the driver in a safe response.

Drivers and officers must also take into consideration other responding vehicles. Extreme care should be exercised when two responding vehicles approach an intersection at the same time. The driver and officer should be familiar with SOPs that prescribe actions to follow when such an incident occurs.

Limitations of Emergency Warning Devices

Most state laws require warning devices to be in operation whenever the apparatus is responding to an emergency. However, drivers should understand the limitations of both visual and audible warning devices. The use of visual warning devices does not guarantee that all drivers will see the vehicle during daylight

Figure 3–17 *The use of onboard computers and software provides officers with the ability to quickly gather needed information on an incident. In turn, the officer can spend more time assisting the driver in a safe response. Courtesy Motorola—Land Mobile Products Section.*

hours and when approaching vehicles from behind. Studies have shown that visual warning devices mounted on top of apparatus are not as effective as those mounted lower on the apparatus. Other limitations to visual warning devices include:

- Environmental factors (glare, fog, rain)
- Visibility from within the vehicle (tinted, dirty, or blocked windows)
- Geographical factors (buildings, hills, trees)
- Other vehicles (especially large tractor trailers)
- Drivers (sunglasses, visual ability)

Audible warning devices are most effective directly ahead of the apparatus. Vehicles approaching emergency apparatus from the side may not hear audible signals as soon as those directly ahead. Other limitations to audible warning devices include:

- Noise inside vehicles (radio, heater/air conditioner, passengers)
- Soundproofing of newer vehicles (some commercials stress the quiet ride of new models)
- Noise caused by environmental conditions (rain, wind, hail)
- Geographical factors (buildings, hills, trees)
- Other vehicles (especially large tractor trailers)
- Drivers (hearing ability)

Drivers should understand that even when emergency warning devices are heard or seen, other drivers may not react appropriately. Some drivers simply disregard warning devices. Others may become nervous or confused and slam on their brakes or turn in the wrong direction. Still others may speed up attempting to cross an intersection or to make a turn to avoid having to slow down or stop.

On-Scene Considerations

Perhaps the most significant consideration when approaching the incident is that the emergency scene will most likely be congested. The driver must exercise extreme care when approaching the emergency scene. The emergency scene congestion can occur from several sources. First, congestion can occur from pre-existing sources such as parked cars, bystanders, narrow roads, utility poles, and trees. Second, the source of on-scene congestion can be the result of the incident, which could include debris from fallen buildings, wrecked vehicles, downed electrical wires, and backed-up traffic. Finally, on-scene congestion can be increased by emergency operations such as other responding vehicles, emergency personnel, and emergency equipment such as hose lines, rehab locations, and staging locations.

After arriving on the scene, the driver must safely position the apparatus. In most cases, this means following the department's SOPs. Several general considerations for positioning the vehicle include the following:

- Position to reduce the likelihood of the vehicle being struck by traffic.
- Use the vehicle to shield emergency personnel from traffic.
- Position the vehicle to enhance emergency operations such as water supply, suppression, and rescue activities.
- Position the vehicle away from hazards such as liquid spills, overhead electrical lines, and poor surface conditions (soft surface or severe grade).
- Consider wind direction, exposure, and emergency escape routes when positioning emergency vehicles.
- Do not part in the collapse zone.
- Never park the apparatus on railroad tracks.
- Park the vehicle on the side of the incident.

Returning from Emergencies

Returning from an emergency response can be dangerous. The stress of the incident coupled with fatigue can cause drivers to be less alert and have sluggish reactions. The desire to get back to the station should not overshadow safety. Finally, drivers should inspect the apparatus to ensure that equipment is properly stowed and personnel are seated for the return trip.

DRIVER OPERATING TESTING

The evaluation of driver operators should include both a written test and a skills assessment. In general, the written test should cover those items within NFPA 1002 listed as prerequisite knowledge and the skills assessment should include those items listed as prerequisite skills. Several driving exercises are typically used to assess the ability to safely operate and control the vehicle. These exercises include the following (Figure 3–18):

Alley Dock. Assesses the ability to back the vehicle into a restricted area such as a fire station or down an alley

Serpentine. Assesses the ability to drive around obstacles such as parked cars and tight corners

Confined Space Turnaround. Assesses the ability to turn the vehicle around within a confined space such as a narrow street or driveway

Diminishing Clearance. Assesses the ability to drive the vehicle in a straight line such as on a narrow street or road

DIMINISHING CLEARANCE

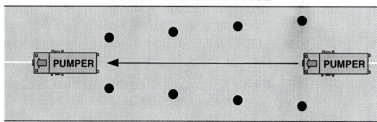

CONFINED SPACE TURNAROUND

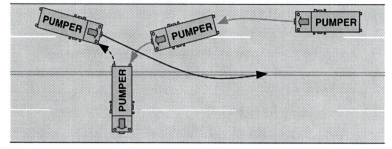

SERPENTINE

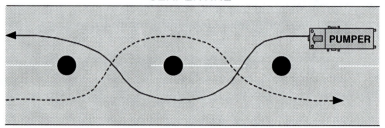

ALLEY DOCK

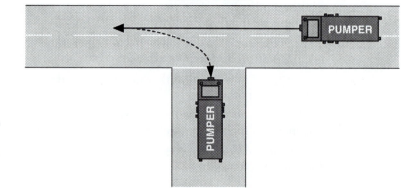

Figure 3–18
Examples of typical driving exercise layouts.

SUMMARY

Apparatus drivers must ensure the safe arrival of the apparatus, equipment, and personnel at the scene. To accomplish this important duty, drivers must be familiar with the laws and standards related to safe driving. In addition, drivers must understand basic driving considerations such as speed and environmental conditions. Finally, drivers must understand the unique requirements for operating their apparatus during emergency response.

REVIEW QUESTIONS

Key Terms and Concepts

On a separate sheet of paper, identify and/or define each of the following.

1. Authorized emergency vehicles
2. Wheel chocks
3. Spotter
4. Total stopping distance
5. Perception distance
6. Reaction distance
7. Braking distance
8. Traction

Multiple Choice and True/False

Select the most appropriate answer.

1. Each year from 1990 to 2000, fire department emergency vehicles were involved in more than 11,000 accidents while responding to or returning from incidents.

   True or False?

2. Which of the following NFPA standards discusses minimum safety-related requirements associated with emergency vehicles, drivers of emergency vehicles, and operation of emergency vehicles?

   a. 1001   **c.** 1500
   b. 1002   **d.** 1901

3. The requirement that emergency vehicle drivers bring the apparatus to a complete stop at red traffic lights or stop signs is contained in which of the following NFPA standards?

   a. 1001   **c.** 1500
   b. 1002   **d.** 1901

4. Which of the following NFPA standards requires drivers to demonstrate the ability to safely maneuver the apparatus over a predetermined route in compliance with all applicable state, local, and department rules and regulations?

   a. 1001   **c.** 1500
   b. 1002   **d.** 1901

5. At a minimum, how many spotters should be used when backing an apparatus?

   a. 1   **c.** 3
   b. 2   **d.** 4

6. When parking an apparatus next to a curb, the rotation of the front tires should be

   a. toward the curb.

   b. away from the curb.

   c. straight.

   d. All of the above are correct.

7. When parking an apparatus with no curb present, the front tires should be positioned to roll the apparatus away from the road.

   True or False?

8. Which of the following should not be used to help regain control while in a skid?

a. take foot off the accelerator

b. depress accelerator slightly

c. apply brakes slightly

d. turn front wheels in the direction of the skid

9. The distance the apparatus travels from the time a hazard is seen until the brain recognizes it as a hazard is known as

a. braking distance.

b. total stopping distance.

c. reaction distance.

d. perception distance.

10. The distance measured from the time a hazard is detected until the vehicle comes to a complete stop is known as

a. braking distance.

b. total stopping distance.

c. reaction distance.

d. perception distance.

11. The distance of travel from the time the brake is depressed until the vehicle comes to a complete stop is known as

a. braking distance.

b. total stopping distance.

c. reaction distance.

d. perception distance.

12. The distance of travel from the time the brain sends the message to depress the brakes until the brakes are depressed is known as

a. braking distance.

b. total stopping distance.

c. reaction distance.

d. perception distance.

13. Posted speed limits are for ideal conditions. Adverse weather may require driving more slowly than posted speed limits.

True or False?

14. Most state laws require warning devices to be in operation whenever the apparatus is responding to an emergency.

True or False?

15. Rarely do accidents occur while returning from an emergency.

True or False?

16. To regain control during a skid,

a. slightly increase speed

b. disengage the clutch

c. Turn wheels in the direction of the skid.

d. Turn wheels in the direction of the skid and depress brake aggressively.

Short Answer

On a separate sheet of paper, answer/explain the following questions.

1. Explain why driving an apparatus is such a critically important duty.

2. List and discuss four common factors associated with apparatus accidents.

3. Explain the purpose of laws and standards for the safe operation of emergency vehicles.

4. Briefly discuss the role of NFPA 1500 and NFPA 1002 in connection with emergency apparatus.

5. List at least four instances when responding apparatus must come to a complete stop.

6. Discuss the attributes of a safe driver.

7. List several examples of defensive driving techniques.

8. Explain the relationship of speed to both steering and stopping apparatus.

9. Discuss how environmental conditions affect the safe handling of apparatus.

10. Explain how stress can both help and hinder a driver while responding.

11. List three general considerations for positioning vehicles on scene.

12. List three basic components contained in most state driving laws.

13. Explain the limitations (conditions) of exemptions extended to emergency vehicle drivers.

14. State laws typically exempt emergency vehicle drivers from several normal driving regulations. List four of these exemptions.

15. Key elements for defensive driving include anticipating and planning for: (list four of these key elements)

16. The key to safely navigating a curve is to maintain _____.

17. List four weather-related conditions that affect traction and, consequently, stopping distance and steering.

18. List four environmental conditions that may affect a driver's vision.

19. List five activities that should be accomplished prior to leaving the station for an emergency.

20. List four limitations to visual warning devices and four limitations to audible warning devices.

ACTIVITIES

1. Obtain a copy of your state law, local law or ordinance, and department SOPs or policies related to emergency response.

2. Evaluate your department's driving-related SOPs or policies in comparison to state and local laws as well as to NFPA 1500 and NFPA 1002.

3. Collect and evaluate data on emergency vehicle accidents at the local or state level.

4. Develop SOPs for (1) emergency driving through intersections and (2) the use of emergency signaling devices during emergency response.

BIBLIOGRAPHY

Most states produce a commercial driver's license manual. Regardless of whether your state requires a commercial driver's license for driving emergency apparatus, this manual provides valuable information on safe-driving techniques for large vehicles.

Teele, Bruce W. *NFPA 1500 Handbook.* Quincy, MA: National Fire Protection Association, 1993.

Emergency Response Guidebook. Washington, DC: United States Government Printing Office, 1993.

Safe Operation of Fire Tankers, FEMA, U.S. Fire Administration Publication # FA 248.

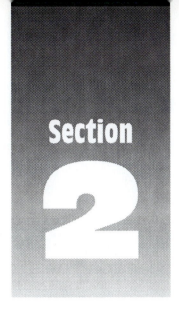

Section

2

PUMP CONSTRUCTION AND PERIPHERALS

The first section of this book discusses basic requirements and concepts for both pump operators and emergency vehicles. Chapter 1 presents the basic qualifications, duties, and responsibilities of the driver/operator referencing NFPA 1002, *Standard for Fire Apparatus Driver/Operator Professional Qualifications*, 2003 Edition, sections 1.4 and 5.1). Chapter 2 discusses preventive maintenance of emergency vehicles (NFPA 1002—4.2 and 5.1.1). Finally, considerations and requirements for driving and operating emergency vehicles are presented in Chapter 3 (NFPA 1002—4.3).

This section of the book covers the requirements of NFPA 1002 presented in sections 5.2.1, 5.2.2, and 5.2.3. It discusses the operating principles, theories, and construction of pumps as well as the systems and components typically used in conjunction with fire pumps. Chapter 4 discusses the need for pumps, types of pumps, and basic pump concepts. The many components used to control and monitor the pump are presented in Chapter 5. Chapter 6 discusses the different types of hose, nozzles, and appliances used in pump operations.

Section 3 continues the discussion of the three duties of the pump operator and the utilization of the tools of the trade in more detail.

Chapter 4

PUMP OPERATING PRINCIPLES AND CONSTRUCTION

Learning Objectives

Upon completion of this chapter, you should be able to:

- Define the term pump.
- Explain the importance of understanding pump operating principles and construction.
- Explain why pumps are needed in the fire service.
- Identify the types and roles of pumps used in the fire service.
- Explain the basic operating principles of positive displacement and dynamic pumps and identify the major construction components of each.
- Discuss the following pump concepts and terms:
    - Intake and discharge sides of a pump
    - Pump efficiency (flow, pressure, speed, slippage)
    - Priming
    - Volume and pressure modes
    - Rated capacity

NFPA 1002
Standard for Fire Apparatus Driver/Operator Professional Qualifications
(2003 Edition)

This chapter addresses the following knowledge elements within sections 5.2.1 and 5.2.2:

Safe operation of the pump

Power transfer from vehicle engine to pump

Operation of pumper pressure control systems

Operation of the volume/pressure transfer valve (multistage pumps only)

INTRODUCTION

pump
a mechanical device that raises and transfers liquids from one point to another

Pumps can be defined as mechanical devices that raise and transfer liquids from one point to another. Note the similarity of this definition to the definition of fire pump operations presented in Chapter 1: "the systematic movement of water from a supply source through a pump to a discharge point."

The ability of a pump to accomplish the task of moving water rests, in part, with the pump itself as well as with the hose and appliances on the intake and discharge sides of the pump. To some extent, the hose lines and appliances can be changed to meet the varying needs of different pumping situations. For example, if higher pressures are needed, smaller hose lines can be used. For greater water flows, larger hose lines can be used. However, one of the major limiting factors of water movement is the type and size of the pump. Different pumps have different capabilities and characteristics for moving water.

■ Note
Knowledge of pump operating principles and construction will help ensure the pump is operated in a safe manner.

Understanding the different types of pumps and their operating principles and construction will assist the pump operator in completing the duties of driving the apparatus, operating the pump, and conducting preventive maintenance. Understanding the limits of a pump will affect the driving and positioning of apparatus for the most effective use of water supplies. In addition, knowledge of the pump will help ensure that the pump is safely operated within its limits to deliver appropriate water quantity and pressure. Providing proper maintenance for safe and efficient operating performance of the pump requires a basic knowledge of pump construction and principles.

Chapter 4 provides a general overview of the pumps typically found on fire apparatus. Discussion centers on the need and types of pumps, pump operating principles, and pump construction.

NEED FOR PUMPS

Water has several characteristics that make it an excellent extinguishing agent for a variety of sizes and types of fires. It is relatively stable at normal temperatures and has a tremendous capacity to absorb heat. It is either readily abundant or at least accessible in most areas. Because of these and other characteristics, the predominant extinguishing agent used in the fire service is water. It makes sense, then, that the fire service would utilize pumps as a means to move water. Indeed, pumps have played an important role in the history and success of the fire service. From human-powered piston pumps to high-capacity pumps to foam systems on modern apparatus, the use of and reliance on pumps in the fire service has steadily increased.

Pumps continue to play an important role in the fire service today for several reasons (Figure 4–1). First, pumps on apparatus play an important role in a municipal water distribution system in that they boost pressure and flows to

Figure 4–1

Emergency vehicles equipped with pumps continue to play an important role in the fire service.

intake
the point at which water enters the pump

discharge
the point at which water leaves the pump

flow
the rate and quantity of water delivered by a pump, typically expressed in gallons per minute (gpm)

pressure
the force exerted by a substance in units of weight per area; the amount of force generated by a pump or the resistance encountered on the discharge side of a pump; typically expressed in pounds per square inch (psi)

speed
the rate at which a pump is operating, typically expressed in revolutions per minute (rpm)

slippage
term used to describe the leaking of water between the surfaces of the internal moving parts of a pump

priming
the process of replacing air in a pump with water

move water from hydrants to the scene. Second, pumps provide some degree of safety from pressure surges. Fire suppression and other emergency activities often require large quantities of water and/or pressure. Either accidents or carelessness can cause dangerous pressure surges during pumping operations. Pressure relief devices protect both personnel and equipment from these powerful pressure surges. Third, pumps are often required to move water supplies maintained on the apparatus to the fire scene. As previously stated, most fire apparatus have at least a minimum water supply on board. Fourth, in areas lacking a municipal water distribution system, pumps are required to move water from static sources such as ponds, lakes, and rivers to the scene. Finally, pumps can provide a continuous supply of water at the proper pressure and quantity to the emergency scene.

BASIC PUMP CONCEPTS AND TERMS

Several concepts and terms are common among all types of pumps. The first concept concerns the terms used to identify the intake and discharge sides of the pump. The **intake** side of a pump is the point at which water enters the pump, also referred to as the "supply side" and the "suction side" of the pump. The use of the term suction side is misleading in that water is not actually sucked into the pump but, rather, flows into the pump under pressure. The **discharge** side of the pump is the location where water leaves the pump. For the remainder of this text, intake and discharge are used to identify the two main sides of the pump.

The second concept concerns the terms used to describe the efficiency of a pump: flow, pressure, speed, and slippage.

- **Flow** refers to the rate and quantity of water delivered by the pump and is expressed in gallons per minute (gpm).
- **Pressure** refers to the amount of force generated by the pump or the resistance encountered on the discharge side of the pump. Pressure is typically expressed in pounds per square inch (psi). Note: Chapter 10 of this textbook discusses flow and pressure in greater detail.
- **Speed** refers to the rate at which the pump is operating and is typically expressed in revolutions per minute (rpm).
- **Slippage** is the term used to describe the leaking of water between the surfaces of the internal moving parts of the pump.

The third concept concerns the priming of pumps. All pumps must be primed to operate. **Priming** the pump is simply getting air out and water in so that it can begin to pump. Positive displacement pumps are typically self-priming while centrifugal pumps must be primed. Often, priming is misunderstood as a suction process, that is, that water is sucked into the pump. The process is actually the creation of a slight negative pressure on the intake side of the pump that allows atmospheric pressure to force water into the pump (Figure 4–2).

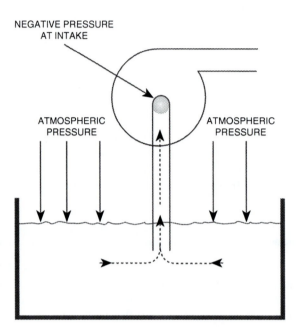

Figure 4–2 *A negative pressure allows atmospheric pressure to force water into the pump.*

NEGATIVE PRESSURE
AT INTAKE

ATMOSPHERIC
PRESSURE

ATMOSPHERIC
PRESSURE

TYPES OF PUMPS

Pump types can be grouped in two ways: by their principle of operation and by their intended use in the fire service. The two broad categories of pumps based on *operating principles* are positive displacement and dynamic. Modern day pumping apparatus utilize pumps from both categories.

 • *Positive displacement pumps* are typically used as priming pumps, high-pressure auxiliary pumps, or portable pumps, and for use in pump and roll situations. These pumps generally provide higher pressures with low flows. Positive displacement pumps are based on **hydrostatic** principles (a branch of hydraulics that deals with liquids at rest and the pressures they exert or transmit). Pressure is generated within the pump by the application of mechanical force.

 • *Dynamic pumps* are typically the main pumps on modern pumping apparatus. These pumps are generally used for large flows with lower pressures. Dynamic pumps are based on **hydrodynamic** principles (a branch of hydraulics that deals with liquids in motion). Pressure is generated by movement and momentum within the pump rather than by force. The common term for dynamic pumps in the fire service is centrifugal pumps. This term is therefore used in the remainder of this textbook.

The three broad categories of pumps based on their *intended use* are main, priming, and auxiliary. Note that this categorization may differ depending on regional preferences.

hydrostatics
the branch of hydraulics that deals with the principles and laws of fluids at rest and the pressures they exert or transmit

hydrodynamics
the branch of hydraulics that deals with the principles and laws of fluids in motion

- The **main pump** is the primary working pump permanently mounted on the apparatus. The size and type of the pump may vary based on the intended purpose of the apparatus. According to NFPA 1901, *Standard for Automotive Fire Apparatus*, 2003 Edition, fire apparatus with the primary purpose of combating structural and associated fires must have a permanently mounted fire pump with a rated capacity of at least 750 gpm. Fire apparatus with the primary purpose of initiating fire suppression efforts and supporting associated fire department operations must have a permanently mounted fire pump with a rated capacity of at least 250 gpm. Specialized vehicles such as tankers, sometimes called water tenders, and high-pressure apparatus may have either centrifugal or positive displacement pumps of various sizes. In most cases however, the term main pump identifies a centrifugal pump.

- **Priming pumps** are positive displacement pumps permanently mounted on an apparatus. Most all apparatus equipped with a centrifugal pump as the main pump will have a priming pump. The sole purpose of these pumps is to prime the main pump when needed.

- **Auxiliary pumps** refer to pumps other than the main pump or priming pump that are either permanently mounted on or carried on an apparatus. The most common of these is the booster pump. Booster pumps do not actually boost the main pump. Rather, they are a smaller pump used to supply small hose lines. Booster pumps can be either positive displacement or centrifugal pumps. Pumps used for **pump-and-roll** operations are included in this category. The pump-and-roll operation is the process of discharging water while the vehicle is in motion. Wildland firefighting apparatus and air crash fire and rescue vehicles often have this capability. High-pressure foam pumps are also included in this category, as are portable pumps, which can be either positive displacement or centrifugal.

POSITIVE DISPLACEMENT PUMPS

Principles of Operation

Two important principles are fundamental to the operation of positive displacement pumps.

Principle 1: Water is virtually noncompressible under normal conditions. This means that forces applied to water will tend to push or move water rather than compress it. This principle can be demonstrated by filling a balloon with water. Pressure exerted on the balloon tends to move the water. The elastic nature of the balloon provides the freedom of the water to move.

Principle 2: When pressure is applied to a confined liquid, the same pressure is transmitted within the liquid equally. Take for example a pump supplying an attack line (Figure 4–3). If the nozzle is closed, the pressure at any point in the line will be the same. Positive displacement pumps utilize these principles to move water.

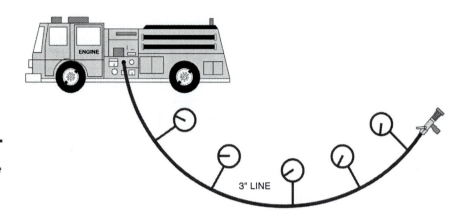

Figure 4–3   *When no water is flowing, the pressure at any point in the line will be the same.*

3" LINE

Operating Characteristics

Positive displacement pumps theoretically discharge (displace) a specific quantity of water for each revolution or cycle of the pump. In reality, slippage occurs that reduces the quantity of water discharged. The greater the slippage, the less efficient the positive displacement pump. Increasing the flow of positive displacement pumps is a simple matter of speeding up the pump. The result of increased pump speed is the same quantity of water per revolution delivered at a faster rate, thus increasing the flow rate.

The relationship between pump intake, discharge, and pressure in positive displacement pumps is, for the most part, straightforward. Each revolution of the pump yields (discharges) a specific quantity of water. Consequently, the pressure on the intake side of the pump is irrelevant. No matter what the intake pressure is, the pump will only discharge its specific quantity per revolution. Increases in pressures may, however, help the pump-driving mechanism function more efficiently in that the drive mechanism will not have to work as hard to pump the water if intake pressure is increased.

Similarly, the water discharged from positive displacement pumps is independent of pressure. Another way of looking at it is that positive displacement pumps simply transfer water. No matter how big the hose or how long the lay on the discharge side, positive displacement pumps will yield the same quantity of water per revolution. Pressure buildup is simply the result of confinement on the discharge side of the pump and does not affect pump discharge. The larger the hose, the longer the pump will take to fill the hose. If the water leaves the hose in the same quantity as the pump delivers it, pressure buildup will be minimal. However, if the hose or associated appliances restrict the amount of water being discharged, pressure increases will occur. If the water is completely restricted as in the case of closing a nozzle, pressure will continue to build up until the pressure forces the pump to stop, the water is released (the nozzle is opened), or something breaks. Because of the potential serious buildup of pressure, some method of relieving excessive pressure on the discharge side of the pump is used. Pressure relief devices are discussed further in Chapter 5.

Positive displacement pumps are typically designed to operate at specific speeds with predetermined pressures to deliver a specific quantity of water. These variables are typically designed into the operation of the positive displacement pump. Priming pumps, for example, often have a simple on-and-off switch to operate the pump.

Construction

The two main types of positive displacement pumps are *piston* (also referred to as reciprocating) and *rotary*. Piston pumps were once the predominate main pump on fire apparatus. The first piston pumps used for firefighting were hand operated (see Figure 4–4). Although rarely used as a main pump today, piston pumps are making a comeback as high-pressure units. Rotary pumps replaced piston pumps as the predominate main pump. However, today they are typically found on modern apparatus in support to the main pump (a centrifugal pump) as either auxiliary or priming pumps. Rotary pumps are further divided into rotary gear, lobe, and vane. Rotary pumps are differentiated from piston pumps based on their circular motion rather than the up-and-down action of piston pumps.

■ Note
Wear and tear in positive displacement pumps can increase slippage, which significantly reduces performance.

Common among positive displacement pumps is the reliance on closely fitting moving parts. Because the parts fit so closely, various liquids and sometimes air can be pumped. The close-fitting moving parts are used to force water or air from the intake side of the pump to the discharge side. The efficiency of positive displacement pumps is related to the close fitting of parts. Consequently, wear and tear of these pumps can increase slippage, which significantly reduces their performance. Improper speeds and debris such as sand in the pump can dramatically increase wear and tear and consequently increase slippage.

Figure 4–4 *Example of an early hand-operated piston pump.*

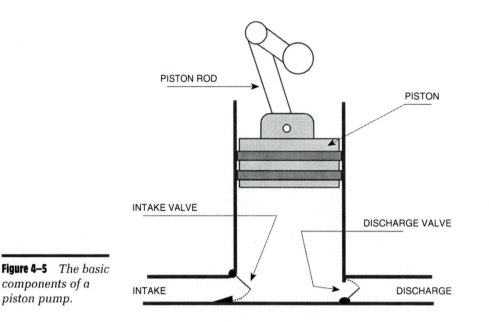

Figure 4–5   *The basic components of a piston pump.*

Piston (Reciprocating) Pumps Piston pumps have a minimum of three moving elements: the piston, the intake valve, and the discharge valve (Figure 4–5). The piston is contained in a cylinder and moves up and down by means of the piston rod. The intake and discharge valves open and close to direct the water.

Figure 4–6 illustrates the operation of a single-action piston pump. When the piston moves upward, the volume between the piston and the cylinder increases.

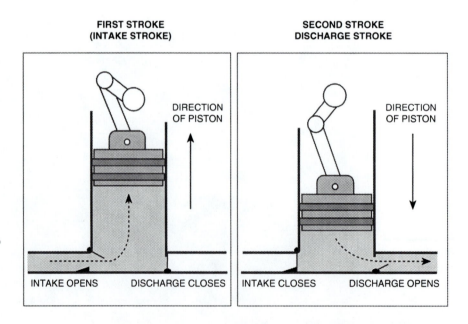

Figure 4–6   *Two-stroke working cycle of a piston pump.*

This action creates a slight vacuum in the chamber that causes the intake valve to open allowing water to enter the chamber. Water will continue to enter until the piston stops. When the piston moves down, the pressure exerted on the water closes the intake valve and opens the discharge valve.

Note that water is discharged only during the downward movement of the piston. This action creates a pulsating effect. A second piston can be added to both reduce the pulsating effect as well as to double the flow of water (Figure 4–7). The pistons simply operate in reverse of each other. While piston A is filling the void space with water, piston B is discharging water. During the next cycle, piston A discharges its water while piston B is filling its void space. Even with two pistons, the pulsating effect still exists. To further reduce the pulsating effect, multiple pistons can be used. Each action of the piston is slightly offset from the others. Air domes have also been used to reduce the pulsating effect. Instead of pumping directly to discharge lines, water is pumped into the air dome. The air dome has

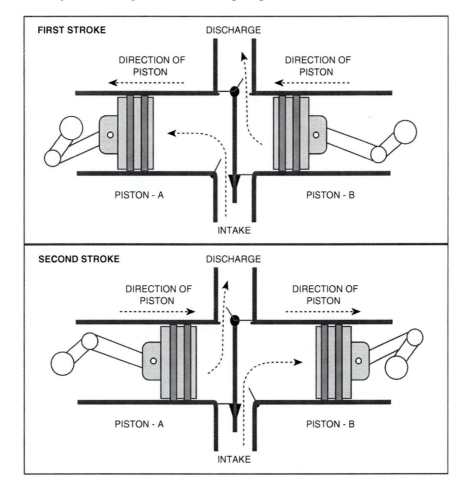

Figure 4–7 *Two-stroke working cycle of a dual piston pump.*

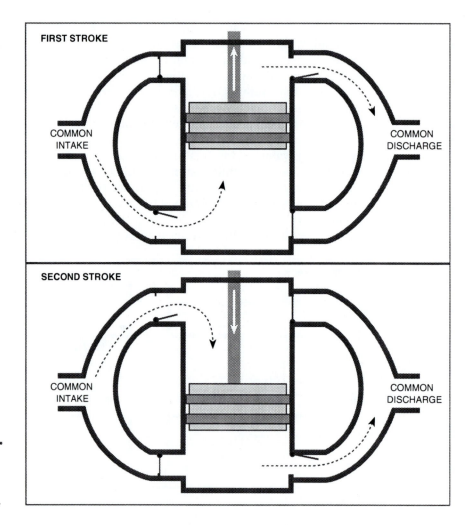

Figure 4–8 *Two-stroke working cycle of a double acting piston pump.*

a pocket of air that acts to cushion the pulsating effect. When the air dome fills to capacity, the internal pressure created from both the water and compressed air forces water out of the dome at a constant pressure.

Piston pumps can also be double acting (Figure 4–8). Double-acting piston pumps discharge water during both the upward and downward motion of the piston. As the piston moves up, water is both drawn in and discharged at the same time. The same is true when the piston is moving down.

Rotary Lobe The moving elements in rotary lobe pumps are two lobes (Figure 4–9). These lobes rotate in opposite directions within a pump casing. The volume between the pump casing and lobe increases alternately between the two lobes. Water enters these cavities and is trapped between the lobes and the pump casing. The lobes move the water along the outside until it reaches the discharge side of

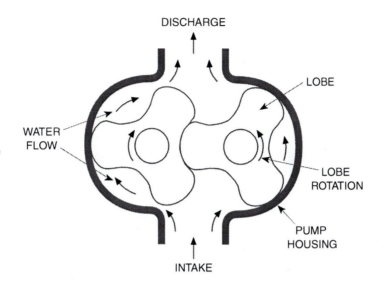

Figure 4–9
Illustration of a rotary lobe pump showing the flow of water from the intake to the discharge side of the pump.

the pump. Note that the close mesh of the lobes prevents water from returning to the intake side of the pump.

Rotary Gear   Rotary gear pumps operate in the same manner as rotary lobes. Instead of lobes, however, closely meshed gears are used to trap water and move it to the discharge side of the pump (Figure 4–10). In some cases, the shape of the rotary gears allows one gear to be powered and in turn power the second gear.

Rotary Vane   The moving elements in a rotary vane pump are a rotor with vanes (Figure 4–11). The rotor is offset (eccentric) from the pump center. This eccentric alignment of the rotor creates a void space of increasing and decreasing volume. The vanes are located in the rotor and are able to move in and out within their slot.

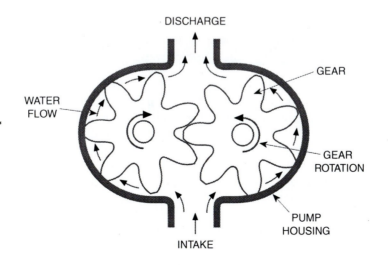

Figure 4–10
Illustration of a rotary gear pump showing the flow of water from the intake to the discharge side of the pump.

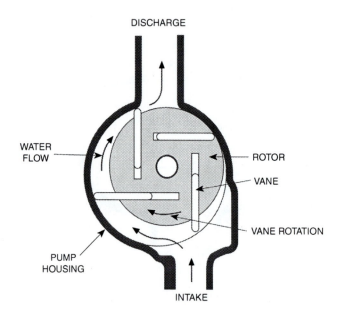

Figure 4–11

Illustration of a rotary vane pump showing the flow of water from the intake to the discharge side of the pump.

The vanes maintain contact with the pump lining by spring tension, centrifugal force, or both. The vanes are typically made of a softer metal than the inside of the pump casing. Although the vanes wear down with use, they are able to maintain contact with the inside of the pump casing. When the vanes wear down to a certain point, they are replaced at a lower cost than replacing a worn pump casing. As the rotor turns, the volume (space) between the pump lining, the vane, and the rotor increases near the intake side of the pump. Water enters this increasing void space and is trapped. The trapped water is forced to the discharge side of the pump along the pump lining. At the discharge side of the pump, the volume (space) decreases in size, and the water is forced out. The close fit of the rotor and pump lining at the discharge side of the pump helps prevent water from returning to the intake.

CENTRIFUGAL PUMPS

■ Note

Centrifugal pumps discharge a quantity of water that is affected by pressure on the discharge side of the pump, and have the ability to change flows and pressures.

Centrifugal pumps have replaced positive displacement pumps as the main pump on modern apparatus. The principles of operation and construction of centrifugal pumps are a little more complicated and confusing than positive displacement pumps.

Recall that positive displacement pumps discharge a specific quantity of water per revolution independent of pressure on the discharge side of the pump. Centrifugal pumps, however, do not discharge a specific quantity of water for each revolution. The quantity of water discharged is affected by pressure on the discharge side of the pump. With centrifugal pumps, flow, pressure, and the speed of the pump (rpm) are interrelated in that a change in one of these factors will change the others. This interrelatedness is one of the useful aspects of centrifugal pumps. The ability to change flows and pressures makes centrifugal pumps very useful for the changing demands at an emergency scene. Calculations for centrifugal pump flows and pressures are discussed further in Section 4 of this textbook.

Principles of Operation

centrifugal force

tendency of a body to move away from the center when rotating in a circular motion

The basic operating principle of centrifugal pumps is centrifugal force. **Centrifugal force** is often defined as the tendency of a body to move away from the center when rotating in a circular motion. Technically, there is no centrifugal force that acts on a body to move it away from the center. Rather, the term is used to explain the *effective force* or behavior of objects in a circular or rotating motion. Centrifugal force is perhaps better thought of as a term used to describe the outward force associated with rotational motion.

Newton's first and third law of motion can be used to help explain the concept of centrifugal force. Newton's first law of motion indicates that a moving body travels in a straight line with constant speed (velocity) unless affected by an outside force. Imagine that you are riding in the passenger seat of a car traveling down the road in a straight line and at a constant speed. As you travel down the road you do not experience or feel any forces acting upon you. If the speed is held constant but the car travels in a circle, you feel or experience what can be described as an outward force as it presses you against the inside of the passenger door. This feeling or experience is referred to as centrifugal force. You are not actually being pushed or forced against the door by centrifugal force. Rather, you are pressed against the passenger door by the inertia of your body as it attempts to keep you moving in a straight line. Another way of thinking about this phenomenon is that the car turns in front of you and your body runs into the passenger door. According to Newton's third law of motion, for every action there is an equal and opposite reaction. You change direction following the path of the car because the passenger door pushes you toward the center of the circle or rotation. This is known as *centripetal force*, a Latin term meaning central (*centrum*) seeking (*petere*). As you travel in a circle, the force you feel is often described as centrifugal force, a Latin term meaning central (*centrum*) fleeing (*fugere*). As indicated earlier, the force you feel is actually caused by inertia. If the door suddenly opened, you would be thrown out of the car and once again follow a straight line. Centrifugal force can also be thought of as the lack of centripetal force.

Figure 4–12 illustrates this concept as it relates to centrifugal pumps. Water placed on the spinning disk would tend to move away from the center in all

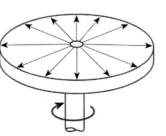

Figure 4–12   *The centrifugal force (outward motion) generated by a spinning disk.*

directions because centripetal force is not exerted on the water. The distance water would travel is related to the speed of rotation and the diameter of the disk. If the water is trapped or contained, pressure will increase relative to the speed of the turning disk. The operation of centrifugal pumps is based on this principle.

Operating Characteristics

The operating characteristics of centrifugal pumps are quite different than those of positive displacement pumps. Centrifugal pump performance is contingent on three interrelated factors: flow (quantity of water discharged), pressure of the water discharged, and speed of the pump. With one factor remaining constant, a change in one of the remaining factors will change the other factors.

Speed: If the speed of the pump is held constant and the flow of water is increased, pressure will drop. Remember, pressure is the result of resistance on the discharge side of the pump. The less resistance, the less pressure. If more water is allowed to flow while the speed of the pump remains the same, the pressure will be reduced.

Flow: If the flow of water is held constant and the speed of the pump is increased, pressure will increase. In this case, the same amount of water is being discharged yet the pump is attempting to discharge more water. The result is an increase in pressure.

Pressure: If the pressure is held constant and the speed of the pump is increased, flow will increase. In this case, the increased speed of the pump will increase the flow. The pressure is maintained constant by increasing or reducing the resistance on the discharge side of the pump.

■ Note

Three important operating characteristics of centrifugal pumps are the use of intake pressure, slippage, and the need for priming.

Three other important operating characteristics of centrifugal pumps are the use of intake pressure, slippage, and the need for priming. Unlike positive displacement pumps, centrifugal pumps take advantage of positive pressure on the intake side of the pump. For example, if the desired discharge pressure is 150 psi and the intake pressure is 50 psi, the pump will only have to produce 100 psi. Slippage is not a factor in the efficiency of a centrifugal pump. Centrifugal pumps have an open path from the intake to the discharge side of the pump. If the discharge side of the pump is closed, water simply spins within the pump. When this occurs, the pump is considered to be experiencing 100% slippage. Because centrifugal pumps rely on the movement of water to operate, the lack of water creates a serious problem. Centrifugal pumps must be primed in order to operate.

Construction

impeller

a disk mounted on a shaft that spins within the pump casing

Centrifugal pumps essentially have one moving part: the impeller, (Figure 4–13). The **impeller** is a disk mounted on a shaft that spins within the pump casing (Figure 4–14). The eye of the impeller is the location where water from the intake enters the pump. Some impellers have an opening on one side while others have openings on both sides. The latter is typically called a double-intake impeller. After entering

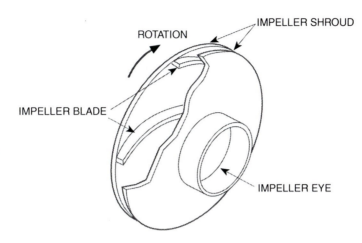

Figure 4–13 *A centrifugal pump impeller. Courtesy W. S. Darely & Company.*

Figure 4–14 *Cutaway of a single stage centrifugal pump showing the impeller on a shaft. Courtesy Waterous Company.*

the eye, water enters the void space between the curved impeller vanes (sometimes called blades) and the sides of the impeller, called the shroud. When the impeller spins, water is forced to the outer edge of the impeller veins by centrifugal force, creating velocity. The vanes also guide the water toward the discharge.

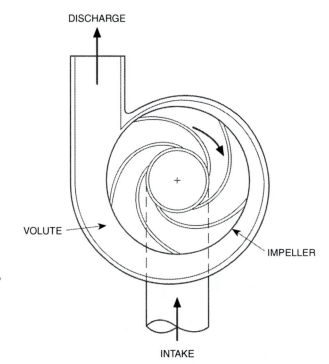

DISCHARGE

VOLUTE

IMPELLER

INTAKE

Figure 4–15
Centrifugal pump impeller and volute. Courtesy W. S. Darely & Company.

volute

an increasing void space in a pump that converts velocity into pressure and directs water from the impeller to the discharge

The impeller is typically mounted eccentric to the pump casing. The void space created by this design is called the **volute** (Figure 4–15). The volute has several functions in a centrifugal pump. First, it converts the velocity created by the impeller into pressure. Second, the increasing void space in the volute allows for increases in water flow created by the impeller. Third, the volute directs water from the impeller to the discharge. Fourth, the volute helps to streamline water, which reduces pressure loss due to turbulence.

The path of a single drop of water through a centrifugal pump is illustrated in Figure 4–16. The drop enters the eye of the spinning impeller. The centrifugal force generated by the impeller forces the water to the outer edge of the impeller vein. As the drop moves away from the eye toward the volute, it picks up velocity. The velocity increases, in part, because the velocity of the impeller increases with the increase in distance from the eye. When the drop leaves the impeller, it is channeled by the volute and directed to the discharge.

Centrifugal pumps can be either single-stage or multistage in operation. A single impeller within a pump casing is called a single-stage centrifugal pump. The speed of the impeller controls the performance of single-stage pumps. Multistage pumps have two or more impellers (stages) enclosed with individual volutes. These pumps provide greater flexibility of flows and pressures. Multistage pumps operate in either volume (parallel) or pressure (series) modes (see Figure 4–17). A transfer valve enables the pump operator to switch between the two modes. The majority of multistage pumps are two stages (meaning they have two impellers)

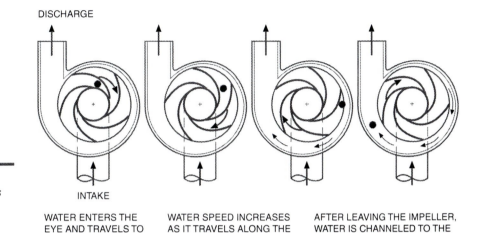

DISCHARGE

INTAKE

WATER ENTERS THE
EYE AND TRAVELS TO
THE IMPELLER

WATER SPEED INCREASES
AS IT TRAVELS ALONG THE
IMPELLER VANE

AFTER LEAVING THE IMPELLER,
WATER IS CHANNELED TO THE
DISCHARGE BY THE VOLUTE

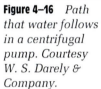

Figure 4–16   *Path that water follows in a centrifugal pump. Courtesy W. S. Darely & Company.*

Figure 4–17   *Cutaway of a dual-stage centrifugal pump showing the impellers on a shaft.*

and are called two-stage centrifugal pumps or series-parallel pumps. The current trend appears to be high-volume single-stage pumps.

Volume Mode   Operating in the volume mode is similar to that of a single-stage operation. The main difference is that water enters both impellers from a common intake and leaves from a common discharge (Figure 4–18). Because both impellers are on the same drive shaft, they spin at the same speed and hence provide the same quantity of water. That means half the total discharge is produced by each of the impellers. If the total discharge of a pump is 1,000 gpm at 100 psi, each impeller is discharging 500 gpm at 100 psi. Multistage pumps are operated in the volume mode when large quantities of water are required at lower pressures, typically when exceeding 50% of the rated pump capacity for longer operations.

Pressure Mode When operating in the pressure mode, the transfer valve redirects the discharge from one impeller to the intake of the second impeller (Figure 4–19). The second impeller can only discharge the quantity given to it by the first. However, the pressure generated by the first impeller can be used by the second. Again, because the impellers are rotating on the same drive shaft, they will each produce a specific quantity and pressure. The second impeller takes the pressure given to it by the first impeller and adds its pressure. For example, if the total discharge of the pump is 500 gpm at 250 psi, then the first impeller is discharging 500 gpm at 125 psi. The discharge of the first impeller is then directed to the intake of the second impeller. The second impeller is only given 500 gpm to pump. However, it takes the 125 psi intake pressure and adds its 125 psi for a total pressure of 250 psi. Multistage pumps are operated in the pressure mode when higher pressures with lower volumes are required.

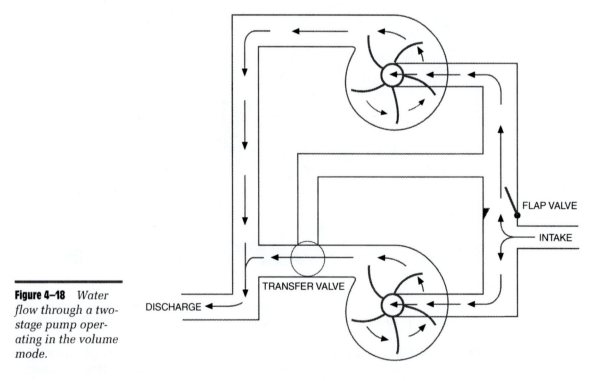

Figure 4–18 *Water flow through a two-stage pump operating in the volume mode.*

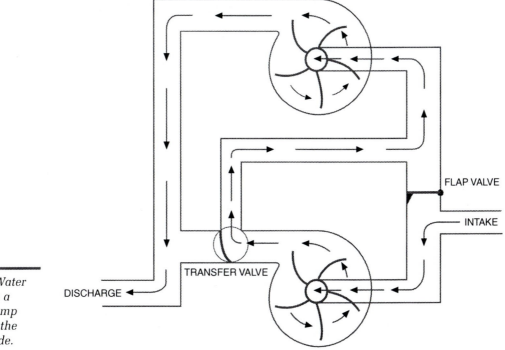

FLAP VALVE

INTAKE

TRANSFER VALVE

DISCHARGE

Figure 4–19 *Water flow through a two-stage pump operating in the pressure mode.*

Rated Capacity and Performance

According to NFPA 1901, *Automotive Fire Apparatus*, pumps used for extended emergency operations must have standard pump capacities of 750, 1,000, 1,250, 1,500, 1,750, 2,000, 2,250, 2,500, or 3,000 gpm. In addition, they must pump their **rated capacity** as follows: 100% at 150 psi, 70% at 200 psi, and 50% at 250 psi. If the pump is rated at 1,000 gpm, then it should deliver 1,000 gpm at 150 psi, 700 gpm at 200 psi, and 500 gpm at 250 psi. See Table 4–1 for the rated capacity of several pumps.

rated capacity
the flow of water at specific pressures a pump is expected to provide

Table 4–1 *Sample rated capacities (in gpm) for several common sizes of fire pumps.*

100% @150 psi	70% @200 psi	50% @250 psi
500	350	250
750	525	375
1,000	700	500
1,500	1,050	750

PUMP LOCATIONS AND DRIVES

The traditional location for mounting pumps is either on the front or in the middle of the apparatus. Today, pumps are mounted on the rear of the apparatus as well, but the majority of pumps are mounted in the middle. Pumps can be powered by a separate engine or by the same engine that drives the apparatus, the latter being the most common.

Pumps can be connected to the engine in several ways (Figure 4–20). One way is to connect the pump directly to the engine crankshaft. This type of connection allows the pump to be utilized in a stationary position as well as while the apparatus is in motion. Front-mounted pumps are typically connected to the engine in this manner. Rear-mounted engines allow rear-mounted pumps to be connected in this fashion as well.

Another method of connecting the pump to the engine is through a power take-off (PTO) from the transmission. This type of connection limits the size of the pump. Pumps connected to the engine through a PTO can operate in a stationary mode or while the apparatus is in motion.

Finally, pumps can be connected to the engine through a split-shaft transmission. The split shaft is located between the transmission and rear axle of the drive line. Midship-mounted pumps are typically connected to the engine in this manner. However, rear-mounted pumps can also be mounted using the split-shaft transmission. This connecting arrangement provides stationary pumping only.

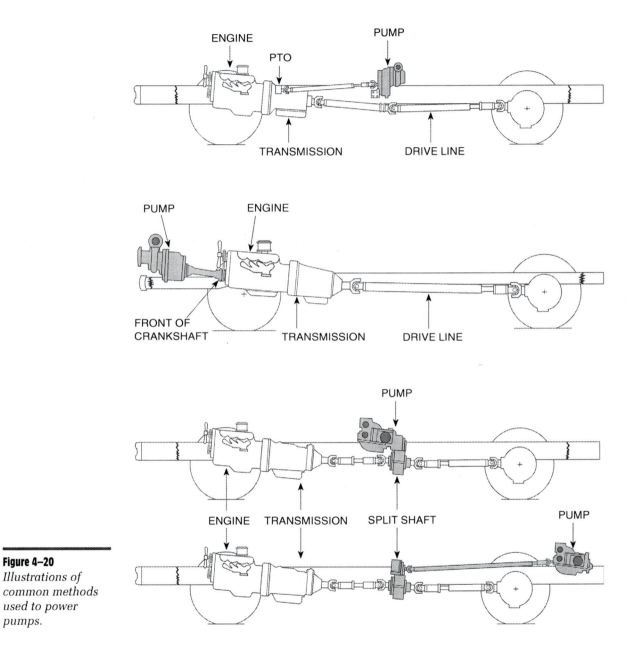

Figure 4–20
Illustrations of common methods used to power pumps.

SUMMARY

The reliance on water as the main extinguishing agent for fire suppression efforts dictates the reliance on pumps in the fire service. To operate pumps in a safe and efficient manner, pump operators must understand the basic operating principles and construction of pumps. The two main types of pumps based on operating principles are positive displacement and centrifugal pumps. The three categories of pumps based on intended use are main, priming, and auxiliary pumps. Pumps can be mounted almost anyplace on the apparatus and can be powered by a separate engine or the drive engine through a direct connection to the engine crankshaft, through a PTO from the transmitter, or through a split-shaft transmission.

REVIEW QUESTIONS

Key Terms and Concepts

On a separate sheet of paper, identify and/or define each of the following.

1. Pump
2. Hydrostatics
3. Hydrodynamics
4. Main pump
5. Priming pump
6. Auxiliary pump
7. Intake
8. Discharge
9. Flow
10. Pressure
11. Pump speed
12. Priming
13. Slippage
14. Centrifugal force
15. Volute
16. Impeller
17. Rated capacity

Multiple Choice and True/False

Select the most appropriate answer.

1. Pumps can be categorized in two ways: by their principle of operation and by their intended use in the fire service.

   True or False?

2. Positive displacement pumps generally provide
   a. lower pressures with higher flows.
   b. lower pressures with lower flows.
   c. higher pressures with lower flows.
   d. higher pressures with higher flows.

3. Centrifugal pumps are based on principles from the branch of hydraulics known as
   a. hydrodynamics.
   b. hydrostatics.
   c. hydromotion.
   d. hydrophysics.

4. According to NFPA 1901, an apparatus intended for sustained operations at structural fires must have a centrifugal pump rated at, at least
   a. 250 gpm.   c. 750 gpm.
   b. 500 gpm.   d. 1000 gpm.

5. The intake side of a pump is the point at which water enters the pump, while the discharge side is the point at which water leaves the pump.

   True or False?

6. Slippage is the term used to describe the extent to which water slides over surfaces providing a measure for friction loss.

   True or False?

7. All pumps must be primed in order to operate.

   True or False?

8. Priming is a suction process that moves water from a static source to the suction side of the pump.

   True or False?

9. Positive displacement pumps theoretically discharge a varying quantity of water inversely related to each revolution or cycle of the pump.

   True or False?

10. Which of the following is incorrect concerning positive displacement pumps?

    a. Each revolution of the pump yields a specific quantity of water.

    b. The pressure on the intake side of the pump does not affect the discharge.

    c. The water being pumped is independent of discharge pressure.

    d. The quantity of water discharged per revolution will increase as pump speed increases.

11. Each of the following is an example of a positive displacement rotary pump except

    a. gear.   c. lobe.

    b. vane.   d. piston.

12. Look carefully at the direction of piston movement, intake and discharge valve positions, and direction of water travel in Figure 4–7 (see page 83). The proper action of a piston pump is illustrated.

    True or False?

13. Rotary gear pumps are similar in operation to rotary lobe pumps, whereas rotary vane pumps are similar in operation to centrifugal pumps.

    True or False?

14. If the speed of a centrifugal pump is held constant and the flow of water is increased (say, for example, a larger line is attached), the pressure will

    a. decrease.

    b. increase.

    c. not change.

    d. increase at first but then decrease.

15. If the flow of water in a centrifugal pump is held constant (same size hose and nozzle) and the speed of the pump is increased, pressure will

    a. decrease.

    b. increase.

    c. will not change.

    d. increase at first but then decrease.

16. 100% slippage can occur in centrifugal pumps because they have an open path from the intake to the discharge side of the pump.

    True or False?

17. The void space created by the impeller being mounted eccentric to the pump casing is called

    a. centrifuge.

    b. countervailance.

    c. hydro-symmetry.

    d. volute.

18. According to NFPA 1901, pumps must deliver their rated capacity for each of the following except

    a. 100% at 150 psi.

    b. 70% at 200 psi.

    c. 50% at 250 psi.

    d. 25% at 300 psi.

19. Pumps connected to the drive engine through a power take-off (PTO) from the transmission can be used for both stationary and pump-and-roll operations.

    True or False?

20. Pumps can be mounted anywhere on the apparatus except toward the back.

    True or False?

Short Answer

On a separate sheet of paper, answer/explain the following questions.

1. Explain why it is important to understand pump operating principles and construction.

2. Explain the difference between a positive displacement pump and a centrifugal pump.

3. Explain the uses of positive displacement and centrifugal pumps in the fire service.

4. Is slippage an important factor in positive displacement and centrifugal pumps? Explain.

5. List and discuss four reasons why pumps are needed in the fire service.

6. List and discuss the terms used to describe the efficiency of a pump.

7. Briefly discuss the priming process.

8. Explain why positive displacement pump discharge is independent of discharge pressure.

9. Piston pumps have a minimum of three moving parts. Draw a simple piston pump and label the three moving parts.

10. Provide the following for the positive displacement pump illustrated in Figure 4–21.

   a. Identify the type of positive displacement pump.

   b. Draw arrows to indicate the path of travel the water will take as it moves through the pump.

   c. Indicate the direction of travel for both lobes.

11. What keeps the vanes in a rotary vane pump in contact with the pump lining?

12. If the pressure of a centrifugal pump is held constant and the speed of the pump is increased, flow will _____. Why does this occur?

13. Indicate the direction of travel for the impeller in Figure 4–22.

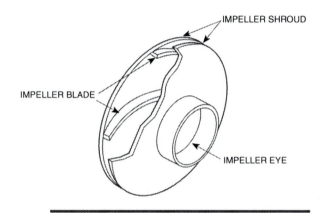

Figure 4–22 *Centrifugal pump impeller. Courtesy W. S. Darley & Company.*

14. Explain what happens when water enters the eye of the impeller.

15. If the total discharge of a two-stage pump operating in the pressure mode is 1,000 gpm at 150 psi, what is the flow and pressure for each of the impellers?

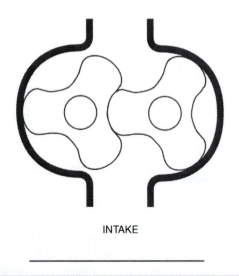

Figure 4–21 *Positive displacement pump.*

ACTIVITIES

1. For each of the following pumps, provide a rough drawing of the pump and identify major components and indicate the path of water flow in each: rotary (either gear or vane), piston, and centrifugal.

2. Develop a rated capacity table for all standard pump capacities. Use the following headings across the top of the table: @ 150 psi, @ 200 psi, and @ 250 psi.

PRACTICE PROBLEMS

1. You are operating in the volume mode of a two-stage 1,250 gpm centrifugal pump flowing capacity @ 150 psi. Diagram water flow, indicating pressure and gpm. Be sure to indicate pressure and gpm for each stage of the pump as well as total flow and pressure.

2. You are operating in the pressure mode of a two-stage 1,250 gpm centrifugal pump flow-ing 50% of rated capacity @ 250 psi. Diagram water flow indicating pressure and gpm. Be sure to indicate pressure and gpm for each stage of the pump as well as for the total flow and pressure at the intake and discharge of each stage.

BIBLIOGRAPHY

Dickenson, T. C. *Pumping Manual.* Oxford, UK: Elsevier Advanced Technology, 1995.
    In-depth text on a wide range of pumps, construction and operation.

Hickey, Harry E. *Hydraulics for Fire Protection.* Quincy, MA: National Fire Protection Association, 1980.
    Excellent review of hydraulic principles related to the fire service.

Chapter

5

PUMP PERIPHERALS

Learning Objectives

Upon completion of this chapter, you should be able to:

- Identify pump peripherals typically found on pump panels.
- Describe the strengths and weakness of different pump panel locations.
- Explain the purpose, construction, and operation of instrumentation, control valves, and systems typically installed on pump panels.

NFPA 1002
Standard for Fire Apparatus Driver/Operator Professional Qualifications
(2003 Edition)

This chapter addresses the following knowledge elements within section 5.2.3:

Proportioning rates and concentrations

Equipment assembly procedures

Foam system limitation, and manufacturer's specifications

The ability to operate foam proportioning equipment and connect foam stream equipment

The following knowledge elements within sections 5.2.1, 5.2.2, and 5.2.4 are also addressed:

Safe operation of the pump

Power transfer from vehicle engine to pump

Operate pumper pressure control systems

Operate the volume/pressure transfer valve (multistage pumps only)

Operate auxiliary cooling systems

INTRODUCTION

pump peripherals
those components directly or indirectly attached to the pump that are used to control and monitor the pump and related components

■ **Note**

Pump peripherals can be grouped into three categories: instrumentation, control valves, and systems.

The ability to move water from one location to another is a function of not only the pump, but also of the peripheral components associated with a pump. As stated earlier, understanding the construction and operating principles of a pump will help the pump operator to:

- Position the pump for best use of available water supply
- Safely operate the pump to deliver proper flows and pressures
- Conduct proper preventive maintenance and inspections on the pump

Understanding the construction and operating principles of pump peripherals will likewise assist the pump operator in those activities.

Pump peripherals are those components directly or indirectly attached to the pump that are used to control and monitor the pump and the engine. Pump peripherals can be grouped into three categories: instrumentation, control valves, and systems. *Instrumentation*, such as gauges, flow meters, and indicators, is used to ensure that the pump is operating efficiently while providing appropriate pressures and flows. *Control valves* include those devices used to initiate, restrict, or direct water flow. *Systems* are those components that operate as a support function to the pump. Examples include priming systems, pressure relief systems, cooling systems, and foam systems. The majority of pump peripherals are located on, and operated from, the pump control panel.

In Chapter 4, pump construction and operating principles were discussed. In this chapter we look at the pump peripherals found on modern pumping apparatus, their construction, and operating principles. Pump peripherals are important for several reasons. First, they allow the pump operator to control the pump by, for example, increasing pump pressure, opening discharge and intake lines, and priming the pump. Second, they help the pump operator ensure that the pump is operating in a safe and efficient manner through the use of gauges and indicator lights and by operating components such as transfer valves and cooling systems. Finally, pump peripherals provide a margin of safety such as intake and discharge pressure relief systems to both the pump and personnel.

PUMP PANEL

pump panel
the central location for controlling and monitoring the pump and related components

The **pump panel** is the mission-control center for pumping operations. It is the one location where the pump operator can control and monitor the pump as well as evaluate its efficiency (Figure 5–1). Pump control, monitoring, and evaluation are accomplished through pump peripherals (instrumentation, control valves, and sys-

Figure 5–1 *Pump operators control and monitor the pump from the pump panel. Courtesy Pierce Manufacturing, Inc.*

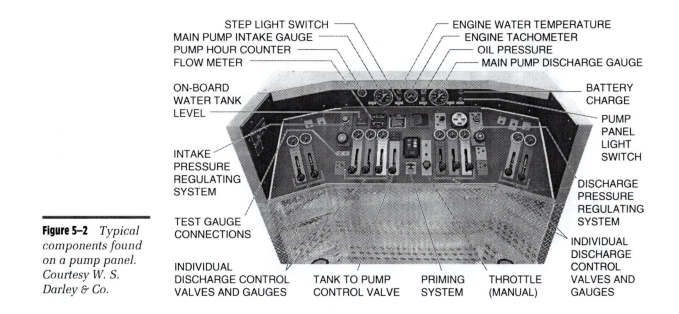

STEP LIGHT SWITCH
MAIN PUMP INTAKE GAUGE
PUMP HOUR COUNTER
FLOW METER

ENGINE WATER TEMPERATURE
ENGINE TACHOMETER
OIL PRESSURE
MAIN PUMP DISCHARGE GAUGE

ON-BOARD
WATER TANK
LEVEL

BATTERY
CHARGE

PUMP
PANEL
LIGHT
SWITCH

INTAKE
PRESSURE
REGULATING
SYSTEM

DISCHARGE
PRESSURE
REGULATING
SYSTEM

TEST GAUGE
CONNECTIONS

INDIVIDUAL
DISCHARGE
CONTROL
VALVES AND
GAUGES

INDIVIDUAL
DISCHARGE CONTROL
VALVES AND GAUGES

TANK TO PUMP
CONTROL VALVE

PRIMING
SYSTEM

THROTTLE
(MANUAL)

Figure 5–2 *Typical components found on a pump panel. Courtesy W. S. Darley & Co.*

tems) located on the pump panel. Figure 5–2 identifies typical pump peripherals found on pump panels. Instrumentation components provide the ability to monitor and evaluate pressures, flows, and pump configuration. Engine operating parameters can also be monitored from instrumentation located on the control panel. Control valves are necessary to physically change the pump configuration such as operating discharge valves, transfer valves, and pump-to-tank valves. Systems on the pump are those components that operate in conjunction with the pump. For example, while the pump is operating, pressure relief systems and cooling systems can be used to help ensure pressures and temperatures remain within safe operating limits. Without a central location for pump peripherals, pump operators would have a difficult time operating the pump.

Physical Characteristics

All pump panels are not created equal. The number, type, and location of pump peripherals on a pump panel are affected by several factors. First, the *size and type of pump* will affect the number of peripherals on a pump panel. Large capacity pumps will often have more instrumentation and control valves, while smaller pumps have fewer peripherals (Figure 5–3). Second, the *intended use* of the pump will affect the types of peripherals on a pump panel. For example, pumps intended to supply foam will have a foam system on the pump panel (note the foam system on the smaller pump panel in Figure 5–3). Third, types and location of peripheral components on a pump panel will be different based on *manufacturer preferences*. Finally, NFPA 1901, *Standard for Automotive Fire Apparatus*, can affect pump panels. For example, according to NFPA 1901, only 2½-inch or

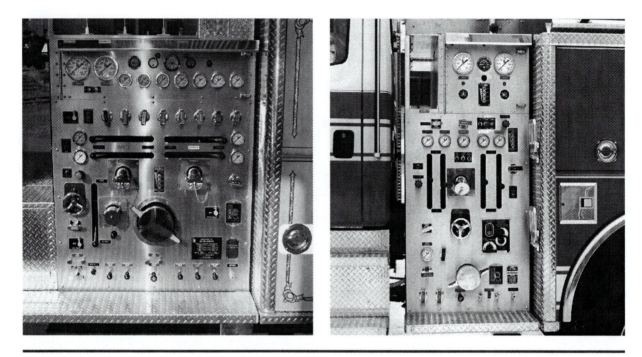

Figure 5–3   *Pump panels differ, based in part on the size of the pump. Courtesy KME Fire Apparatus.*

smaller discharges can be located on the pump panel. In addition, NFPA 1901 requires the following warning on pump panels:

> **WARNING:** Death or serious injury might occur if proper operating procedures are not followed. The pump operator as well as individuals connecting supply or discharge hoses to the apparatus must be familiar with water hydraulics hazards and component limitations.

Compare the pump panels shown in Figure 5–3 with the pump panels in Figure 5–4. Note the different sizes and configurations of the panels.

The traditional appearance of pump panels is in a state of change. One reason for this change is the use of electronic and microprocessor-driven components. Increasingly, flow meters are augmenting pressure devices, and small electric actuating motors are installed on pump panels. Figure 5–5 illustrates the dramatic change in the appearance of a modern electronic pump panel over traditional panels. A second reason that pump panels are in a state of change is the impact that national standards have on them. In the case of the pump panel, the change is toward standardization. For example, NFPA 1901 sets minimum gauge sizes, requires intake relief valves, and suggests a standard color code to match discharges with their gauges (see Table 5–1). Finally, pump panels are changing to

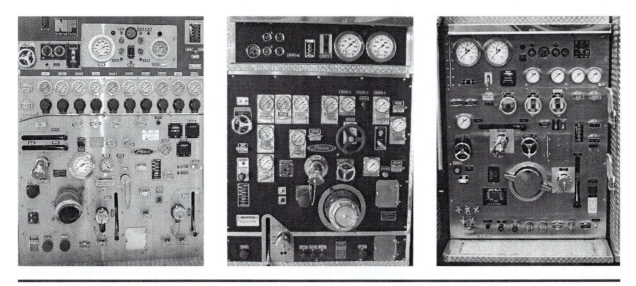

Figure 5–4   *A variety of pump panel configurations exist. Photographs courtesy of Joe Reed, Bob Laeng, and KME Fire Apparatus respectively.*

Figure 5–5   *A modern electronic pump panel called a "diagrammatic" pump panel. All valves are activated by Elkhart electric activators. Courtesy Saulsbury Fire Apparatus.*

Table 5–1   *NFPA 1901 suggested color coding for matching discharge gates with their respective gauges.*

Discharge Number	Color
Preconnect #1 (front bumper jump line)	Orange
Preconnect #2	Red
Preconnect or discharge #3	Yellow
Preconnect or discharge #4	White
Discharge #5	Blue
Discharge #6	Black
Discharge #7	Green
Deluge/deck gun	Silver
Water tower	Purple
Large-diameter hose	Yellow with white border
Foam line(s)	Red with white border
Booster reel(s)	Grey
Inlets	Burgundy

provide a wider margin of safety to pump operators, for example, by the removal of intake and discharge hose connections from the pump panel (Figure 5–6). In doing so, the pump operator is less likely to be injured from lines rupturing or from simply tripping over them.

Figure 5–6   *This apparatus has no intake or discharge connections on the pump panel. Courtesy Saulsbury Fire Apparatus.*

Although pump panels can be different in a number of ways, they are often more alike than different:

- Basic pump peripheral components can be found on most pump panels. Components such as engine throttle, discharge valves and gauges, main intake and discharge gauges, and tank-to-pump and pump-to-tank valves are present on most pump panels.
- Pump discharge control valves and their gauges are typically grouped together on most pump panels (see Figure 5–7).
- The general location of peripherals is similarly located on most pump panels. For example, the main intake and main discharge gauges are typically located close together in the upper section of the panel while the pump intake is typically located in the lower section of the pump panel.
- The size of pump panels are about the same allowing for illumination and ease of access to components.
- Pump panels are typically constructed of stainless steel with some having a vinyl coating.

Compare the pump panels in the previous figures for their similarities.

Pump Panel Location

The location of pump panels on the apparatus has been restricted, in part, by the constraints of mechanical operating components and by tradition. However, the introduction of electronic automation allows pump panels to be located almost anywhere on the apparatus. In addition, tradition appears to be giving way to *safety and efficiency issues* related to pump configuration and location.

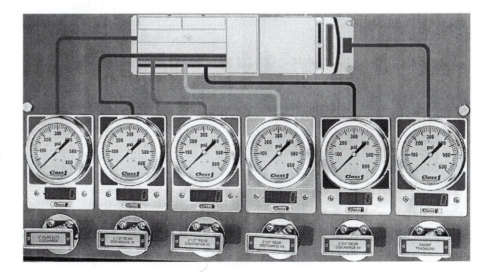

Figure 5–7
Discharge control valves and their gauges are typically located next to one another. Courtesy Class 1.

One safety issue centers on the connecting intake and discharge lines to the pump panel. The concern is the possibility of tripping over, or the rupturing of, intake and discharge lines. Another safety issue focuses on the location of the pump operator while at the pump panel. The concern is the possibility of his being hit by moving vehicles. An issue related to efficiency includes the ability to get pump operations initiated quickly while at the same time staying clear of equipment being removed from the apparatus. Other issues include the ability to observe emergency scene operations, noise levels, and the ability to communicate.

Different locations of pump panels on apparatus have both advantages and disadvantages. In some cases, trade-offs are made in moving the pump panel from one location to another.

Traditional Location The traditional location of the pump panel (Figure 5–8) is just behind the crew cab entrance on the driver's side. One advantage of this location is that pump operators are practically at the pump panel when they exit the cab. In addition to quick access to the pump panel, this position keeps the pump operator out of the way while equipment is removed from the apparatus.

Safety appears to be the greatest disadvantage of this location. First, intake and discharge lines typically connected to the pump panel subject the pump operator to hazards from tripping, as well as the potential rupture of hose at the panel. Second, this position typically exposes the pump operator to the possibility of being hit by passing vehicles. Finally, this position offers pump operators limited visibility of the emergency scene.

Front-Mounted Pump panels mounted on the front of apparatus are not common. When pump panels are mounted on the front of the apparatus, the pump is typically mounted there as well (Figure 5–9). Front-mounted pumps are usually smaller resulting in fewer peripherals. Consequently, the pump panel is relatively small with only basic components (Figure 5–10). One advantage to this position is that

Figure 5–8 *The traditional location of the pump panel is just behind the crew cab on the driver's side. Courtesy Pierce Manufacturing, Inc.*

Figure 5–9
*Apparatus with
front-mounted
pump panel located
next to front-
mounted pump.
Courtesy Saulsbury
Fire Apparatus.*

Figure 5–10   *Close-
up view of a front-
mounted pump
panel. Courtesy
Saulsbury Fire
Apparatus.*

pump operators have a relatively good view of activities on the emergency scene. Another advantage is positioning for drafting evolutions. The overriding concern with this location is the potential for inadvertent movement of the apparatus while the pump operator is standing in front of it. Tripping hazards and breakage of lines are also problems at this location.

Top-Mounted Pump panels found on the top of apparatus are typically mounted across the apparatus so the pump operator faces the rear while operating the pump (Figure 5–11). The major reason for having the pump panel on top of the apparatus is to reduce the likelihood of pump operators being hit by traffic. In addition to being safe from moving vehicles, this position often provides a good view of fireground activities. Another advantage is that pump operators have fewer obstructions to deal with while working at the pump panel since intakes and discharges are typically located elsewhere on the apparatus.

There are two main disadvantages to a top-mounted pump panel: the potential for leaving the pump panel unattended while connecting or changing hose lines, and the increased risk of slipping and falling when the pump operator ascends and descends the apparatus.

Rear-Mounted Rear-mounted pump panels (Figure 5–12), can be located on the end or toward the end on the sides of the apparatus. One advantage of pump panels located on the back of apparatus is that discharge lines can be preconnected to the panel. In addition, other hose lines are readily accessible to the pump operator. This position also allows the pump operator to assist with pulling lines. One disadvantage of this location is that the pump operator may get in the way of equipment being taken off the apparatus. In addition, the distance of the pump panel from the cab requires further travel by the pump operator, and, occasionally, emergency

Figure 5–11
Apparatus with top-mounted pump panel. Courtesy KME Fire Apparatus.

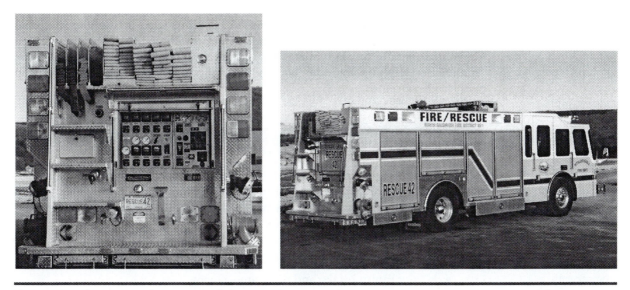

Figure 5–12   *Rear-mounted pump panel showing close-up and position on apparatus. Courtesy Saulsbury Fire Apparatus.*

apparatus are struck in the rear by vehicles while operating on the emergency scene.

Inside Location   Pump panels are also located inside the cab/crew compartment (Figure 5–13). This position has several advantages. First, the pump operator is off the street thus reducing the likelihood of being hit by traffic. Second, the working conditions are more favorable. Heating and air conditioning as well as reduced noise levels allow the pump operator to concentrate on operating and monitoring. One

Figure 5–13

Apparatus with the pump panel located inside the crew cab. Courtesy KME Fire Apparatus.

disadvantage of this location is the limited view provided inside the cab. Small windows, fog, and equipment may obscure the view of the fireground. In addition, the pump operator must go in and out of the cab to hook up hose lines, resulting in increased slipping potential, leaving the pump unattended, and loss of contact with the fireground.

INSTRUMENTATION

instrumentation
devices such as pressure gauges, flow meters, and indicators used to monitor and evaluate the pump and related components

All modern pumping apparatus have **instrumentation** located on their pump panels. Instrumentation is vital for setting up and monitoring pumping operations as well as for evaluating the efficiency of the pump and its engine. In order to safety operate and monitor the pump and its peripherals, the pump operator must be familiar with all instrumentation located on the pump panel. According to NFPA 1901, the following instrumentation must be located on the pump panel:

- Master pump intake and discharge gauges
- Engine tachometer
- Pumping engine coolant temperature and oil pressure indicators
- Voltmeter
- Water tank level indicator

Instrumentation on pump panels can be grouped into three categories: pressure gauges, flow meters, and indicators.

pressure gauge
device used to measure positive pressure in pounds per square inch (psi) or negative pressure in inches of mercury (in. Hg)

Pressure Gauges

Obviously, the purpose of a **pressure gauge** is to measure pressure. The purpose of pressure gauges on a pump panel is to provide a central location for the collection of information on pressures during pumping operations. On the fireground, pump operators use this information within hydraulic calculations to determine flow. Additional information on calculated flow is presented in Section A of this text. Pressure gauges on pump panels may measure positive and negative pressure. Positive pressure gauges measure pressure in pounds-per-square-inch (psi) above atmospheric pressure. Negative pressure gauges measure pressure in inches of mercury (in. Hg). A **compound gauge** reads both positive pressure (psi) and negative pressure (in. Hg).

compound gauge
a pressure gauge that reads both positive pressure (psi) above atmospheric pressure and negative pressure (in. Hg)

bourdon tube gauge
the most common pressure gauge found on an apparatus, consisting of a small curved tube linked to an indicating needle

The **bourdon tube gauge** is the most common pressure gauge found on pump panels. It consists of a small curved tube filled with liquid (Figure 5–14). One end of the tube is connected to the indicating needle through a linkage system. When the liquid in the tube is under pressure, the tube straightens itself, causing the indicating needle to turn. When the pressure of the liquid decreases, the tube begins to return to its original curved position, again causing the indicating needle to turn.

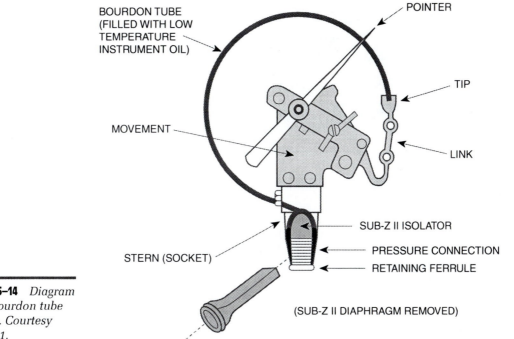

BOURDON TUBE
(FILLED WITH LOW
TEMPERATURE
INSTRUMENT OIL)

POINTER

TIP

MOVEMENT

LINK

SUB-Z II ISOLATOR

PRESSURE CONNECTION

STERN (SOCKET)

RETAINING FERRULE

(SUB-Z II DIAPHRAGM REMOVED)

Figure 5–14   *Diagram of a bourdon tube gauge. Courtesy Class 1.*

Vibrations of the engine and pump and varying water pressures can cause the indicating needle to fluctuate or bounce. To compensate for this movement, some manufacturers fill the gauge with a heavy clear liquid, such as silicone, to help keep the needle from excessive fluctuation. Another means used to reduce needle bounce is a dampening device. These devices place pressure on the needle to reduce the fluctuation. However, if too much pressure is placed on the needle, the gauge will not function correctly.

Bourdon tube gauges are used because they provide the degree of accuracy and durability required on mobile apparatus. However, the often harsh working conditions and environments these gauges operate in can, over time, affect their accuracy. Therefore, all gauges should be calibrated at regular intervals to ensure that they provide accurate readings. Unfortunately, calibration of pump gauges is often not performed. All the efforts of a pump operator to provide adequate pressures and flows will be negated if pump panel gauges do not provide accurate readings. Pump operators should take the initiative to ensure that pump panel gauges are calibrated periodically by qualified personnel.

Digital pressure gauges are increasingly finding their way onto pump panels. A sensing device on the discharge outlet measures the pressure and sends an electrical signal to the pump panel. The signal is received, translated, and digitally displayed on the pump panel (Figure 5–15). When digital pressure gauges are

■ **Note**

All gauges should be calibrated at regular intervals to ensure that they provide accurate readings.

used, NFPA 1901 requires the device to display pressure in increments of 10 psi (70 kPa) or less.

Pressure gauges found on most pump panels can be grouped into main pump gauges and individual discharge gauges. The main pump gauges are usually located close to each other on the pump panel and are typically larger than the individual discharge gauges.

Main Pump Gauges Most all pump panels have two main pump gauges, sometimes referred to as master gauges. According to NFPA 1901, these gauges are to be labeled "Pump Intake" and "Pump Discharge." The *pump intake gauge*, is a compound gauge attached to the intake side of the pump. This gauge measures negative and positive pressures from 30 in. Hg. to at least 300 psi. Negative pressure readings are important for monitoring drafting (a process for priming the pump) operations. Positive pressure readings are important for monitoring pressurized water supplies, such as from hydrants or relay operations. The other main gauge, called the *pump discharge gauge*, is a positive pressure gauge mounted on the discharge side of the pump. The main pump discharge gauge measures the highest positive pressure generated on the discharge side of the pump. NFPA 1901 requires the intake and discharge gauges to be mounted within 8 inches of each other with the intake gauge installed to the left or below the discharge gauge.

Individual Discharge Gauges Individual gauges are commonly provided for each 1½-inch and larger discharge outlet located on the pump. These gauges measure positive pressure in psi and are typically smaller than the main gauges. Individual discharge gauges measure positive pressure for the specific outlet to which they are attached and are usually located next to their respective discharge control valves.

Knowing the discharge pressure of each outlet is important to ensure that excessive pressures are not supplied to attached discharge lines. In addition, by knowing the pressure of the discharge outlet, the pump operator can calculate the amount of water flowing through the line. This is a critical point in that the effectiveness of extinguishing efforts rests, in part, with the quantity of water flowing. (More information on the relationship between water flow and extinguishment and the calculation of water flow is provided in Section 4.)

Figure 5–16 *Flow meters are increasingly being installed on pump panels. Courtesy Bob Laeng.*

Flow Meters

flow meter

a device used to measure the quantity and rate of water flow in gallons per minute (gpm)

Flow meters are placed on pump panels to augment or replace pressure gauges (Figure 5–16). The purpose of **flow meters** is to measure the quantity and rate of water flow in gallons per minute (gpm). Unlike pressure gauges that require hydraulic calculations to determine gpm, flow meters measure gpm directly. The result is reduced hydraulic calculations and increased accuracy of flow rates.

Flow meters are comprised of several components. A sensor mounted on the discharge pipe provides a reading of the flow. In most cases, a paddlewheel is used to measure flow (see Figure 5–17). The sensor transmits the signal to the pump panel where a microprocessor chip translates the signal into the flow rate that is displayed on the digital readout. Because of the "smart" electronics of flow meters, calibration is a simple process that does not require special instruments. Flow

Figure 5–17
Illustration of a paddlewheel sensor. Courtesy Fire Research Corp.

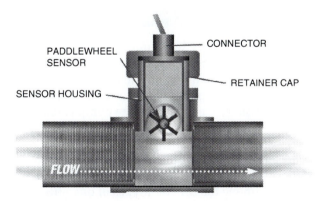

PADDLEWHEEL
SENSOR

SENSOR HOUSING

CONNECTOR

RETAINER CAP

FLOW

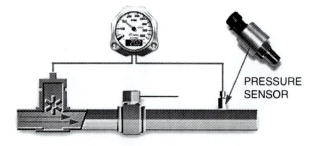

Figure 5–18 *Both a flow meter and pressure sensor sending electronic signals to a combination flow/pressure instrument. Courtesy Fire Research Corp.*

meters can provide current flow and flow accumulation for an operation, and some flow meters provide pressure readings as well. Discharge flow meters, according to NFPA 1901, must have display digits of at least ¼ inch and display flow in increments of 10 gpm or less. The current trend is to include both pressure and flow information on pump panels (see Figure 5–18).

Indicators

indicators

devices other than pressure gauges and flow meters (such as tachometer, oil pressure, pressure regulator, and on-board water level) used to monitor and evaluate a pump and related components

Indicators are loosely categorized as all instrumentation on the pump panel other than pressure gauges or flow meters. They can be simple or complex devices that provide information on a component. Several indicators can be found on the pump panel to provide information on the pump engine. The most common engine indicators are the tachometer, oil pressure, and coolant temperature. Digital readouts that combine engine information into a single display are finding their way onto pump panels (Figure 5–19). The onboard tank usually has an indicator on the pump panel that provides the status of the water level contained in the tank. Often a pump engagement indicator is provided on the pump panel to let the pump operator know when the pump is properly engaged. Finally, a pressure regulator indicator is typically found on pump panels. This indicator lets the pump operator know when the pressure regulator has been activated.

CONTROL VALVES

control valves

devices used by a pump operator to open, close, and direct water flow

Control valves are used by the pump operator to open, close, and direct water flow. Control valves consist of a valve and an operating mechanism (Figure 5–20). The *valve* is the component that physically directs water flow. The most common types of valves found on pump panels are ball valves and butterfly valves. Piston and gated valves can also be found on the pump panel or on other intake/discharge locations on the apparatus.

The *operating mechanism* is the device that controls the valve. These can be manual mechanisms such as push–pull (commonly referred to as "T-handles"), quarter turn, and crank control. Electric, pneumatic, and hydraulic actuators are also used. Common among the different types of operating mechanisms is the requirement to hold the valve in position, when set, so that it will not change while

Figure 5–19 *These digital displays provide information on engine rpm, oil pressure, engine temperature, voltage, and can display engine alert messages. Courtesy Class 1 and Fire Research Corp. respectively.*

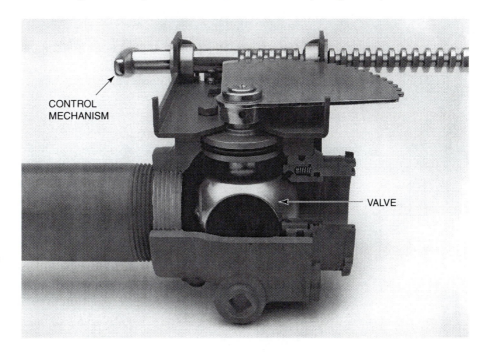

Figure 5–20
Example of a control mechanism and valve. Courtesy Waterous Company.

Figure 5–21 *Example of slow-acting control mechanism in compliance with NFPA 1901. Courtesy Elkhart Brass Manufacturing Company, Inc.*

operating at the maximum flows and pressures of the pump. In addition, control valves are designed to operate smoothly within normal operating pressures of the pump. According to NFPA 1901, intakes and discharges of 3 inches or larger must have slow-acting (cannot go from full close to full open in less than 3 seconds) control mechanisms (Figure 5–21).

Electric Control Valves

Electrically activated control valves are increasingly being installed on pump panels. Several reasons for their rise in popularity include the following:

- An increase in pump panel design opportunities (control valves are operated remotely rather than tied to a mechanical linkage from the pump panel)
- A reduced water hammer potential (electric motors control the speed of opening and closing the valve)
- Ease of operation even under high pressure (an electric motor does all the work)

Figure 5–22 is a schematic of an Elkhart electric control valve. The pump operator controls the valve through the switch/indicator that sends information through the wiring harness to and from the control valve. The switch activates the motor to open and close the valve. The indicator lights (the left one is green, the middle one is amber, and the right one is red) provide information on the position of the valve:

Indicating Signal	*Meaning*
Steady burning red	Valve ball in fully closed position
Steady burning green	Valve ball in fully open position
Steady burning amber	Valve ball in an intermediate position
Steady burning amber with flashing green	Valve ball moving toward open position
Steady burning amber with flashing red	Valve ball moving toward closed position

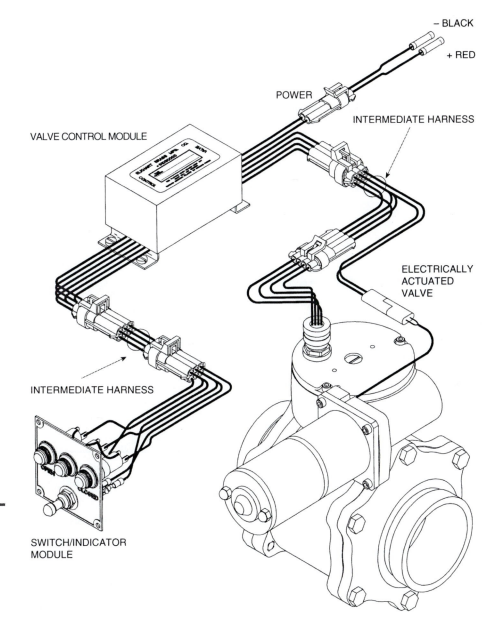

Figure 5–22

Schematic of an Elkhart electrically actuated valve. Courtesy Elkhart Brass Manufacturing Company, Inc.

Intake Control Valves

Intakes provide for the connection of external supply sources to the pump. Intake control valves allow the pump operator to regulate external supply sources to the pump. Pumps usually have several intakes located in various positions on the apparatus. NFPA 1901 has several requirements for intakes. Table 5–2 indicates the minimum number and size of intakes per rated capacity of the pump. Not

Table 5–2 *Suction hose size, number of suction lines, and lift for fire pumps.*

Rated Capacity		Maximum Suction Hose Size		Maximum Number of Suction Lines[a]	Maximum Lift	
gpm	L/min	in.	mm		ft	m
250	1,000	3	75	1	10	3
300	1,100	3	75	1	10	3
350	1,300	4	100	1	10	3
500	2,000	4	100	1	10	3
750	3,000	4½	110	1	10	3
1,000	4,000	6	150	1	10	3
1,250	5,000	6	150	1	10	3
1,500	6,000	6	150	2	10	3
1,750	7,000	6	150	2	8	2.4
2,000	8,000	6	150	2	8	2.4
2,000	8,000	8	200	1	6	1.8
2,250	9,000	6	150	3	6	1.8
2,250	9,000	8	200	1	6	1.8
2,500	10,000	6	150	3	6	1.8
2,500	10,000	8	200	1	6	1.8
3,000	12,000	6	150	4	6	1.8
3,000	12,000	8	200	2	6	1.8

Source: NFPA 1901, Standard an *Automotive Fire Apparatus*, 2003 edition.

[a]Where more than one suction line is used, all suction lines do not have to be the same hose size.

all intakes have control valves. However, NFPA 1901 requires that at least one auxiliary gated 2½-inch intake be operable from the pump panel. All intakes are required to have a removable or accessible strainer to keep large debris from entering the pump. All 3½-inch or larger intakes with a control valve must also have an automatic pressure relief device. Figure 5–23 shows the intake relief valve in operation. Note the water being discharged under the apparatus.

Discharge Control Valves

Unlike intakes, each individual discharge requires a control valve. Pumps typically have at least two 2½-inch discharges. However, the pump usually has as many 2½-inch or larger discharges as are needed to discharge the rated capacity of the pump. As a rule of thumb, a pump will have one 2½-inch discharge for each 250

Figure 5–23
Example of pressure relief valve in operation. Note the water discharging under the apparatus.

gpm of rated capacity. For example, if the pump is rated at 1,250 gpm, then most likely the pump will have a minimum of five 2½-inch discharges. Pumps typically have at least two 1½-inch preconnect discharges as well. Discharge control valves provide a great deal of flexibility over individual hose lines, allowing multiple hose lines of different pressures and flows to be deployed at the emergency scene.

Other Control Valves

Several other control valves are located on the pump panel. One control valve, called the *tank-to-pump*, allows water to flow from the onboard water supply to the intake side of the pump. Another control valve located on the pump panel is called the *pump-to-tank* or *tank fill*. This valve allows water to flow from the discharge side of the pump to the tank. These two control valves can be used together to help keep the pump from overheating by circulating water from the pump to the tank and then back to the pump again. The *transfer valve* is a control valve found on multistage pumps that redirects the water flow in the pump between the pressure mode and volume mode.

systems
those components that directly or indirectly assist in the operation of the pump (e.g., priming systems, pressure regulating systems, and cooling systems)

SYSTEMS

Pump peripheral **systems** are those components that directly or indirectly assist in the operation of the pump. The common systems found on pump panels include priming, pressure regulating, cooling, and foam systems.

Priming Systems

Priming systems are used by the pump operator to prime the pump. Recall that priming a centrifugal pump is the process of removing air from the intake side of the pump and replacing it with water. Priming systems accomplish this task. The Waterous priming system is used to illustrate a typical priming system. This system is comprised of a priming pump, an oil reservoir, and a priming valve activated on the pump panel (Figure 5–24).

The priming process is initiated when the pump operator activates the priming valve. Referring to Figure 5–25, when this occurs, a sliding plunger (A) in the priming valve accomplishes two tasks. One task is to open a passageway between the intake side of the priming pump (B) and the intake side of the pump being primed (C). The second task is to depress the priming pump motor switch (D).

With the priming motor activated, the priming pump begins to pump air from the intake side of the main pump and oil from the priming oil tank reservoir (refer to Figure 5–24). Oil is included in the priming process to help lubricate the priming pump as well as to help provide a tighter seal between the moving parts. Recall

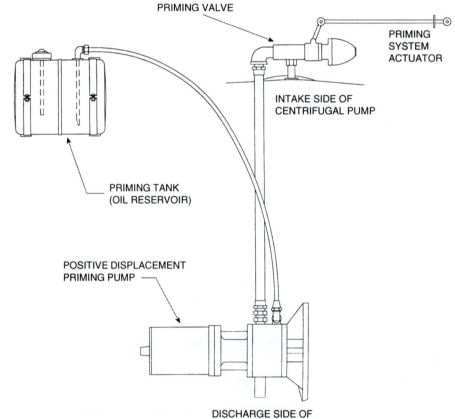

Figure 5–24

Components of a typical priming system. Courtesy Waterous Company.

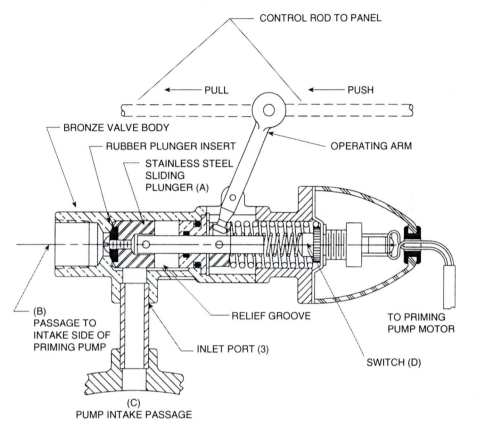

CONTROL ROD TO PANEL

← PULL   ← PUSH

BRONZE VALVE BODY

RUBBER PLUNGER INSERT

STAINLESS STEEL
SLIDING
PLUNGER (A)

OPERATING ARM

(B)
PASSAGE TO
INTAKE SIDE OF
PRIMING PUMP

RELIEF GROOVE

INLET PORT (3)

TO PRIMING
PUMP MOTOR

SWITCH (D)

(C)
PUMP INTAKE PASSAGE

Figure 5–25
*Waterous priming
valve. Courtesy
Waterous Company.*

that the ability to move air and water rests, in part, with the close fit of moving parts in the positive displacement priming pump. The air and oil mixture is typically discharged under the apparatus. Because this water/oil mixture is discharged, environmentally friendly priming oil is now being used with some priming devices.

As the air is removed by the priming system, a negative pressure is created on the intake side of the pump. When this happens, atmospheric pressure forces the water into the pump (refer back to Figure 4–2). When the mixture being discharged from the priming pump turns from an air/oil mixture to a water/oil mixture, the priming system has accomplished its task of removing air from the pump and replacing it with water. Other indicators that the pump is primed are:

- A positive reading on the pressure gauge.
- The priming motor sounds as if it is slowing.
- The main pump will sound as if it is under load (begins to pump water).

Priming devices on modern pumping apparatus are typically rotary vane or rotary gear positive displacement pumps. Older apparatus may use exhaust or vacuum primers. Exhaust primers redirect exhaust gases through a chamber connected to the intake side of the centrifugal pump. As the rapidly moving gases pass through the chamber, they create a negative pressure helping to draw out air from the pump. Higher engine pressures are needed for this device to work efficiently. Vacuum primers use the vacuum created in the manifold of gasoline-powered engines. A float is typically used to help ensure water does not travel from the pump into the engine manifold.

Pressure Regulating Systems

pressure regulating systems
devices used to control sudden and excessive pressure build-up during pumping operations

All modern pumping apparatus have a pressure regulating system attached to the centrifugal pump. **Pressure regulating systems** provide safety to personnel and equipment and assist with the efficient delivery of discharge flows. *Safety* is attained by controlling sudden and excessive pressure buildup during pumping operations. *Efficiency* of discharge flows is assisted by compensating for fluctuating pressures, which helps keep the pump and engine under a constant load. Without pressure regulating devices, pump operators would have an impossible task of maintaining safe and efficient pressures and flows.

Pressure regulators are needed because pressures can fluctuate during any pumping operation. These fluctuating pressures can be severe enough to damage equipment and injure personnel. In addition, they can occur rapidly and without warning. Even the most astute pump operator cannot adequately correct and control these pressure fluctuations without the aid of a pressure regulator.

Pressure increases can result from simply closing a discharge line. Recall that pressure buildup results, in part, from restrictions on the discharge side of the pump. Take for example a pump flowing a total of 500 gpm through two discharge lines. Each discharge line flows half of the volume, or 250 gpm. If one line is closed, the pump attempts to flow the 500 gpm through the one line. A sudden pressure increase occurs due to the restrictions of the one remaining open discharge. Pressure regulators can be set to automatically compensate for such an increase in pressure.

The two pressure regulators commonly found on pumps are *pressure relief devices* and *pressure governors*. Pressure relief devices control pressure buildup by sending excess water pressure back to the intake side of the pump. Governors control pressure buildup by controlling the speed of the engine, which in turn controls the speed of the pump. Pump speed is increased and decreased to handle pressure fluctuations. According to NFPA 1901, the pressure regulating system must operate within 3 to 10 seconds after an increase of 30 psi above the set discharge pressure. In addition, the pressure regulating system must be operational from the pump panel and must indicate when the system is in operation.

pressure relief device
a pressure regulating system that protects against excessive pressure build-up by diverting excess water flow from the discharge side of the pump back to the intake side of the pump or to the atmosphere

Pressure Relief Devices To protect against excessive pressure buildup by diverting excess water flow from the discharge side of the pump back to the intake side of the pump, **pressure relief devices** are used (see Figure 5–26). In the previous example, the excess 250 gpm caused by closing one line would be pumped back into the intake side of the pump. The pump, in essence, would continue to flow

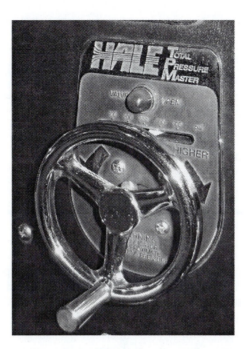

Figure 5–26 *Pressure relief device.*

500 gpm—250 gpm through the open line and 250 gpm back to the intake side of the pump.

> ■ **Note**
>
> **The pressure exerted by a liquid on a surface is proportional to the area of the surface.**

A variety of pressure relief devices are installed on pumps; however, the operating principle behind each of them is virtually the same. Pressure relief devices operate, in part, based on a hydraulic principle stating that the pressure exerted by a liquid on a surface is proportional to the area of the surface. In essence, a small quantity of water can be used to force a piston against higher pressures. Relief valves that operate automatically utilize this principle.

Pressure relief devices consist of two main components: a control mechanism and a relief valve. The *control mechanism* is used to set the relief device to operate automatically at specific pressures, allowing excess pressure to be diverted back to the intake side of the pump.

> ■ **Note**
>
> **Pressure relief devices consist of two main components, a control mechanism and a relief valve.**

The *relief valve* is a piston contained in a cylinder that opens and closes a passage between the discharge side of the pump to the intake side of the pump, (Figure 5–27). Both sides of the piston receive pressure from the discharge side of the pump. However, the smaller side of the valve receives pressure directly from the discharge side of the pump while the larger side receives pressure that is regulated from the control mechanism. When the pressure from the control mechanism is equal to or greater than the discharge pressure, the relief valve is closed. When pressure increases on the discharge side of the pump, the piston is forced open and excess pressure is diverted to the intake side of the pump. When discharge pressure decreases, the relief valve is forced to close by the regulated pressure.

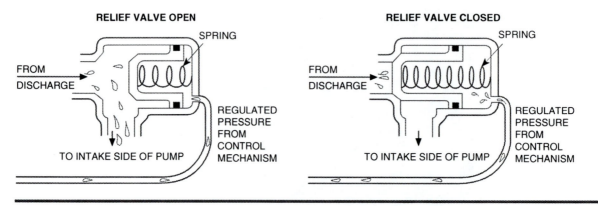

Figure 5–27   *Illustration of a pressure relief valve operation.*

Two operating characteristics affect the efficiency of pressure relief devices. First, because excess pressure is diverted to the intake side of the pump, high intake pressures can negate the operation of the relief valve. Increasingly, intake relief devices are also used to ensure that intake pressure is maintained at a safe level. Second, pressure relief devices decrease pressure by diverting excess water flow back to the intake. They are not designed to increase pressure. Consequently, when discharge pressure is reduced, the engine speed must be increased to maintain desired discharge pressures.

Governors To protect against excessive pressure buildup by controlling the speed of the pump engine to maintain a steady pump pressure, **pressure governors** are used. When discharge pressures increase, the governor reduces engine speed and the desired pressure is maintained. When pump discharge pressures decrease, the governor increases engine speed to maintain the desired pressure.

Older pressure governors utilize a mechanical linkage to control engine speed. When discharge pressures rise above the governor setting, a device connected to the engine accelerator by mechanical linkage reduces engine speed. When discharge pressures fall below the governor setting, the device increases engine speed. Two problems are associated with these older governors. First, if intake pressure is reduced or lost, the discharge pressure would decrease, causing the governor to attempt to increase pressure by increasing engine speed. Because the discharge pressure cannot attain the governor pressure setting, the governor continues to increase engine speed to its maximum rpm. Second, problems with the mechanical linkages could cause a slow response by the governor, resulting in a continuous increase and decrease of the engine speed by the governor as it attempts to reach its pressure setting.

Newer pressure governors utilize microprocessors to control engine speed. Information on intake and discharge pressures is sent to the microprocessor (see

pressure governor
a pressure regulating system that protects against excessive pressure buildup by controlling the speed of the pump engine to maintain a steady pump pressure

Figure 5–28
Electronic auto-matic pressure control governors that also monitor essential engine vital signs. Courtesy Class 1 and Fire Research Corp. respectively.

Figure 5–28). The processor analyzes the information and increases or decreases engine speed as needed. Utilizing intake and discharge pressure information guards against high engine speed as the result of lost intake pressure or cavitation. In addition, electronic devices tend to react quickly, reducing the increase and decrease of engine speed as the governor attempts to maintain the pressure setting.

Cooling Systems

■ **Note**
Two types of cooling systems can be found on pump panels: auxiliary cooling systems and radiator fill systems.

All combustion engines are equipped with a cooling system. This cooling system maintains the engine within designed operating temperatures under normal environmental and working conditions. Engines used to drive pumps are often subjected to intense working conditions as well as harsh environments. In addition, combustion engines are designed to maintain operating temperatures while in motion. Engines used to drive pumps work hard while in a stationary position. Because of this, additional cooling systems are required to maintain engine temperatures. Two types of cooling systems can be found on pump panels: auxiliary cooling systems and radiator fill systems.

auxiliary cooling system
a system used to maintain the engine temperature within operating limits during pumping operations

Auxiliary Cooling Systems Sometimes called heat exchange systems, **auxiliary cooling systems** are used to maintain engine temperatures within operating limits during pumping operations. Such a system consists of tubing running from the discharge side of the pump through a heat exchange system and back to the intake side of the pump. Water is pumped from the discharge side through the heat exchanger and back to the intake side. The heat exchange system allows the transfer of heat from the engine coolant to the water in the auxiliary cooling system. The engine coolant and the auxiliary cooling system water never actually mix. The heated water in the auxiliary cooling system is then pumped back to the intake side. This system helps keep the engine coolant system cool enough to maintain engine temperatures within acceptable operating limits. The auxiliary cooling system is controlled by a valve on the pump panel.

Radiator Fill Systems On older pumps, a radiator fill system can be found. This system helps keep the engine within acceptable temperature limits by pumping water from the discharge side of the pump directly into the engine cooling system. Although an effective means of preventing engine overheating, extreme care must be taken because water coming from the discharge can be under high pressure. In addition, this pressurized water is added to a system that is typically already under pressure. One way to guard against overpressurizing the system is to remove the radiator cap. However, this practice can also be extremely dangerous. Another reason why caution should be taken with this system is that it is capable of cooling the engine too fast. The end result might be a cracked head or block. Also, the gate controlling the water flow should be opened slowly. Adding fresh water into the radiator has the drawback of displacing the antifreeze/coolant in the system. This type of system should be used only as a last resort to keep the engine from overheating.

Foam Systems

The use of foam as an extinguishing agent has increased in popularity because it appears to increase the efficiency and safety of fire suppression efforts. Systems that produce foam (both Class A and B foams) are being added to pumps, with the controls typically located on the pump panel. Such systems include premixed systems, in-line eductor systems, around-the-pump proportioning systems, balanced pressure systems, and direct injection/compressed-air systems. The basic differences between the systems lie in how and where the foam and water come together.

Premixed System Premixed systems consist of a tank in which foam concentrate and water are added at appropriate proportions. Often, the onboard water tank is used, and the foam concentrate is simply added to the tank. When the tank-to-pump control valve is opened, the foam/water mixture enters the intake side and is pumped to the discharge lines.

> **Advantages:** no special components, such as metering valves or special gauges, are needed and the correct amount of foam can be added for effective foam/water streams.

Disadvantages: a limited foam supply is available (when the tank is empty, the foam supply is exhausted), and the potential harm caused by the foam to the tank, valves, and fittings.

In-Line Eductor Systems   In-line eductor systems utilize eductors to add foam to water in appropriate proportions. Water passes through the eductor, which causes foam to mix with the water. In addition to the eductor, a foam tank (or supply) is required. This system can be either external or internal to the apparatus. Internal systems have a foam concentrate valve and metering control installed on the pump panel (Figure 5–29). The eductors are typically installed in-line between

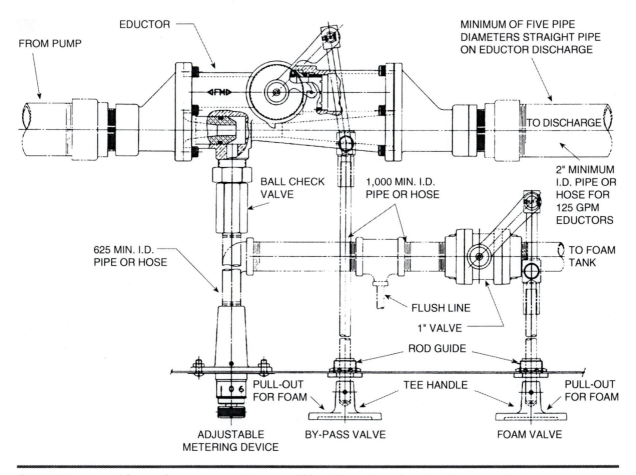

Figure 5–29   *Internal foam eductor system. Courtesy Elkhart Brass Manufacturing Company, Inc.*

the pump discharge and the discharge outlet on the pump panel. In addition, a foam tank is usually installed on the apparatus. External systems place the eductor, with built-in metering control, in-line between the discharge outlet on the pump panel and the nozzle. The foam is typically supplied in small containers.

Advantages: the ease of operation and that foam can be resupplied to this system for extended foam operations.

Disadvantages: the narrow operating pressure requirements of the eductor (because eductors require a specific pressure for maximum operating efficiency, hose line length and size can be limited), and extremely cold operating temperatures may limit the operating efficiency of eductors.

More information on eductor theory is presented in Chapter 6.

Around-the-Pump Proportioning Systems An around-the-pump proportioning system also uses an eductor to mix foam with water. In this system, the eductor is located between the discharge and intake sides of the pump. When the system is activated, water from the discharge side of the pump is allowed to flow through the eductor. After picking up foam, the solution then travels to the intake side of the pump and can be pumped through any and all discharge outlets attached to the pump. One limitation to this system is that water and foam cannot be pumped at the same time. In addition, because the foam/water mixture enters the intake side of the pump, much the same way as a pressure regulator relieves excess water, excessive intake pressures can hinder the system's efficiency.

Balanced Pressure Systems Balanced pressure systems mix foam with water by means of pressure. Two types of balance pressure systems are used on pumps. One system utilizes discharge pressure to force foam from a bladder contained in a vessel. As discharge water enters the vessel, the pressure buildup forces foam from the bladder. This type of a system is also known as a *pressure proportioning system.* The balancing part of the system comes into play when foam is forced from the bladder in proportion to the pressure under which water enters the vessel. The foam, under pressure, passes through a metering system and is then mixed with a water stream. One advantage of this system is that foam enters the system under pressure; therefore, this system is efficient over a variety of pressures and flows.

The second balanced pressure system consists of a separate foam pump and foam tank. Foam from the tank is pumped under pressure through a metering system and is mixed with a water stream. The two common types are by-pass and demand systems. In the former, foam pressure is controlled through a *by-pass system* that diverts excess foam concentrate back to the foam tank. With the *demand system* (Figure 5–30), the discharge pressure of the main pump is measured and used to regulate the pressure at which the foam pump will discharge foam.

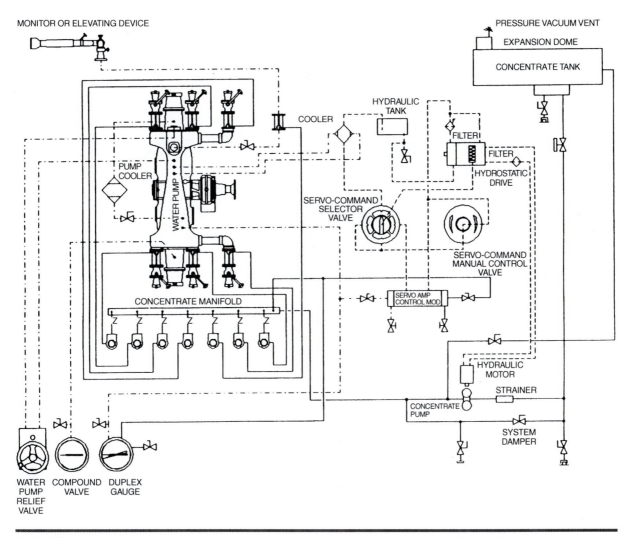

Figure 5–30   *Balanced pressure demand-type foam proportioning system. Courtesy National Foam.*

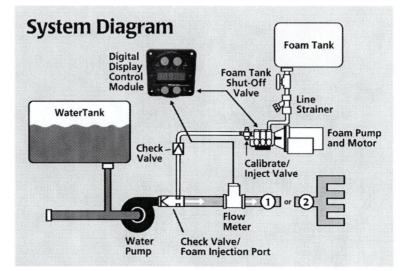

System Diagram

Figure 5–31 *Diagram of a direct injection foam system with digital control and display. Courtesy Hypro Corporation.*

Direct Injection/Compressed-Air Foam Systems Direct injection systems consist of a separate foam pump and foam tank. Foam from the tank is pumped directly into discharge lines. The rate at which foam is injected into discharge lines is controlled by a microprocessor that receives a signal from a discharge flow meter (see Figure 5–31). Compressed-air foam systems (CAFS) add one more step to the process. After the foam is injected into the discharge line, and prior to leaving the discharge outlet, compressed air is added to the foam/water (Figure 5–32). This process creates a unique lightweight foam of fluffy consistency.

MISCELLANEOUS

There are a number of miscellaneous components typically found on pump panels. Common to most all pump panels is the *pump engine hand throttle*. This hand throttle controls the speed of the pump engine and can set and hold the speed necessary for appropriate flows. Several *auxiliary outlets* can also be found on pump panels. One auxiliary outlet allows the engine speed to be measured independent of the tachometer mounted on the pump panel. Two other outlets commonly found on pump panels allow pressure test gauges to be connected, one to the intake and the other to the discharge (Figure 5–33). *Drain valves* are commonly found on pump panels (Figure 5–34). Drain valves, also called bleed-off valves, allow pressure to be relieved from hose connected to the pump or to drain the pump and its associated piping. Finally, water tank level indicators can be found on pump panels. Several examples of modern tank level indicators are provided in Figure 5–35.

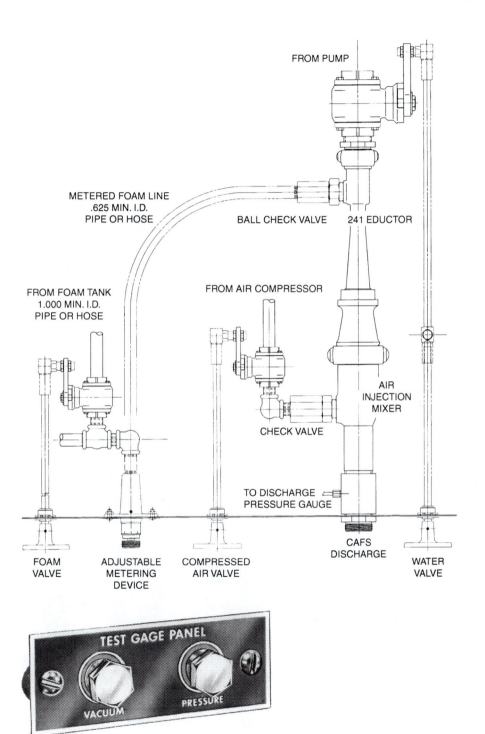

Figure 5–32 *Typical Compressed Air Foam System (CAFS). Courtesy Elkhart Brass Manufacturing Company, Inc.*

FROM PUMP

METERED FOAM LINE
.625 MIN. I.D.
PIPE OR HOSE

BALL CHECK VALVE   241 EDUCTOR

FROM FOAM TANK
1.000 MIN. I.D.
PIPE OR HOSE

FROM AIR COMPRESSOR

AIR
INJECTION
MIXER

CHECK VALVE

TO DISCHARGE
PRESSURE GAUGE

FOAM
VALVE

ADJUSTABLE
METERING
DEVICE

COMPRESSED
AIR VALVE

CAFS
DISCHARGE

WATER
VALVE

Figure 5–33 *Test gauge outlets. Courtesy Akron Brass Company.*

TEST GAGE PANEL

VACUUM   PRESSURE

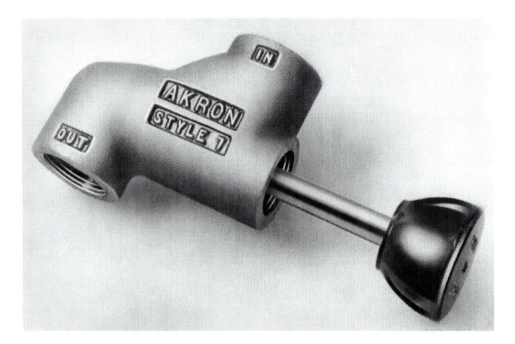

Figure 5–34 *Drain valve. Courtesy Akron Brass Company.*

Figure 5–35 *Examples of modern tank level indicators. Courtesy Class 1, Fire Research Corp., and Class 1 respectively.*

SUMMARY

The ability to move water from one location to another is a function of the pump and its attached peripheral components. The pump panel is the mission-control center of pump operations, allowing the pump operator to control and monitor pump efficiency from one location. The peripheral components on the pump panel help control the pump, ensure the efficiency of pump operations, and provide safety to personnel and equipment during pump operations.

Pump peripherals can be grouped into three categories: instrumentation, control valves, and systems. Instrumentation allows the pump operator to gather information on pressures, flows, and components. Control valves operate to initiate, restrict, and direct water flow through the pump and associated piping. Systems support the function of the pump and include priming, pressure regulating, cooling, and foam systems. Miscellaneous components typically found on pump panels include the pump engine hand throttle, auxiliary outlets, and drain valves.

REVIEW QUESTIONS

Key Terms and Concepts

On a separate sheet of paper, identify and/or define each of the following.

1. Pump peripherals
2. Pump panel
3. Instrumentation
4. Pressure gauges
5. Compound gauge
6. Bourdon tube gauge
7. Flow meter
8. Indicators
9. Control valves
10. Pressure regulating systems
11. Pressure relief device
12. Pressure governor
13. Auxiliary cooling system

Multiple Choice and True/False

Select the most appropriate answer.

1. The majority of pump peripherals are located
   a. in the cab of the apparatus.
   b. on the pump panel.
   c. at various locations on the apparatus.
   d. under the apparatus.

2. Which NFPA standard suggests a standard color code to match discharges with their gauges?
   a. 1500   c. 1911
   b. 1901   d. 1961

3. Instrumentation on pump panels can be grouped into three categories. Which of the following is not one of the categories?
   a. emergency warning lights
   b. pressure gauges
   c. flow meters
   d. indicators

4. The bourdon tube gauge is the most common pressure gauge found on pump panels.

   True or False?

5. Calibrating pressure gauges used on pump panels is unnecessary because of their durability and accuracy.

   True or False?

6. Most all pump panels have two main pump gauges, one for the intake and one for the discharge side of the pump.

   True or False?

7. Which of the following is not considered an indicator on the pump panel?

   a. tachometer

   b. water level instrument

   c. pump engagement light

   d. flow meter

8. Each of the following is an example of a valve except

   a. ball.   **c.** ram.

   b. butterfly.   **d.** gated.

9. According to NFPA 1901, intake relief valves are not required unless the main pump is rated at 2,000 gpm or higher.

   True or False?

10. The two pressure regulators commonly found on pumps are pressure relief devices and pressure governors.

    True or False?

Short Answer

On a separate sheet of paper, answer/explain the following questions.

1. Identify the pump peripherals identified in the figure below.

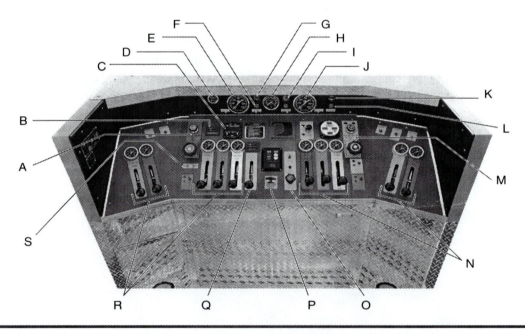

Figure 5–36 *Typical components found on a pump panel. Courtesy W. S. Darley & Co.*

2. Explain how pump panels are both similar and different.

3. List the different locations where pump panels can be found on apparatus and explain the advantages and disadvantages of each location.

4. How can the fluctuations or bounces of indicating needles be controlled?

5. What is the function of the two main gauges and individual discharge gauges located on pump panels?

6. What is the main advantage of using flow meters on pumps?

7. As a rule of thumb, a pump will have one 2½-inch discharge for each _____ gpm of rated capacity.

8. Explain the function for each of the following:

   Tank-to-pump

   Pump-to-tank (recirculating valve)

   Transfer valve

9. What is the purpose of oil in the priming system?

10. How can you tell when the priming process has actually primed the pump?

11. List the different types of foam systems and identify the advantages and disadvantages of each.

12. Explain the difference between pressure gauges and flow meters.

13. What are control valves? Provide several examples of common control valves found on pump panels.

14. Why are pressure regulators so important to pump operations?

15. Explain how air is replaced with water during the priming process.

ACTIVITIES

1. Draw a schematic, or take a picture, of two different pump panels. First, identify all the components found on the pump panels. Next, discuss the differences and similarities between the two pump panels.

2. After careful investigation and testing, your department has decided to take the plunge and begin using foam during suppression efforts. The next big question is, what systems should your department use to generate foam streams? Because of your excellent research talents, your chief has asked you to contact manufacturers and vendors to determine what types of foam-generating systems are available. Contact the manufacturers and vendors and prepare a report to your chief highlighting the different systems available. Be sure to indicate which system you think is the best for your department and why. Also, include the information you received during your research.

PRACTICE PROBLEM

1. You are the lead apparatus operator for your department. Because of your experience and knowledge of pump operations, your chief has asked you for your advice on pump panel location and configuration. Respond to your chief's request in the form of a memo. Be sure to provide justification for each of your recommendations.

BIBLIOGRAPHY

Pump manufacturers' literature is an excellent source for information on pumps and pump peripherals construction, operation, and maintenance.

Dickenson, T. C. *Pumping Manual.* Oxford, UK: Elsevier Advanced Technology, 1995.
An in-depth text on a wide range of pumps, construction, and operation.

Chapter 6

HOSE, APPLIANCES, AND NOZZLES

Learning Objectives

Upon completion of this chapter, you should be able to:

- Identify the NFPA standards that focus on hose, appliances, and nozzles.
- Explain the role hose plays in pump operations.
- List the basic parts of a hose and explain the three main types of hose construction.
- Discuss the two broad categories of hose and identify several tools used when working with hose.
- List and discuss several appliances used during pump operations.
- Explain the role nozzles play in pump operations.
- Explain the relationship of flow, pressure, shape, and nozzle reaction in the design and operation of a nozzle.

NFPA 1002
Standard for Fire Apparatus Driver/Operator Professional Qualifications
(2003 Edition)

This chapter addresses parts of the following knowledge elements within sections 5.2.1, 5.2.2, and 5.2.4:

Operate at a fire hydrant and at a static water source

Assemble hose lines, nozzles, valves, and appliances

INTRODUCTION

Why study such items as hose, appliances, and nozzles in a textbook on pump operations? First, keep in mind that pump operations is the process of moving water from a supply source through a pump to a discharge point. Getting water from a supply source to the pump and from the pump to a discharge point is the job of hose, appliances, and nozzles. Second, misuse or inappropriate use of hose, appliances, and nozzles can reduce the effectiveness of a pump. Too small a hose on the intake side of the pump may not be able to supply the required flows on the discharge side. Finally, and perhaps most important, lack of knowledge can create hazardous situations for both personnel and equipment. Providing too much pressure can rupture hose or overpower personnel operating nozzles. Understanding the operating principles and construction of hose, appliances, and nozzles will help ensure that pump operations are conducted in a safe, efficient, and effective manner while providing appropriate flows and pressures.

> **Safety**
> Lack of knowledge can create hazardous situations for both personnel and equipment.

Pump operators should be familiar with the NFPA standards that focus on hose, appliances, and nozzles. The following standards are referenced in this chapter:

- NFPA 1961 *Standard on Fire Hose.* This standard establishes requirements for the design, construction, inspection, and testing for new fire hose.

- NFPA 1962 *Standard for the Inspection, Care, and Use of Fire Hose, Couplings, and Nozzles and the Service Testing of Hose.* This standard applies to all fire hose, couplings, and nozzles currently in operation or that will be placed in operation.

- NFPA 1963 *Standard for Fire Hose Connections.* This standard defines the minimum requirements for new fire hose connections.

- NFPA 1964 *Standard for Spray Nozzles.* This standard provides the minimum performance requirements for new adjustable-pattern spray nozzles for the following firefighting applications: general use, marine and offshore platform use, standpipe system use.

- NFPA 1965 *Standard for Fire Hose Appliances.* This standard provides the minimum performance, operation, testing requirements for appliances with up to 6-inch connections.

HOSE

Hose is an essential, yet often misused and abused, component in fire pump operations. The ability to move water from a supply source to the pump as well as from the pump to a discharge point would be impossible without hose. NFPA 1961 defines fire hose as a "flexible conduit used to convey water." Because of its importance in transporting water, pump operators should be familiar with the different classifications of hose, hose tools, hose construction, and hose care principles.

Classification of Hose

The two broad classifications of hose are intake or supply hose and discharge or attack hose. Intake hose moves water from a supply source to the pump, while discharge hose moves water from a pump to the discharge point. In some cases, the same type of hose can be used for either intake or discharge water movement. NFPA 1961 identifies specific requirements for supply and suction hose (intake hose) and attack hose (discharge hose). This standard also identifies **large diameter hose** (LDH) as any hose with a diameter of $3\frac{1}{2}$ inches (90 mm) or larger.

large diameter hose
hose with a diameter of $3\frac{1}{2}$ inches or larger

Intake Hose Intake hose, often referred to as *supply* lines, moves water from a source to the pump. Typically, this means getting as much water to the pump as possible. Depending on the water supply and the discharge flow requirements, several types of hose can be used. **Supply hose,** according to NFPA 1961, is designed to move water from a pressurized water source such as a hydrant to a pump. Supply hose have a minimum trade size diameter of $3\frac{1}{2}$ inches (90 mm), which means they are also classified as LDH. NFPA 1962 requires that supply hose not be operated at pressures exceeding 185 psi. Supply hose can transfer water over longer distances with minimum loss of pressure due to friction because of the large diameter size of the hose. Two other types of intake hose are used to transfer water over shorter distances. **Soft sleeve** is typically a shorter section of supply hose with female couplings on both ends used when the pump is close to a pressurized water supply such as a hydrant. Soft sleeve hose ranges in size from $2\frac{1}{2}$ inches to 6 inches in diameter. **Suction hose,** also referred to as hard suction hose, is defined by NFPA 1961 as a "hose designed to prevent collapse under vacuum conditions so that it can be used for drafting water from below the pump (lakes, rivers, wells, etc.)." The size of hard suction hose ranges from $2\frac{1}{2}$ inches to 6 inches in diameter and is typically 10 feet in length. Note the confusing term "hard suction" given to this hose. Recall that drafting operations raise water to the pump by atmospheric pressure rather than by a suction process.

supply hose
used with pressurized water sources and operated at a maximum pressure of 185 psi

soft sleeve
shorter section of hose used when the pump is close to a pressurized water source such as a hydrant

suction hose
special noncollapsible hose used for drafting operations

Discharge Hose Discharge hose, referred to as attack hose in NFPA 1961 and NFPA 1962, is used to move water from the pump to a discharge point. The discharge point could be a nozzle used for suppression efforts or for covering exposures. In addition, the discharge point could be a different pump that uses the discharge hose as a supply line. Often, hoses used to supply a nozzle are called *attack lines* while those used to supply other pumps or devices are called *supply lines*. NFPA 1961 defined **attack hose** as "hose designed to . . . combat fires beyond the incipient stage." Attack hose can be either woven-jacket or rubber-covered, range in size from 1 inch to 3 inches, and have a normal highest operating pressure of 275 psi. The typical sizes used for attack hose include $1\frac{1}{2}$ inches, $1\frac{3}{4}$ inches, 2 inches, $2\frac{1}{2}$ inches, and 3 inches.

attack hose
$1\frac{1}{2}$" to 3" hose used to combat fires beyond the incipient stage

Construction

The basic parts of a hose include a reinforced inner liner and an outer protective shell with couplings attached at both ends. The inner liner, sometimes referred to as the inner tube, keeps the water contained in the hose. According to NFPA 1961 liners are made from one of the following materials:

■ **Note**

The basic parts of a hose include a reinforced inner liner and an outer protective shell with couplings attached at both ends.

- Rubber compound
- Thermoplastic material
- Blends of rubber compounds and thermoplastic material
- Natural rubber-latex-coated fabric

An extrusion method, a process of forcing heated rubber or plastic through a die to create a uniform component, is typically used to create a continuous and seamless liner. In addition to keeping the hose leakproof, the liner provides a smooth surface to reduce the friction of water as it passes over the surface. Hose liners by themselves cannot withstand the pressures often required during pump operations. Therefore, liners are reinforced with natural fiber, synthetic fiber, or a combination of the two. The purpose of the outer shell is to protect the reinforced inner liner from the many hazards that may damage or destroy the integrity of the hose, such as abrasion, cuts, chemicals, and heat. The outer shell or cover can be made of the same selection of material as is available to the liner.

■ **Note**

The two main types of hose construction are woven jacket and rubber covered. The two main types of couplings are threaded and sexless.

Although NFPA 1961 no longer specifies the required length of hose, a section of hose is commonly 50 feet, some times 100 feet, in length with couplings attached to both ends. Hose and coupling diameters range from 1 inch to 6 inches. The two main types of hose construction are woven jacket and rubber covered. The two main types of couplings are threaded and sexless.

Woven-Jacket Hose Woven-jacket hose has been the hose of choice in the fire service for many years. The inner liner is usually made from a synthetic rubber compound or special thermoplastic. The reinforcement material is typically a synthetic fiber weave called a *jacket*. Synthetic fiber has, for the most part, replaced cotton in the construction of woven-jacket hose. The inner liner and reinforcement jacket are bonded together by a number of different processes. A liner with a reinforced woven jacket is commonly referred to as a *single-jacketed lined hose*. Typically,

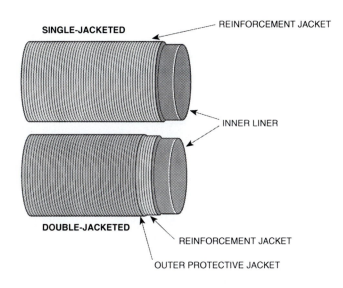

Figure 6–1
Illustration of a single-jacketed and a double-jacketed hose.

a second jacket, or outer shell, is used to provide added strength and resistance to the harsh working environments of hose. A woven hose with this second jacket is called a *double-jacketed lined hose* and is the most durable and popular of the two. Figure 6–1 shows both a single- and double-jacketed hose. To increase the durability of the hose, a variety of treatments are used on the outer shell to increase abrasion, heat, and chemical resistance.

Rubber-Covered Hose   Today, rubber-covered hose is popular in the fire service because it is lightweight and durable. As the term implies, the hose has a rubber outer cover (Figure 6–2). Several methods are used to construct rubber-covered hose. In one method, the liner is attached to a reinforcing synthetic fiber jacket as is the case with woven-jacket hose. However, instead of using a second woven jacket as the outer shell, a rubber outer cover is extruded over the reinforced liner. In another method, the rubber liner and outer shell are extruded over the reinforced woven fiber in one process. Both processes produce a unified hose appearance.

Couplings   NFPA 1963 defines couplings as a "set or pair of connection devices attached to a fire hose that allow the hose to be interconnected to additional lengths of hose or adapters and other firefighting appliances." The two most

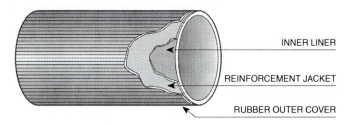

Figure 6–2 *Cutaway illustration of a rubber-covered hose.*

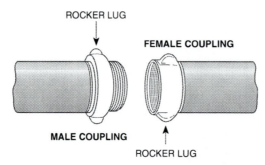

ROCKER LUG

FEMALE COUPLING

MALE COUPLING

ROCKER LUG

Figure 6–3 *Example of a male and female threaded coupling.*

popular materials used to make couplings are brass and lightweight aluminum alloys. Couplings used on hose, appliances, and nozzles are typically either threaded or sexless. Couplings must be easy to connect yet able to withstand high pressures and the demands of fireground operations. NFPA 1963 requires couplings pass a variety of rigorous tests to include being dropped from 6 feet without damage and being able to withstand the service test pressure of the hose without leakage or failure after being connected 3,000 times.

Threaded couplings utilize two different couplings, both having a common thread. Of the two different couplings, one is called a male and the other a female (Figure 6–3). The thread used in nearly all threaded coupling construction is referred to in NFPA 1963 as "American National Fire Hose Connection Screw Threads" and is abbreviated with the thread symbol **NH**. Sometimes this thread type is referred to as national standard thread (NST). However, some municipalities still utilize nonstandard threads. In such cases, adapters should be readily available to connect NH couplings and appliances. Threaded couplings have lugs or handles to assist with connecting and disconnecting the couplings. The most common lug on threaded couplings is the rocker lug (see Figure 6–3).

Sexless couplings, sometimes referred to as nonthreaded couplings, are so named because the two connecting couplings are identical to each other. Storz couplings (Figure 6–4) are the most common sexless couplings used in the fire service. Storz couplings are connected by joining the two identical couplings and twisting (one-third turn) to lock them.

Hose Tools

Pump operators have a variety of hose tools from which to choose when working with hose. **Spanner wrenches** come in many sizes and shapes to help connect and disconnect couplings of various types, sizes, and lug construction (Figure 6–5). **Hose clamps** are critical for pump operators in that they control the flow of water in the supply hose. The three main types of hose clamps are: press, screw, and hydraulic. Figure 6–6 shows a press hose clamp in action. Care should be exercised when using clamps as they can cause injury and/or damage to hose and personnel. **Hose bridges** (Figure 6–7) are used to allow vehicles to move across hose

NH
a common thread used in the fire service to attach hose couplings and appliances

spanner wrench
tool used to connect and disconnect hose and appliance couplings

hose clamp
device used to control the flow of water in a hose

hose bridge
device used to allow vehicles to move across a hose without damaging the hose

■ **Note**
The three main types of hose clamps are press, screw, and hydraulic.

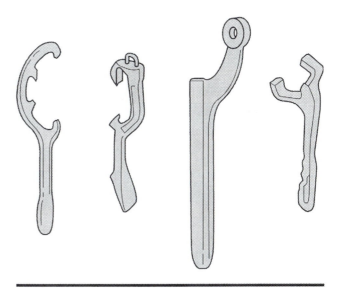

Figure 6–4 *Storz couplings are sexless in that the couplings are identical, that is, there are no male and female couplings.*

Figure 6–5 *Spanner wrenches are available in a variety of styles to fit the different types of hose couplings.*

Figure 6–6 *Example of a press hose clamp. Courtesy Greg Burrows.*

Figure 6–7 *Hose bridges help protect hose from damage when driven over. Courtesy Ziamatic Corp.*

hose jacket
device used to temporarily minimize flow loss from a leaking hose or coupling

without causing damage to the hose. **Hose jackets** provide a quick method of minimizing flow loss from a leaking hose (Figure 6–8).

There are many more components used in conjunction with hose. Some are discussed in the following section, "Appliances." Others are discussed later in the text when presenting specific pump operations.

Hose Care

Because of the important role hose plays in pump operations, appropriate attention should be given to its care. Care of hose begins with preventing it from getting damaged. This includes proper use as well as the recognition of situations that may help prevent damage to hose such as

- Minimizing chafing
- Reloading so that hose folds are in different locations each time
- Keeping vehicles from running over hose
- Minimizing heat and cold exposure
- Allowing air circulation in hose loads to reduce mildew growth and rust/corrosion

Care of hose continues with three important activities including inspecting, cleaning, and record keeping.

Inspecting All hose should be inspected prior to being placed in service and after each use. According to NFPA 1962, hose should be inspected for the following when placed in service:

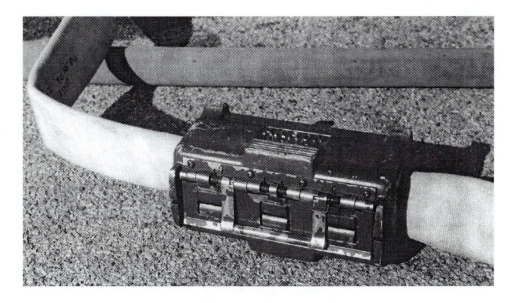

Figure 6–8 *Hose jackets can be used to temporarily stop flow from a leaking hose or to temporarily connect mismatched or damaged couplings. Courtesy Greg Burrows.*

- Not vandalized and free of debris
- No evidence of mildew or rot
- No damage by chemicals, burns, cuts, abrasions, or vermin
- Service test of hose is current
- Liner shows no sign of delamination

In addition, hose should be tested annually and after any repairs have been made. The inspection of in-service hose after it is used should include the following components, as a minimum:

- Hose outer cover (inspecting for its condition, discoloration, abrasions, cuts, fraying)
- Coupling (looking for damage, proper operation, condition of threads, slippage from hose)
- Gasket (checking its presence and condition)
- Liner (no sign of delaminating)
- Hydrostatic test date (ensuring hose is within annual test date)

Should any section of hose fail this inspection it should be removed from service and either repaired or condemned.

■ Note
Hose should be tested annually and after any repairs have been made.

Cleaning Hose should be properly cleaned after each use. For woven-jacket hose this means a good cleaning and thorough drying. The main reason for cleaning and drying is that dirt, grime, and chemicals can work their way into the woven jacket, attacking and deteriorating the fibers. For rubber-covered hose, this means a good wipe down depending on the amount of dirt and grime on the hose. The cleaning process can be manually accomplished by laying hose section out and then scrubbing with a mild soap or detergent. Some departments purchase or build a mechanical hose washer. In most cases, the inspection of hose can be accomplished during this cleaning process.

Record Keeping According to NFPA 1962, fire departments must establish and maintain an accurate record of each hose section. This requires that each hose section be assigned an identification number. The identification number must be stenciled on the jacket/cover or stamped on the boal or swivel of the coupling. In addition, NFPA 1961 requires that each hose section be indelibly marked on both ends using 1-inch letters with the following information:

- Manufacturer's identification
- Month and year of manufacture
- The statement "Service Test to ____ per NFPA 1962"

LDH must use 2-inch letters and be marked as either "supply hose" or "attack hose."

In the past, most hose record-keeping systems were paper based. Today, many departments develop or purchase electronic hose record-keeping systems. Typical information maintained includes the following:

- Identification number
- Manufacturer and vendor
- Hose size and length
- Type and construction of hose
- Date placed in service
- Hose location (station and/or apparatus)
- Date of each service test
- History of damage and repairs

The collection and analysis of hose information/data is important for the safe use of the hose. Analysis of hose data can help determine the cost-effectiveness of hose and how well or poor a hose performs over time. For example, a trend in high repair rates for a hose may either be due to poor hose construction or improper use. In either case, corrective action can be taken to ensure the safe use of hose.

Hose Testing

Most new attack hose have a service test of 250 psi while supply hose have a service test pressure of 200 psi. Hose that comply with NFPA 1962 will have the service test pressure stenciled on each section of hose. The testing of fire hose requires an appropriate location with a water source. When possible, the location should be level with a smooth surface and adequate space to lay out the hose being tested. The water supply can be a hydrant, tank, or static source, such as a pond or lake. The quantity of water sufficient to fill hose being tested is the primary consideration for a water supply. In general, the fire hose is layed out, filled with water, and the pump discharge pressure increased to the service test pressure. NFPA 1962 provides excellent guidance on conducting fire hose service tests and includes important safety considerations and sample forms to document the test. Several considerations for conducting annual service tests of fire hose include the following.

Preparation (prior to starting the test) Identify the correct test pressure for the hose. More than one hose size and section can be tested at one time, but all hose should have the same test pressure. Conduct an inspection of the hose making sure to check gaskets, threads, and obvious damage. A maximum of 300 foot hose sections should be tested. Ensure testing instrumentation calibration is within 12 months.

Conducting the Test The basic steps for conducting hose tests are as follows:

Step 1: connect water supply and test hose sections.
Step 2: open the discharge control valves to which test hose sections are connected.
Step 3: slowly increase the discharge pressure to 45 psi.
Step 4: remove all air from within the hose, check for leaks, and mark hose near coupling to determine coupling slippage.
Step 5: increase pressure slowly; NFPA 1962 suggests a rate no faster than 15 psi per second, to the service test pressure.
Step 6: conduct the test for three minutes periodically checking for leaks, keep all non-essential personnel away from testing area.
Step 7: Record the results of the test per fire department requirements.

Post Test Any hose that fails the test will be removed from service and repaired or condemned. Repaired hose must pass a retest prior to being placed back in service. All hose passing the pressure test should be cleaned and dried as per normal operating procedures.

APPLIANCES

appliances
accessories and components used to support varying hose configurations

■ **Note**
The most common appliances used in pump operations include wyes and siamese, adapters, and several specialized devices.

During pump operations, a variety of hose configurations are used to move water from a supply to a discharge point. The accessories and components used to support these varying hose configurations are called **appliances**. Appliances can be used on either the intake or discharge side of hose configurations. In most cases, friction loss occurs when water flows through an appliance. Pump operators must include appliance friction loss factors when performing hydraulic calculations. The most common appliances used in pump operations include wyes and siamese, adapters, and several specialized devices such as manifolds, water thieves, and eductors. NFPA 1965 establishes requirements for appliances up to 6 inches (150 mm) in diameter. According to NFPA 1965, appliances must have a maximum operating pressure of 200 psi (13.8 bar) or greater, and the maximum operating pressure along with manufacture name and product/model identification must be permanently marked on the appliance.

Wyes and Siamese

wye
appliance used to
divide one hose line
into two or more lines

A **wye** (Figure 6–9) is used to divide one hose line into two or more lines. The inlet side of the wye has a female coupling and the discharge outlets are male couplings. A plain wye cannot control flow in the divided lines as it has no control or clapper valves. However, the majority of wyes have control valves, typically a ball valve, and are called *gated wyes.* According to NFPA 1965, appliances with lever-operated handles must indicate a closed position when the handle is perpendicular to the hose line. A *reducing wye* has a larger coupling on the intake side and smaller couplings on the discharge outlets, for example, a 2½-inch inlet and several 1½-inch outlets. Wyes come in a variety of sizes and configurations (Figure 6–10).

Figure 6–9   *This plain wye has one 2½-inch female NH inlet and two 2½-inch male NH outlets. Courtesy Elkhart Brass Manufacturing Company, Inc.*

Figure 6–10   *A variety of wye sizes and configurations are used in the fire service. Courtesy Elkhart Brass Manufacturing Company, Inc.*

siamese

appliance used to combine two or more lines into a single line

A **siamese** is used to combine two or more lines into a single line. The inlet sides of a siamese have female couplings while the discharge outlet has a male coupling. As with wyes, plain siamese have no control or clapper valves (Figure 6–11). Some siamese have clapper valves. Clapper valves on a siamese prevent higher pressure from one of the intake hose lines from entering other connected intake hose of lower pressure. The majority of siamese will have either the same size couplings on both the inlets and the outlet, or a larger outlet than the inlets. Like wyes, siamese come in a variety of sizes and configurations (Figure 6–12).

Figure 6–11 *This plain siamese has two 2½-inch female NH inlets and one 2½-inch male NH outlet. Courtesy Elkhart Brass Manufacturing Company, Inc.*

Figure 6–12 *A variety of siamese sizes and configurations are used in the fire service. Courtesy Elkhart Brass Manufacturing Company, Inc.*

Adapters

When laying out hose configurations for pump operations, mismatched hose ($2\frac{1}{2}$-inch to a $1\frac{1}{2}$-inch) and couplings (two female couplings) may occur. To connect these mismatched lines a variety of **adapters** are available. Double male and female adapters (Figure 6–13) are used to connect threaded couplings of the same size and sex. Increasing and decreasing adapters allow couplings of different size to be connected. Figure 6–14 shows two common decreasers, $2\frac{1}{2}$ inches (female) to $1\frac{1}{2}$ inches (male). Increasers can be made by adding an adapter to a decreaser. For example, an increaser can be made by adding a double $1\frac{1}{2}$-inch female to the $1\frac{1}{2}$-inch male side of the decreaser and a double $2\frac{1}{2}$-inch male added to the female side of the decreaser.

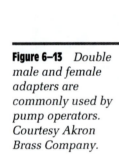

adapter
appliance used to connect mismatched couplings

Figure 6–13 *Double male and female adapters are commonly used by pump operators. Courtesy Akron Brass Company.*

Figure 6–14 *Two examples of decreasers. Courtesy Akron Brass Company.*

Specialized Devices

A number of specialized devices are used to support and enhance pumping operations. **Four-way hydrant valves** are used to increase hydrant pressure without interrupting the flow (Figure 6–15). **Water thieves** are similar to gated wyes in that they are used to connect additional smaller lines from an existing larger line (Figure 6–16). **Manifolds** are larger devices that provide the ability to connect numerous smaller lines from a large supply line (Figure 6–17). Pressure relief devices and gauges are often mounted on manifolds. Manifolds provide a greater degree of flexibility when used for both forward and reverse hose lays. Sharp curves and twisting in the hose lines connected to manifolds should be avoided when possible especially with longer hose lays. Excessive curves and twisting can cause the manifold to move and even flip. Use of manifolds and hose lays are discussed further in Chapter 7. Care must be exercised when using manifolds.

Eductors are specialized devices used in foam operations. The basic components of an eductor include a metering valve and pickup hose (Figure 6–18). The *metering valve* controls the percentage of foam drawn into the eductor. The *pickup hose* is a noncollapsible tube used to move the foam to the eductor. The eductor utilizes the **venturi principle** to draw foam into the water stream (Figure 6–19). Water pressure increases as the eductor tube decreases in size. The pressure is converted to velocity as it enters the induction chamber and then travels out the eductor. This process creates a low-pressure area in the induction chamber that allows foam to be drawn into the eductor and mixed with the water stream.

four-way hydrant valve
appliance used to increase hydrant pressure without interrupting the flow

water thieves
similar to gated wyes, water thieves are used to connect additional smaller lines from an existing larger line

manifolds
devices that provide the ability to connect numerous smaller lines from a large supply line

eductor
a specialized device used in foam operations that utilizes the venturi principle to draw the foam into the water stream

venturi principle
process that creates a low-pressure area in the induction chamber of an eductor to allow foam to be drawn into and mixed with the water stream

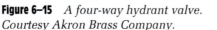

Figure 6–15 *A four-way hydrant valve. Courtesy Akron Brass Company.*

Figure 6–16 *A water thief. Courtesy Akron Brass Company.*

Figure 6–17 *A distribution manifold. Courtesy Angus Fire.*

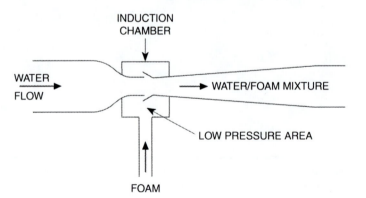

Figure 6–18
Eductors are used to pick up and mix foam during foam operations. Courtesy Scott Sheets.

Figure 6–19
Eductors use the venturi principle to draw foam into water streams.

NOZZLES

Nozzles are an important component in fire suppression efforts. Without nozzles, fire suppression activities would be minimal at best. In addition, nozzles that are supplied with inefficient flows and pressures can reduce extinguishment effectiveness as well as increase hazards to personnel. You might argue, then, that the primary purpose of pump operations is to supply nozzles with appropriate flows and pressures so that efficient and effective suppression efforts can be made. Therefore, pump operators should have a good working knowledge of the operating principles and classification of nozzles.

Operating Principles

nozzle pressure
the designed operating pressure for a particular nozzle

All nozzles are designed to provide a given flow, or range of flows in the case of automatic nozzles, at a specific pressure. Recall that flow is the quantity of water delivered at a specific rate, typically expressed in gallons per minute (gpm). The specific pressure, called **nozzle pressure**, is the designed operating pressure for a particular nozzle. Most nozzles are designed for an operating nozzle pressure of either 50, 75, 80, or 100 psi. When proper flow and pressure are provided, the three nozzle functions of controlling flow, reach, and stream shape will be the most effective for suppression efforts. In addition, the nozzle reaction will most likely be within acceptable limits for those operating the nozzle. Conversely, if the specific flow and pressure are not provided to the nozzle, the nozzle will not operate as designed, resulting in poor stream quality and reduced suppression capability, as well as increased risks to those operating the nozzle.

■ Note
When proper flow and pressure are provided, the three nozzle functions of controlling flow, reach, and stream shape will be the most effective for suppression efforts.

Nozzle Flow The amount of water flowing from a nozzle (**nozzle flow**) is a function of the discharge orifice of the nozzle. The larger the orifice, the more water the nozzle is capable of flowing. Some nozzles have a fixed orifice size in which a specific flow is provided. Others have the ability to select from several orifice sizes to provide the ability to vary the flow. Finally, some nozzles can automatically change the orifice size while maintaining proper nozzle pressure.

nozzle flow
the amount of water flowing from a nozzle; also used to indicate the rated flow or flows of a nozzle

Reach *Webster's Ninth New Collegiate Dictionary* defines a nozzle as a "constriction used . . . to speed up or direct a flow of fluid." Note the terms *constriction* and *speed*. Restrictions in a hose will cause pressure to increase. One inherent design feature of a nozzle is to restrict the flow of water which, in turn, increases pressure (Figure 6–20). This increased pressure is converted to velocity, or speed, as water leaves the nozzle. The end result is an increase in the distance water travels when it leaves the nozzle. The distance water travels after leaving a nozzle is called **nozzle reach**.

nozzle reach
the distance water travels after leaving a nozzle

The concept of restriction and reach can be illustrated using a garden hose. The distance water will travel when flowing from the open end of a garden

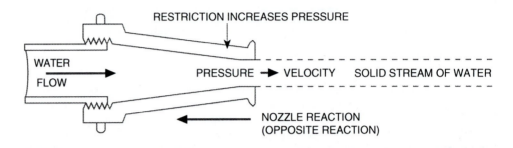

Figure 6–20   *Illustration of a smooth-bore nozzle. The nozzle acts to restrict water flow, which increases pressure. Pressure is converted to velocity as it leaves the nozzle. Nozzle reaction is the tendency of the nozzle to move in the opposite direction of water flow.*

hose without a nozzle is minimal. However, when a thumb is placed over the opening, causing a restriction, the distance water will travel is significantly increased. Nozzles used for fire suppression utilize this principle to provide maximum usable reach of the nozzle at a specific pressure. Gravity, wind, and the stream's shape all affect the distance water travels.

stream shape
the configuration of water droplets (shape of the stream) after leaving a nozzle

Stream Shape   The shape of a stream of water (**stream shape**) after it leaves a nozzle is determined by the type and function of the nozzle. Nozzles produce a variety of stream shapes ranging from a compact water stream, referred to as a solid or straight stream, to a disperse stream consisting of small water drops, referred to as a fog stream. A straight stream is a compact column of water used to attain greater reach. A fog stream is a wide pattern of water used to provide a large quantity of water droplets over a wide area immediately in front of the nozzle. The most effective shape the nozzle is capable of producing will be attained when the correct flow and pressure are provided.

nozzle reaction
the tendency of a nozzle to move in the direction opposite of water flow

Nozzle Reaction   In addition to providing the correct flow and pressure required by a nozzle, pump operators must also consider **nozzle reaction**. Newton's third law of motion states that "for every action there is an equal and opposite reaction." Relating this law of motion to that of a modern fire suppression nozzle flowing water, the first action would be water leaving the nozzle in a forward direction while the second action would be the tendency of the nozzle to move in the reverse direction (see Figure 6–20). This tendency to move in the opposite direction of water flow is called nozzle reaction.

The person operating the nozzle constantly battles the nozzle's opposite reaction to the water being discharged. Note that this opposite reaction is equal to the first action. Consequently, when an increase in flow or nozzle pressure is experienced at the nozzle, a corresponding increased nozzle reaction will occur. The concern to pump operators, and especially those operating the nozzle, is to provide the required pressures and flows, which in turn, help maintain manageable nozzle reaction.

Classification

Although a wide variety of nozzles exists for fire suppression activities, the vast majority of nozzles can be grouped into either smooth-bore or combination nozzles.

smooth-bore nozzle
nozzles designed to produce a compact solid stream of water with extended reach

Smooth-Bore Nozzles Also referred to as straight or solid-stream nozzles, **smooth-bore nozzles** are the least complex of nozzles. The nozzle illustrated in Figure 6–20 is an example of a smooth-bore nozzle. These nozzles are designed to produce a compact, solid stream of water with extended reach. The nozzle diameter determines the quantity of water the nozzle is designed to flow. Smooth-bore nozzle diameters, also called *tips*, range from ¼ inch to 2½ inches. Some smooth-bore nozzles have stacked or interchangeable tips that provide the ability to change the quantity of water depending on flow requirements (Figure 6–21). Smooth-bore nozzles are used on hand lines and master stream applications. When used as hand lines, tip sizes up to 1⅛ inch are typically used with a nozzle pressure of 50 psi. When used as master stream devices, tip sizes of 1¼ inches or larger are used with a nozzle pressure of 80 psi.

combination nozzle
a nozzle designed to provide both a straight stream and a wide fog pattern; most widely type used in the fire service

Combination Nozzles More complex than smooth-bore nozzles in both design and operation, **combination nozzles** (Figure 6–22) are designed to provide both a straight stream and a wide fog pattern. They are the type of nozzle most widely used in the fire service. The vast majority of combination nozzles are designed to provide their rated flow at 100 psi nozzle pressure. However, some combination nozzles, called *low pressure nozzles*, provide their rated flow at less than 100 psi, typically 75 psi, nozzle pressure.

Figure 6–21 *Smooth-bore nozzle with stacked tips. Courtesy Greg Burrows.*

Figure 6–22 *Example of a combination nozzle. Courtesy Greg Burrows.*

Combination nozzles can be divided into three groups: fixed-flow, selectable flow, and automatic flow nozzles. *Fixed-flow nozzles*, also referred to as constant flow or fixed-gallonage nozzles, are designed to provide a specific flow at 100 psi nozzle pressure. These nozzles deliver the same flow or gpm regardless of the stream pattern. For example, a nozzle designed to deliver 125 gpm will continue to deliver 125 gpm when flowing a fog stream or a straight stream as long as the nozzle pressure remains at 100 psi. Some special application fixed-flow nozzles are designed to deliver their rated flow at 75 psi. It is important to note that the pump operator must maintain a 100-psi nozzle pressure to obtain maximum flow rates and stream patterns. Nozzle pressures less than 100 psi will result in less flow and reach as well as poor pattern development. Nozzle pressures higher than 100 psi will result in poor pattern development and excessive nozzle reaction. Attempting to control excessive nozzle pressure at the nozzle, by partially closing the nozzle, will result in decreased flow as well as in effective patterns.

Selectable flow nozzles, also referred to as manual adjustable nozzles or selectable gallonage nozzles, provide the ability to change the flow at the nozzle. These nozzles provide some flexibility on the fireground by allowing the flow to be increased or decreased as needed. Typically, a series of flow settings are available on these nozzles that increase or decrease the nozzle discharge opening to deliver the selected flow at 100 psi nozzle pressure (see Figure 6–23); however, when changes are made at the nozzle, adjustments must also be made by the pump operator to maintain the designed operating nozzle pressure of 100 psi. If changes are made at the nozzle without the pump operator's knowledge, the nozzle will not flow the selected rate.

Automatic nozzles, also referred to as constant-pressure nozzles, are designed to maintain a constant nozzle pressure over a wide range of flows. These nozzles have a built-in sensing device that automatically increases or decreases the discharge opening (see Figure 6–24). As pressure increases, the baffle moves forward, increasing the nozzle discharge orifice. When pressure decreases, the spring pulls the baffle back, decreasing the nozzle discharge orifice. The result is an increase or decrease in flow while maintaining 100 psi nozzle pressure (see Figure 6–25). In addition to maintaining a constant nozzle pressure, automatic nozzles also maintain effective flow patterns over nozzle flow range. Because of this, patterns tend to "look good" at both the low and high end of the nozzle flow range. The pump operator must ensure that "acceptable" flows are provided when using automatic

Figure 6–23

Example of a selectable flow nozzle.

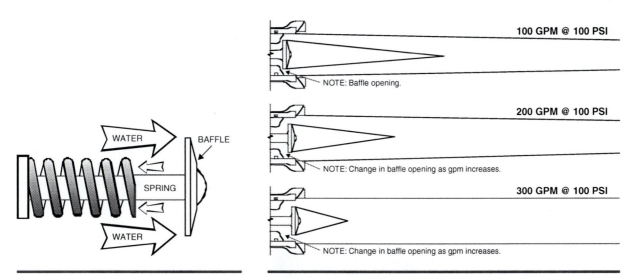

Figure 6–24 *Automatic nozzle sensing device that increases and decreases the nozzles discharge opening. Courtesy Task Force Tips.*

Figure 6–25 *Automatic nozzles maintain 100 psi nozzle pressure while flows increase or decrease. Courtesy Task Force Tips.*

nozzles to ensure firefighter safety during fire suppression activities. A newer feature on some automatic nozzles is the ability to switch between two operating pressures. These dual pressure nozzles can be switched between 100 psi and approximately 60 psi nozzle pressure, allowing for great flows.

Nozzle Maintenance

According to NFPA 1962, nozzles should be inspected after each use and annually. At a minimum, the inspection should include the following:

- Water way is clear of obstructions
- Tip is not damaged
- Controls and adjustments operate properly
- No missing or broken parts
- Gasket is in good condition

According to NFPA 1964, nozzles must be permanently marked with the following information:

- Manufacturer name
- Product or model
- Rated nozzle pressure and flow(s)
- Minimum and maximum discharge (automatic nozzles)
- FLUSH operating position if so equipped
- Straight stream and fog pattern indications

SUMMARY

Hose, appliances, and nozzles are used by pump operators to accomplish the task of moving water from a supply to a discharge point. Hose provides the ability to move water over a distance and is used for different tasks depending on its size and construction. Nozzles provide flow, reach, and shape to extinguish fires. For optimal and safe performance, nozzles must be supplied with the proper flow and pressure. Pump operators utilize the pump and pump peripherals to control flows and pressures to nozzles. Pump operators utilize appliances to provide the variety of hose configurations often required to move water to the point of discharge. Hose, appliances, and nozzles, have certain flow, pressure, and friction-loss characteristics. Hydraulic calculations for providing proper flows and pressures must include these characteristics. (Section 4 of this text discusses hydraulic calculations in detail.)

REVIEW QUESTIONS

Key Terms and Concepts

On a separate sheet of paper, identify and/or define each of the following.

1. NH
2. Supply hose
3. Suction hose
4. Spanner wrench
5. Hose clamp
6. Hose bridge
7. Hose jacket
8. Nozzle pressure
9. Nozzle flow
10. Nozzle reach
11. Stream shape
12. Nozzle reaction
13. Smooth-bore nozzle
14. Combination nozzle
15. Appliances
16. Wye
17. Siamese
18. Adapters
19. Four-way hydrant valve
20. Water thieves
21. Manifolds
22. Eductor
23. Venturi principle

Multiple Choice and True/False

Select the most appropriate answer.

1. The basic parts of a hose include each of the following except
   a. reinforced inner liner.
   b. outer protective shell.
   c. couplings.
   d. safety relief ring.

2. The most common length of a section of hose is
   a. 50 feet.   c. 150 feet.
   b. 100 feet.   d. 200 feet.

3. The two main types of hose construction are woven jacket and rubber covered.
   True or False?

4. Cotton, rather than synthetic fiber, is still the most popular material used in the construction of woven-jacket hose.
   True or False?

5. Which of the following is considered the most common sexless coupling used in the fire service?

a. rocker   **c.** storz

b. lug   **d.** NST

6. Each of the following can be used for intake hose except

   a. LDH.   **c.** hard suction.

   b. soft sleeve.   **d.** SDH.

7. NFPA _____ provides guidelines for the care, use, and maintenance of fire hose.

   a. 1961   **c.** 1963

   b. 1962   **d.** 1964

8. Most nozzles are designed for an operating nozzle pressure of either 50, 75, 80 or 100 psi.

   True or False?

9. A _____ is used to divide one hose line into two or more lines.

   a. wye   **c.** adapter

   b. siamese   **d.** increaser

10. A _____ is used to combine two or more lines into one line.

    a. wye   **c.** adapter

    b. siamese   **d.** increaser

11. Both wyes and siamese have female inlets and male discharge couplings.

    True or False?

12. Eductors utilize the venturi principle to draw foam into a water stream.

    True or False?

13. The correct nozzle pressure for a smooth-bore hand line is:

    a. 50 psi

    b. 100 psi

    c. 150 psi

    d. depends on the hose length

14. NFPA _____ defines minimum requirements for new fire hose connections.

    a. 1961   **c.** 1963

    b. 1962   **d.** 1964

15. NFPA _____ focused on fire hose appliances.

a. 1961   **c.** 1963

b. 1962   **d.** 1964

16. According to NFPA, supply hose have a maximum operating pressure of _____ psi while attack hose have a maximum operating pressure of _____ psi.

    a. 185 and 285   **c.** 150 and 250

    b. 185 and 275   **d.** 95 and 150

17. The thread used in nearly all fire hose threaded coupling construction is referred to as

    a. NST.   **c.** HN.

    b. NH.   **d.** STORZ.

Short Answer

On a separate sheet of paper, answer/explain the following questions.

1. Explain the importance of studying hose, appliances, and nozzles to pump operations.

2. Discuss the basic types of hose construction. Which do you feel is better? Why?

3. Identify four appliances commonly used during pump operations.

4. What are the critical points associated with nozzles operating as designed?

5. List the standard nozzle pressures most often used by fire service nozzles.

6. Explain the operation of a nozzle. Be sure to discuss reach, shape, flow, and nozzle reaction.

7. What are the differences between wye and siamese appliances?

8. List and discuss the purpose of NFPA standards that focus on hose, appliances, and nozzles.

9. List four common types of hose tools used by pump operators.

10. What should you look for when conducting a hose inspection?

11. When proper flow and pressure are provided, the three nozzle functions of controlling

_____, reach, and _____ _____ will be the most effective for suppression efforts.

12. Explain the difference between smooth-bore nozzles and combination nozzles.

13. List and explain the difference between the three major types of combination nozzles.

14. List several types of wyes commonly used in the fire service.

15. List several adapters commonly used in the fire service.

ACTIVITIES

1. Identify and categorize the hose, appliances, and nozzles used by your department.

2. Using NFPA 1962 as a guide, develop a detailed procedure for testing fire hose to include safety considerations and records documentation.

PRACTICE PROBLEM

1. You have been asked to identify all the pump-related items needed for a new 1250-gpm pumper that the chief will propose in the new budget. List the items you feel should be provided with the new pumper.

BIBLIOGRAPHY

David P. Fornell. *Fire Stream Management Handbook.* Saddle Brook, NJ: Fire Engineering, 1991. Excellent discussion on nozzles, hose, and streams.

Section

3

PUMP PROCEDURES

The first section of this text discusses basic requirements and concepts for both the pump operator and emergency vehicles. The second section discusses pumps and the many components connected to, or used with, pump operations.

This section of the text discusses the concepts and procedures for the three interrelated fire pump operation activities: securing a water supply, operating the pump, and maintaining discharge pressures referencing NFPA 1002, *Standard for Fire Apparatus Driver/Operator Professional Qualifications*, 2003 edition, Sections 5.2.1, 5.2.2, and 5.2.4. Chapter 7 presents the various water supply sources and considerations for their use, Chapter 8 discusses various procedures for operating the pump, and Chapter 9 discusses how to initiate and maintain discharge hose lines.

The next section presents detailed explanations of hydraulic theories and principles as well as fireground flow considerations and pump discharge calculations.

Chapter

7

WATER SUPPLIES

Learning Objectives

Upon completion of this chapter, you should be able to:

■ List and explain basic considerations for preplanning and selecting water supplies.

■ Explain each of the five water supply systems typically used in fire pump operations.

■ Discuss the unique considerations for using each of the five water supply systems.

■ Describe the hydrant coding system as suggested by NFPA 291.

■ Explain how relay and tanker shuttle operations are designed.

NFPA 1002
Standard for Fire Apparatus Driver/Operator
Professional Qualifications
(2003 Edition)

This chapter addresses parts of the following knowledge elements within sections 5.2.1, 5.2.2, and 5.2.4:

Problems related to small-diameter or dead-end mains

Low pressure and private water supply systems

Hydrant coding systems

Reliability of static sources

Position a fire department pumper to operate at a fire hydrant and at a static water source

INTRODUCTION

When an alarm is received, the first duty of a pump operator is to ensure the safe arrival of the apparatus, equipment, and personnel at the scene. The next duty is to initiate fire pump operations. As stated earlier, fire pump operations consist of three interrelated activities:

1. Securing a water supply
2. Operating the pump
3. Maintaining discharge pressures

The critical first step of a pumping operation is to secure a water supply. Securing a water supply is the first step in the process of moving water from a source to the intake side of a pump. This particular activity is perhaps the most challenging and difficult of the three in that controlling the pressure and flow of a supply is typically limited.

This chapter presents information as though the pump operator makes the decision concerning which water supply to secure. In reality, this is not always the case. Depending on a department's standard operating procedures (SOPs), the determination of which water supply to secure can rest with any of the following:

- Pump operator
- Assigned officer
- Pump operator and the officer (they share the responsibility)
- Incident commander, water supply officer, or other officer within the command system

Regardless of who actually makes the decision, the pump operator should be capable of evaluating and securing water supplies. Therefore, pump operators should be familiar with the strengths and limitations of available water supplies.

Recall from Chapter 1 that water supplies come from three kinds of sources: the apparatus, static sources, and pressurized sources. *Apparatus water supplies* include water tanks carried on the apparatus and are secured through onboard tank operations. *Static sources* include ponds, lakes, and rivers and are secured through drafting operations. *Pressurized sources* include elevated towers and municipal water supplies. Municipal water supplies are the most common pressurized source used in the fire service and are secured through hydrant operations. Private water supplies, common in large industrial complexes, are similar in nature to municipal systems. All three sources can be used in combination through relay operations and tanker shuttle operations to provide initial and sustained water supply to the scene.

Obviously, when only one source is available, pump operators have no choices. However, when more than one source is available, pump operators should be able to choose, or recommend, the water supply that maximizes the efficient and effective use of both the supply source and the apparatus to meet the flow requirements of an incident.

This chapter discusses the pump operator's task of securing a water supply, starting with basic water supply considerations. Then common supply systems typically used in fire pump operations are presented to include onboard supplies, municipal private supplies, static supplies, relay operations, and tanker shuttle operations.

BASIC WATER SUPPLY SYSTEM CONSIDERATIONS

The first few minutes of an emergency are typically accompanied by stress, anxiety, and confusion. In most cases, *preplanning* water supplies will help reduce this stressful time of an incident. In addition, preplanning will help the pump operator complete the task of securing a water supply in an efficient and effective manner. The following considerations should be included in the preplanning and selection of a water supply.

required flow
the estimated flow of water needed for a specific incident

• *Required Flow.*   Determining the **required flow**, an estimated flow of water needed for a specific incident, is one of the first considerations when selecting a water supply. Considerations for calculating required flow are discussed in Section 4. In some cases, multiple pumpers are needed to provide the required flow for an incident. In other cases, the required flow may not be known. In either case, a water supply should be selected that can provide enough water to flow the capacity of the pump. In doing so, the pump operator is in the best position to assist with providing the required flow. The bottom line is that when securing a water supply, the amount of water the pump operator is expected to provide should be known.

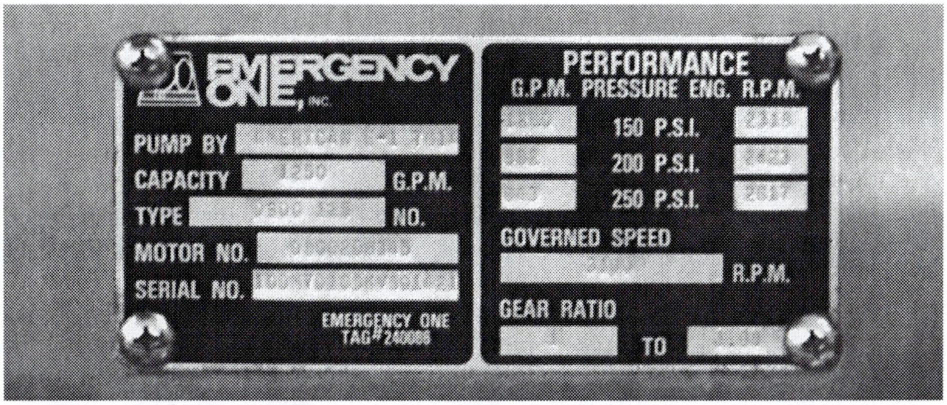

Figure 7–1 *The rated capacity and performance of a pump is usually identified on or near the pump panel.*

■ **Note**

A water supply should be selected that can provide enough water to flow the capacity of the pump.

• *Pump Capacity.*   Another important consideration when securing a water supply is the capacity of the pump (Figure 7–1). In general, supplies should be chosen that are capable of providing enough water to allow the pump to flow capacity. Larger capacity pumps may rule out some water supplies, while smaller capacity pumps will have a wider selection of supplies to choose from. Keep in mind that pumps, at draft, are expected to flow 100% capacity at 150 psi. When pump pressures in the operation reach 200 psi, the pump can only be expected to deliver 70% of its rated capacity. When pressures reach 250 psi, only 50% of its rated capacity should be expected. In essence, increased pump pressures reduce the quantity of water the pump is able to flow. (See "Rated Capacity and Performance" in Chapter 4.)

• *Supply Hose Capacity.*   The size of supply hose is another consideration when selecting a supply source. Apparatus equipped with 2½-inch supply hose may rule out some supplies, while those equipped with large diameter hose (LDH) may have a wider selection of supplies to choose from. In general, larger diameter supply hose will provide less friction loss over longer distances.

• *Water Availability.*   Another important consideration is the availability of the source water. Several factors contribute to the availability of water. One is the *quantity* of water available. For example, a swimming pool may not provide as much water as a large lake. Other factors are the *flow* (gpm) and *pressure* (psi) at which water is available from the supply. Municipal supplies, for example, may provide water at a variety of flows and pressures. Finally, the *physical location* of the water is a factor. A marginal hydrant distant from the incident may not be the best water supply if a good static source like a lake or pond is located closer to the incident. **Water availability**, then, relates to the quantity, flow, pressure, and accessability of a water supply.

water availability
the quantity, flow, pressure, and accessibility of a water supply

supply reliability
the extent to which the supply will consistently provide water

• *Supply Reliability.*   The reliability of a source should also be considered when selecting a water supply. The **supply reliability** is the extent to which the supply will consistently provide water. Another way of looking at it is the extent to which the supply fluctuates in flow, pressure, and quantity. For example, a river or pond may not be a reliable supply if the water level changes frequently or is frozen and tidal water supplies may change dramatically within a short period of time.

• *Supply Layout.*   Finally, the supply layout hose should also be considered when selecting a water supply. **Supply layout** is the required supply hose configuration necessary to efficiently and effectively secure the water supply. The layout of supply hose may be affected by the type of source, the size of hose, the hose appliances available, and the number and size of intakes on the pump. Therefore, a variety of supply hose configurations can often be used. Keep in mind, though, the more elaborate the supply configuration, the longer it will take to set it up. For example, connecting a supply hose to a pump located near the hydrant will not take as long to secure as a relay or tanker shuttle operation. In general, the supply configuration should be sufficient to provide the flow needed for an operation.

Although pump operators should be prepared to establish each kind of water supply system, *preplanning* will make the task a little smoother and less stressful. This is especially true when considering the compressed time in which decisions must be made during an emergency. Pump operators should be familiar with the different types of water supply systems, considerations for their use, and considerations for securing them.

ONBOARD WATER SUPPLIES

The **onboard supply** is simply the water carried in a tank on the apparatus (Figure 7–2). The onboard water supply is used for several reasons. First, it is used when no other water supply is available. In locations where this is a common occurrence, apparatus typically have larger onboard tanks. Second, it is used when an incident requires only a small quantity of water, such as car fires and small brush fires. This saves time and energy when a supply source is not readi-

supply layout
the required supply hose configuration necessary to efficiently and effectively secure the water supply

■ Note
Preplanning will make the task of securing a water supply a little smoother and less stressful.

onboard supply
the water carried in a tank on the apparatus

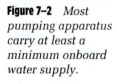

Figure 7–2 *Most pumping apparatus carry at least a minimum onboard water supply.*

ENGINE

ly available. Third, it is used when an immediate water supply is deemed more critical than the time it would take to secure a supply with greater flows. In this situation, an alternate supply is secured while utilizing the onboard source. Finally, onboard water supplies can be used as a backup or emergency water supplies in the event other water supply is interrupted.

Water Availability

An immediate and readily available supply of water is the main advantage of the onboard source. Although readily available, the onboard supply is the most limited in terms of quantity of water. Most pumping apparatus have at least a small tank of water on board. NFPA 1901 specifies minimum tank capacity for fire apparatus as follows:

initial attack apparatus:	200 gallons per minute
pumper fire apparatus:	300 gallons
mobile water apparatus/tanker:	1,000 gallons

Tank capacities of 500 to 1,000 gallons are common on many pumper fire apparatus and many mobile water apparatus have tank capacities over 2,000 gallons. NFPA 1901 also requires that the piping between the tank and the pump be capable of flowing at least 250 gpm for tank capacities less than 500 gallons and must be able to flow 500 gpm for tank capacities of 500 gallons or larger. Greater flows from the tank can be achieved by specifying a larger diameter pipe between the tank and the pump.

The length of time water can be supplied when utilizing the onboard water depends on the size of the tank and the flow of discharge lines. For example, a 1¾-inch preconnect flowing 125 gpm can be sustained for 8 minutes with a 1,000 gallon tank, 6 minutes with a 750 gallon tank, and 4 minutes with a 500 gallon tank. In comparison, a 2½-inch line flowing 250 gpm can be maintained for 4, 3, and 2 minutes with 1,000, 750, and 500 gallon tanks, respectively.

Supply Reliability

In general, the onboard water supply is a reliable source. A potential concern, however, is not having water in the tank when on scene. This can occur when the tank is accidentally left empty or when the tank's level indicating device is malfunctioning showing the tank to be full when it is not. Another potential concern is that air pockets may impede pumping the tank water. Air pockets may occur for several reasons: when the tank and piping develop leaks and holes, when air is trapped in the piping and pump, or when the pump is left dry to guard from freezing. Priming the pump to eliminate air pockets is usually all that is needed to begin pumping the tank water.

Supply Layout

The onboard water supply is by far the fastest supply to secure in that the supply line is permanently attached. All that is required to secure the supply is to pull the tank-to-pump control valve. The tank-to-pump control valve opens and closes a passage from the tank to the pump. The tank is usually mounted higher than the pump to allow gravity to move water to the intake side of the pump. A control valve is also typically installed in the piping from the pump to the tank and has several names: "pump-to-tank," "recirculating valve," and "tank fill." This valve opens and closes a passage from the pump to the tank allowing the tank to be refilled from the pump. The two valves working together circulate water between the tank and the pump. In addition, a tank level indicator is usually provided to monitor water levels in the tank. The tank-to-pump and pump-to-tank control valves as well as the tank level indicator are commonly mounted on the pump panel (Figure 7–3). Often, the onboard water supply can be pumped using either the booster pump or the main pump.

In most cases, only small discharge lines are used in conjunction with the onboard tank. Typically, the discharge lines are preconnected and are either 1-inch booster lines, 1½-inch or 1¾-inch attack lines. When using the onboard tank, a good habit is to plan ahead for an alternate water supply should additional water be required. When an alternate supply is secured or when the incident is over, the pump operator should consider refilling the tank as soon as possible.

■ **Note**
The pump operator should consider refilling the tank as soon as possible.

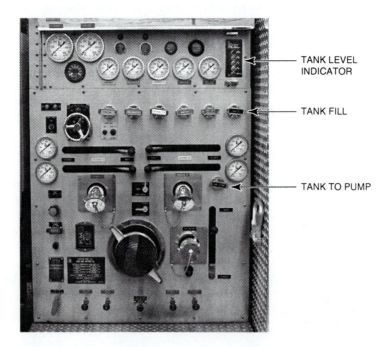

Figure 7–3 *Onboard water supply control valves are typically located on the pump panel. Courtesy of KME Fire Apparatus.*

TANK LEVEL INDICATOR

TANK FILL

TANK TO PUMP

Often, the onboard supply is used to provide immediate water to the incident. A difficult decision is how quickly water is needed at the scene. The quickest delivery of onboard water is to simply drive the apparatus to the scene and start pumping. The major concern with this approach is the ability to secure another source before the onboard supply runs out. The second quickest way is to lay a line from the hydrant on the way to the scene. Although this approach takes a little longer to start flowing the onboard water, an alternate water supply can usually be secured within a reasonable time frame. This timing is critical when the total quantity of water needed exceeds tank capacity.

MUNICIPAL SYSTEMS

Most densely populated areas in the United States utilize a municipal water supply system. In many cases, a city, county, or special water district utilize a public works or water department to design, maintain, and test the municipal water system. Often, the difficult part is to not only maintain existing systems but to expand systems in population growth areas as well as to upgrade systems as they wear out or become obsolete.

municipal supply
a water supply distribution system provided by a local government consisting of mains and hydrants

    Municipal supplies deliver water to the intake side of the pump under pressure. They are used when water demands exceed onboard water supply ability or when the distance from the source to the incident is great. Hydrants are connected via a water-main distribution system. Both hydrants and their water mains are components of a municipal water supply system. Pump operators should be familiar with the municipal water system, its distribution system, and the types of hydrants used in their service area.

Municipal Water Systems

Municipal water supplies commonly provide water for two purposes. First, they provide water for normal consumption such as household and industrial uses. Second, they provide water for emergency use to hydrants and fixed fire-protection systems. In some locations, two completely separate systems are used, one to provide water for normal consumption and one to provide water for emergency use. More commonly, the same system provides water for both domestic consumption and emergency use. When this is the case, fluctuating hydrant pressures and flows can be expected with the changes in domestic consumption.

    The basic components of municipal water systems include a water supply course, a distribution system, and hydrants.

Municipal Water Supply Sources

The water supply source for a municipal system is obtained from either surface water such as lakes and ponds or groundwater such as wells or springs. In some cases, a municipal system will use water from both sources. The water can be moved from the source to the distribution system by means of a pumping system, gravity system, or a combination of the two. Pumping systems utilize pumps of sufficient size and number to meet the consumption demands of the service area. Typically, backup electrical power in the form of diesel generators and redundant pumps are maintained in case the primary electrical power or pumps fail. Gravity systems use elevation as a means to move water from the supply to the distribution system. A mountain lake providing water to a city in the valley is one example of a gravity system. Another example of a gravity system is an elevated storage tank (see Figure 7–4). Elevated storage tanks are common in many areas and can either be located on a hill or mountain or the tank itself may be elevated. Sometimes a combination pumping system and gravity system are used. In combination systems, the elevated storage tanks are used to assist pumping systems with meeting peak demand needs or as an emergency or backup water supply. In many cases, the elevated storage tanks are filled at night when demands are lower and pressures are higher. During the day, elevated storage tanks use gravity to assist pumps in meeting increased water demands.

In many cases, the water passes through a processing facility to filter and treat water. This occurs most often when the system provides potable water for domestic consumption as well as for fire suppression. Typically, these facilities are

Figure 7–4 *Elevated storage tanks are sometimes located on higher elevations to further increase gravity.*

designed to process water with sufficient capacity to handle peak domestic consumption and fire suppression needs; however, areas of rapid growth place a heavy burden on processing facilities to the extent that they are no longer able to provide required flow to the community and/or for fire suppression efforts. Natural disasters, equipment malfunctions, and terrorist activities may also impact the ability of the processing facility to provide adequate water flows. Preplanning is therefore important to determine alternate water supplies. See Figure 7–5 for an illustration of the three systems.

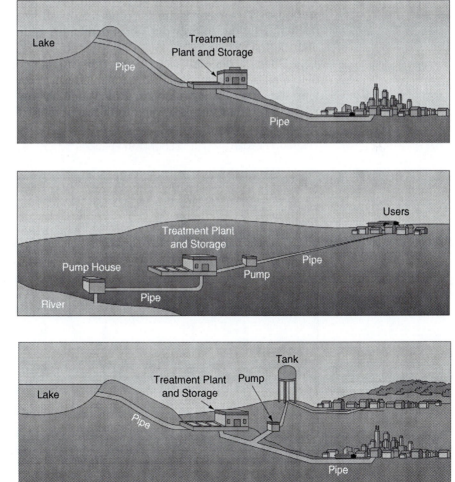

Figure 7–5

Illustrations of (A) a gravity-fed water system, (B) a pumping system, and (C) a combination system.

Distribution System

Municipal water distribution systems consist of a series of decreasing sized pipes, call mains, and control valves that move water from the source or treatment facility to individual hydrants and commercial and residential occupancies. The distribution system typically starts with large-sized pipes, 16 inch or larger, called feeder mains, which carry water from the source or treatment facility to various locations within the distribution system. Larger-sized pipes are used to reduce the loss of pressure caused by friction and can carry water long distances. Because of this, feeder mains are typically spaced farther apart. Secondary feeder mains of intermediate size, 12 to 14 inches in diameter, carry water to a great number of locations within the distribution system. Finally, distributors, the smallest size pipe in the system at 6 to 8 inches in diameter, complete the system and connect directly to hydrants and residential/commercial customers (see Figure 7–6). Actual pipe sizes for a distribution system can differ considerably between one system and other. The major factor for pipe size in all systems is the ability to provide the required flow. Simply put, larger demands require larger-sized pipes within the distribution system.

Often, feeder mains and distributors are interconnected, allowing water to be delivered to the same location through alternate routes. This interconnected system, sometimes called a loop or grid system has several important features. First, when the system supplies water from two or more directions, it helps reduce fric-

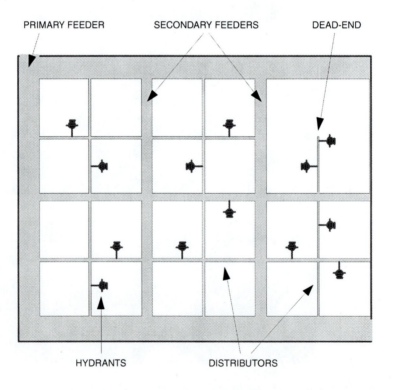

Figure 7–6 *A grid distribution system provides better water supplies because several mains can feed one hydrant.*

tion loss. Second, damage or maintenance on one part of the system can be isolated using control valves so that the distribution system remains in operation. Third, a high demand in one area will not adversely affect another area as much as if a grid system were not in place. Basically, water can flow from multiple locations to compensate for a high consumption area. The ability to provide water from multiple locations can also be used to help maintain adequate water flow and pressure to high risk areas.

A hydrant supplied from only direction is called a dead-end main and hydrants located on the main are known as dead-end hydrants (see Figure 7–6). Should the dead-end main become damaged, all subsequent hydrants from that point to the end of the main will become inoperable. Also, if a pumper begins pumping large quantities of water from one of the first hydrants on a dead-end main, all subsequent hydrants will realize a significant reduction in water pressure and flow. The more water pumped from the hydrant, the less water is available for the remaining hydrants on the dead-end main.

Control valves are used to direct water within a distribution system. These valves can be opened and closed to redirect water within the system to isolate sections of the grid or to enhance flow and pressure to sections of the grid. Therefore, control valves should be located in strategic locations within the grid to provide maximum flexibility. In some locations, only the public works or water department can operate control valves within the distribution system. Other locations allow fire department personnel to operate control valves. Regardless, preplanning is essential, and pump operators should be familiar with all aspects of the distribution system, including control valve locations.

Indicating and nonindicating control valves are the two broad types of control valves used in distribution systems. As the name implies, indicating valves provide a visual indication on the status of valves—whether open, closed, or in between—and are more commonly found in private distribution systems. The two common types of indicating valves are the outside screw and yoke valve, commonly referred to as an OS&Y valve, and the post-indicating valve, referred to as a PIV (see Figure 7–7). The OS&Y valve, typically used in sprinkler systems, is made up of a threaded stem connected to a gate. The location of the stem within the yoke indicates the location of the gate. The PIV consists of a stem within a post that is attached to the valve. When the valve is fully open or closed the words "open" or "shut" appear in the PIV. Nonindicating control valves are more commonly found in municipal distribution systems and are usually either buried or installed within manholes. Gated valves and butterfly valves are the two most common types used in distribution systems.

Hydrants

The two basic types of hydrants used by the fire service are wet and dry barrel hydrants, the latter being the most common. **Wet barrel hydrants** are typically used where freezing is not a concern. This type of hydrant is usually operated by individual control valves for each outlet (Figure 7–8). As the name implies, water

■ Note
The two basic types of hydrant used by the fire service are wet and dry barrel hydrants.

wet barrel hydrant
a hydrant operated by individual control valves that contain water within the barrel at all times; typically used where freezing is not a concern

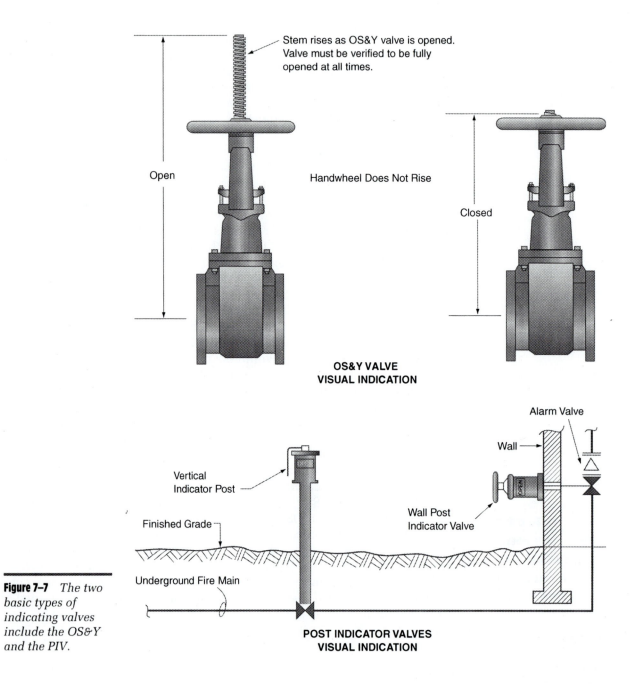

Stem rises as OS&Y valve is opened.
Valve must be verified to be fully
opened at all times.

Open

Handwheel Does Not Rise

Closed

**OS&Y VALVE
VISUAL INDICATION**

Alarm Valve

Wall

Vertical
Indicator Post

Wall Post
Indicator Valve

Finished Grade

Underground Fire Main

**POST INDICATOR VALVES
VISUAL INDICATION**

Figure 7–7 *The two
basic types of
indicating valves
include the OS&Y
and the PIV.*

is typically maintained in the hydrant at all times. Because each outlet is individually operated, supply hose can be connected and charged independently.

dry barrel hydrant

a hydrant operated by a single control valve in which the barrel does not normally contain water; typically used in areas where freezing is a concern

Dry barrel hydrants are typically used in areas where freezing is a concern. This type of hydrant is operated by turning the stem nut located on the bonnet of the hydrant, which opens the main valve at the base (Figure 7–9). The main valve is located at the bottom of the hydrant below the frost line. When the main valve opens, water enters the barrel for use through all the outlets. Because of this, supply hose should be attached to outlets prior to opening the main valve. In addition, gated valves should be attached to unused outlets (Figure 7–10). In doing so, the outlets will be available for use after the main valve is opened. When the main valve

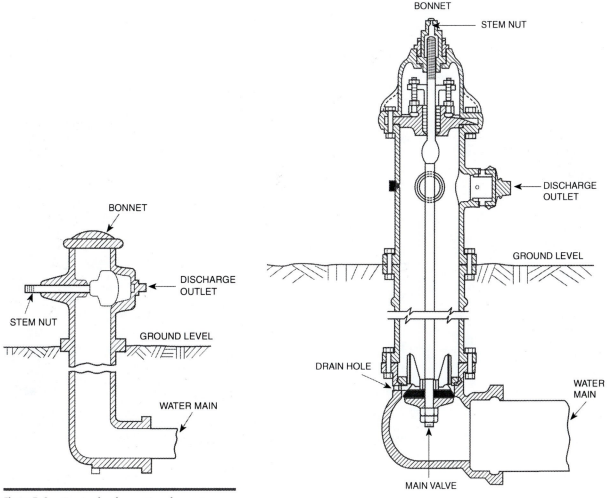

Figure 7–8   *Typical schematic of a wet barrel hydrant.*

Figure 7–9   *Typical schematic of a dry barrel hydrant.*

closes, the drain, also known as the weep valve, opens, allowing water to escape and returning the barrel to its normal state. One potential concern is that if the hydrant is not fully closed or opened when in use, the drain valve may remain slightly open allowing water to flow from the hydrant and causing erosion and potential damage.

Both wet and dry barrel hydrants have a variety of outlet configurations and can be found in a variety of positions. The normal configuration is one large outlet (4 inches or greater) and two $2\frac{1}{2}$-inch outlets (Figure 7–10). However, hydrants with only two $2\frac{1}{2}$-inch outlets are also common (Figure 7–11). Most of these outlets use American National Fire Hose connection screw threads (NH) couplings, although some locations still have special threads. In these cases, adapters are required to connect supply hose to the hydrant. Typically, hydrants are installed with the 4-inch outlet facing the street and the $2\frac{1}{2}$-inch outlets parallel with the street. In addition, the outlets should be installed so they are not obstructed. However, they are not always properly installed. Outlets can face almost any direction and can be obstructed in any number of ways. Some of the more common obstructions occur when hydrants are improperly installed (Figure 7–12).

Figure 7–10 *Additional lines can be attached to a dry barrel hydrant while in use if gated valves are attached to unused outlets before the hydrant is charged. Courtesy Greg Burrows.*

Figure 7–11 *Some hydrants have only two $2\frac{1}{2}$-inch outlets.*

Figure 7–12
Improper installation can reduce the effective use of a hydrant.

Hydrants can be located almost anywhere and are often hidden from view (Figure 7–13). Therefore, fire departments utilize a variety of methods to assist in the quick identification of hydrant locations. One common method is to use fluorescent paint when painting the hydrant. Another common method is to place three reflective bands on street poles, signs, or other tall objects located near a

Figure 7–13 *A hydrant is blocked from view by the first mail box on the right. Note the hydrant marker located in the middle of the street (arrow).*

Table 7–1 *NFPA 291 suggests the following for classification and color coding hydrants.*

Class	Rated Capacity	Color of Bonnets and Nozzle Caps
AA	1,500 gpm or greater	Light blue
A	1,000–1,499 gpm	Green
B	500–999 gpm	Orange
C	Less than 500 gpm	Red

hydrant. The reflective strips can also match the color code of the hydrants (see Table 7–1). Another method gaining popularity is to place traffic reflective markers (Figure 7–14) in the middle of the road adjacent to a hydrant (Figure 7–15). (A traffic reflective marker can also be seen in Figure 7–13). To identify on which side of the road the hydrant is located, the reflector can be offset on the side of the hydrant. Another method for identifying on which side of the road a hydrant is located is to use two different colors on either side of the reflector, one color to indicate the left side and another color to indicate right side. Road reflectors don't work well in areas where snow plows routinely work the streets in the winter. In

Figure 7–14 *Hydrant traffic reflectors can help pump operators locate hydrants from a distance, especially at night.*

Figure 7–15 *Traffic reflectors are usually located near the middle of the road adjacent to the hydrant.*

these areas, a tall reflective flag or other indicating device is used. The device must be tall enough to be seen over snow drifts and snow banks.

Private Water Systems

Industrial, commercial, and large complexes or facilities often maintain their own private water supply system. In most cases, these private water systems are similar to municipal water systems. They can either receive their water from a municipal water system or use surface or groundwater sources. In addition, private water supply systems can use a pumping system, gravity, or a combination of the two to move water from the source to the hydrant, standpipe system, or sprinkler system. Typically, private water supply systems provide water for industrial/manufacturing use, employee use, and fire protection. Private fire protection systems are typically maintained separate from other water systems. As with municipal systems, private fire protection systems consist of a series of decreasing-sized pipes, called mains, and control valves that move water from the source to individual hydrants and/or sprinkler and standpipe systems.

Water Availability

Municipal supplies, as a water source, have three main advantages. First, hydrants provide a readily available supply of water over an expanded geographical area. How available this water supply is depends on the extent to which hydrants are distributed within an area. Second, hydrant flow capacities can be determined in advance through flow tests. NFPA 291, *Fire Flow Testing and Marking of Hydrants*, identifies testing procedures and suggests that hydrants be classified and color coded based on their rated capacity (Table 7–1). Finally, municipal systems can provide a sustainable supply for several hours. It is important to realize that municipal water supplies are typically designed to provide required fire flows for a specific time frame, usually between 2 to 10 hours. When pumping less than the required flow, hydrant operations can be sustained for longer periods. If pumps are used in the water system to move water, several types of pumps are installed. Some of the pumps run constantly, others turn on when additional pressure and volume is required, and still others are maintained as emergency backup pumps complete with emergency electrical power sources.

Hydrant Flow Test

Hydrant flow testing provides the means to determine the available pressure and flow within municipal and private distribution systems. The information collected during flow tests is critical for identifying and selecting hydrants that can provide the estimated fire flows calculated during prefire planning activities. In some locations, the public works or water department is responsible for conducting hydrant flow tests. In others, the pump operator conducts or assists with hydrant flow testing. Regardless, pump operators should be familiar with hydrant flow test procedures.

Essentially, the flow test consists of measuring static and residual pressures from one hydrant and measuring flow from one or more other hydrants. The data collected during tests provide the means to calculate available hydrant flow at specific residual pressure. NFPA 291, *Recommended Practices for Fire Flow Testing and Marking of Hydrants*, provides excellent information on all aspects of conducting flow tests to include the following.

Preparation (prior to conducting the flow test) Several considerations and actions to complete before conducting the flow test include the following:

- Hydrant flow testing during periods of ordinary water demand will yield more accurate results based on realistic conditions.
- Test hydrant location consideration may help to ensure minimum impact on traffic and to surface areas.
- Use of hydrant diffusers can reduce the impact of erosion.
- Select a test hydrant, sometimes referred to as the residual hydrant, and one or more flow hydrants.
- Place a cap with pressure gauge onto the $2\frac{1}{2}$-inch test hydrant outlet.
- To increase the accuracy of reading, ensure individuals at flow hydrants have received training and have practice taking pitot gauge readings.

General Procedure The general procedures for conducting hydrant flow test are as follows:

Step 1: the test begins with a static pressure reading at the test hydrant. The reading is taken after fully opening the hydrant valve and removing the air.

Step 2: open flow hydrant(s) one at a time until a 25 percent drop in residual pressure is achieved.

Step 3: after a sufficient drop is noted, continue flowing to clear debris and foreign substances.

Step 4: take all readings at the same time; a residual reading at the test hydrant and flow readings using the pitot gauges at each of the flow hydrants.

Step 5: record the exact interior size, in inches, of each outlet flowed.

Step 6: after recording all readings, slowly shut down hydrants one at a time to reduce the likelihood of a water hammer or surges within the system.

Equipment The following equipment is often required during hydrant flow testing.

- pressure gauge mounted on an outlet cap, the pressure gauge calibrated within the past twelve months

- pitot gauge for each hydrant; note: NFPA 291 requires gauges calibrated within the past twelve months
- a hydrant diffuser can be used to reduce damage caused by large volume flows from hydrants
- hydrant wrenches
- portable radios

Calculating Results First, calculate the discharge from hydrants used for flow during the test. The flow from hydrant pitot readings can be determined by either a chart, see Table 4.10.1(a) within NFPA 291, or by the following formula:

$$Q = 29.84 \times c \times d^2\sqrt{p} \tag{7-1}$$

where c = coefficient of discharge
    0.90 for smooth and rounded outlets
    0.80 for square and sharp outlets
    0.70 for square outlets projecting into the barrel
  d = diameter of the outlet in inches
  p = pitot pressure in psi

When multiple hydrants are used in the flow test, calculate (or look up) each flow and then add them together.

Next, calculate the discharge at the specified residual pressure and/or desired pressure drop using the formula:

$$Q_R = Q_F \times (h_r^{0.54}/h_f^{0.54}) \tag{7-2}$$

where Q_R = predicted flow at specified residual pressure
  Q_F = total flow from hydrants
  h_r = pressure drop to desired residual pressure (initial static pressure reading—desired residual pressure)
  h_f = pressure drop measured during test (initial static pressure reading—final residual pressure reading)

The formula can be computed using a calculator capable of logarithms or by looking up the values of pressure readings to the 0.54 power in a table. NFPA 291 suggests both a form to use to document flow test data and a form to graph results. A variety of computer software programs are available to assist with hydrant flow calculations and reporting.

Reliability

Municipal systems generally provide a reliable supply of water; however, several factors may reduce the reliability of this supply source. First, the flow from hydrants may decrease over time based on gradual increases in municipal consumption or through the normal deterioration of piping and components. The color coding of hydrants, as suggested by NFPA 291, provides flow rates for a single hydrant. When multiple hydrants are used, individual hydrant flows may change dramatically. In addition, the color code indicates flows during normal municipal consumptions. During peak use hours, hydrant flow rates may again change dramatically. Finally, leaks, preventive maintenance, scheduled outages, and damaged or broken components may all lead to reduced flow capacity of hydrants from time to time.

Supply Layout

Hydrant supply hose configurations can vary depending on the flow and pressure of the hydrant, the position of the apparatus, and the number and size of intakes available on the apparatus. The more complex the configuration, the longer it will take to set up the supply operation. The first step, though, is to select a hydrant. In some cases, there is no choice in that only one hydrant is available. In general, hydrants closer to the incident should be selected to reduce the time to set up the operation as well as to reduce the loss of pressure due to friction; however, in some cases a stronger hydrant (with more volume and pressure) that is farther away may be picked over weaker hydrants closer to the incident. In addition, the number and size of hydrant outlets may influence which hydrant to select. Finally, the pump capacity and the size of supply lines available will influence the selection of a hydrant.

Once the hydrant is selected, the next step is to determine supply line configurations. In general, larger supply lines should be used when possible. Supply lines should be laid to minimize bends and kinks to reduce losses in pressure from friction. Regardless of the configuration, the pump will be positioned either next to the hydrant or some distance from the hydrant.

The pump can be positioned at the hydrant for several reasons. One reason is that the hydrant is relatively close to the scene and attack lines can be advanced from that position. Another reason is when a reverse lay is conducted. A **reverse lay** (see Figure 7–16) is when the apparatus stops at the scene, drops off attack lines, equipment, and personnel, and then advances to the hydrant. Another reason for the pump being located at the hydrant is when it is the first pump in a relay operation. Finally, the pump may be located at a hydrant when increasing hydrant pressure to another pump by use of a four-way hydrant valve (see Figure 6–15).

Because hydrant outlets and pump intakes vary, a wide range of configurations is possible when the pump is located at a hydrant. The quickest way is to use a soft suction hose to connect directly into one of the main intakes. Another way is to use a 50-foot section of LDH. When smaller-diameter-hose is used, two or more lines can be connected from the hydrant. This is typically accomplished using wyes on the hydrant and siamese on the pump intake. Because hydrants are located at different distances from the curb and outlets can face just about any

reverse lay
supply hose line configuration when the apparatus stops at the scene; drops attack lines, equipment, and personnel; and then advances to the hydrant laying a supply line

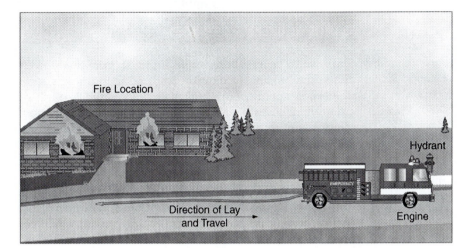

Fire Location

Hydrant

Direction of Lay
and Travel

Engine

Figure 7–16 *A reverse hose lay.*

direction, pump operators must realize that positioning the apparatus at hydrants will vary. When possible, the configuration should provide enough water to pump capacity or deliver the required flow.

If the pump is not positioned at the hydrant, either a forward lay has been conducted or the pump is being supplied as part of a relay operation. A **forward lay** (see Figure 7–17), sometimes referred to as a straight lay, is when the apparatus stops at the hydrant and lays a supply line to the fire. If the apparatus proceeds directly to the scene, a second apparatus can lay a line to or from a hydrant. When a supply line is laid to the pumper, the line is connected to one of the pump's intakes.

forward lay

supply hose line configuration when the apparatus stops at the hydrant and a supply line is laid to the fire

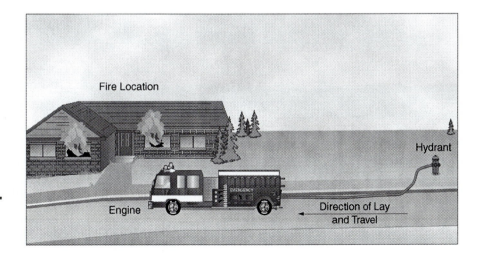

Fire Location

Hydrant

Engine

Direction of Lay
and Travel

Figure 7–17 *A forward or straight hose lay.*

STATIC SOURCES

static source

water supply that generally requires drafting operations, such as ponds, lakes, and rivers

drafting

process of moving or drawing water away from a static source by a pump

Static sources are those supplies such as ponds, lakes, rivers, and swimming pools, that generally require drafting operations (Figure 7–18). **Drafting** is the process of moving or drawing water away from a source by a pump. Static sources are used for several reasons. First, they are used when hydrants are not available. Many communities simply cannot afford or don't need a municipal water system. Second, static sources are used when available hydrants cannot provide the needed flows. Finally, they are used when natural disasters or mechanical failures cause hydrant systems to shut down.

Drafting operations require the use of a special supply hose called hard suction hose (Figure 7–19 see also "Classification of Hose" in Chapter 6). Hard suction hose is used because supply line pressures will be at or below atmospheric pressure within the hose. It is important to match the size of hard suction with the rated capacity of the pump (Table 7-2). Hard suction hose too big or too small may make it difficult or impossible to prime the pump. In addition, the ability to pump capacity may be significantly reduced. Pumpers that comply with NFPA 1901 will have the appropriate size of hard suction hose for the rated capacity of the pump.

To keep debris from the static source from entering the pump, strainers are connected to the end of hard suction hose. Note the strainer attached to the bottom section of hard suction hose in Figure 7–19. Care must be taken to keep the strainer off the bottom of the water source to prevent clogging (Figure 7–20). In addition, if the strainer is too close to the surface, whirlpools may develop allow-

Figure 7–18 *Example of a static water supply.*

Figure 7–19 *Drafting operations require hard suction hose. Pictured are the newer, more maneuverable, sections of hard suction. Courtesy Greg Burrows.*

Table 7–2 *Rated capacity of hard suction hose.*

Rated Capacity	Hard Suction Size
750 gpm	4½-inch
1,000 gpm	5 inch
1,250 gpm and above	6 inch

Figure 7–20 *This strainer also keeps the hard suction off the bottom to prevent clogging. Courtesy Greg Burrows.*

ing air to enter the pump (Figure 7–21). The result may be either inefficient pumping or loss of prime.

Drafting operations typically require apparatus to position fairly close to the source, as shown in Figure 7–21, because the height to which water can be drafted is limited. Recall that water is not sucked or pulled into the pump. Rather, when the priming system reduces the pressure in the pump (below 14.7 psi) at sea level, water is forced into the pump by atmospheric pressure (14.7 psi). Water will rise approximately 2.3 feet for each 1 psi of pressure. If the priming device reduces

Figure 7–21 *The strainer being used in this drafting operation floats on the surface. It is designed to skim debris free surface water while controlling whirlpool production. Courtesy Ziamatic Corporation.*

the pressure inside the pump from 14.7 psi to 9.7 psi (a 5 psi reduction), the atmospheric pressure, now 5 psi greater than the inside the pump, will force water to a height of 11.5 feet (2.3 × 5 = 11.5 feet). If the priming system reduced the pressure to 0 psi, a perfect vacuum, how high will atmospheric pressure force the water? This height is called *theoretical lift* and can be calculated by multiplying 14.7 psi (atmospheric pressure) by 2.3 feet per psi, approximately 33.81 feet. Obviously, priming systems on apparatus come nowhere near creating a perfect theoretical lift. The general rule of thumb is to attempt to pump no more than 22.5 feet, or two-thirds of the theoretical lift.

Capacity tests for pumps are conducted at draft. In general, pumps are required to flow capacity through 20 feet of hard suction with a maximum height of 10 feet. Greater heights and longer supply lines reduce the ability to flow capacity.

Availability

Regional differences determine the availability of static sources. Some areas have many sources, while others may have only a few. Where static sources are available, access to the supply may be limited. For example, the banks of rivers may be too steep or too soft to support the weight of the apparatus (Figure 7–22). The available flow from different static sources also vary as well. The available water from a small pond will not be as great as a large river. Pump operators should be familiar with drafting locations and available flows for their service area

Figure 7–22 *The banks of this static water supply will not support the weight of a pumper.*

Figure 7–23 *Providing an adequate road will enhance the availability of a static source, especially in areas where seasonal conditions impair access.*

static source hydrants
prepiped lines that extend into a static source.

Access and flow for static sources can be enhanced in several ways. One way is to provide an adequate road to the drafting site (Figure 7–23). In doing so, access can be increased during seasonal and weather changes. Another way is to provide a stable area close to the source where drafting operations can take place safely. Dredging the immediate area surrounding the drafting location is another way to enhance the static source by helping to provide unrestricted flows. Finally, **static source hydrants**, sometimes called dry hydrants, can be installed (Figure 7–24). These hydrants are simply prepiped lines that extend into the static source (Figure 7–25). Static hydrants can be used for natural static sources (ponds, lakes, and rivers) as well as artificial sources such as underground water tanks. These hydrants can be beneficial in that they reduce setup times, can increase flows by using large lines, and can increase the number of access points to a static source. A disadvantage is sediment that may form in the piping requiring a "back flush" prior to priming the pump.

Reliability

Static water supply reliability is affected by three factors. The first factor is the supply itself. Static supplies can change based on environmental and seasonal conditions. For example, droughts tend to diminish water supplies, while excessive rains may flood normal drafting locations or soften the ground thus restricting access. In addition, excessive silt and debris may render the source unusable. In

Figure 7–24
Example of a static source hydrant.

the winter, the water may freeze, hindering access to the supply. Tidal water supplies can change dramatically within a short period of time. The second factor is the pump and equipment used to draft the static source. Drafting is demanding for both the pump and the pump operator. Pumps and equipment must be maintained in good working order to efficiently and effectively use static sources as a supply. Finally, the pump operator is a factor. Pump operators must be thoroughly familiar with pump operations, priming operations, and drafting operations to reliably use static sources.

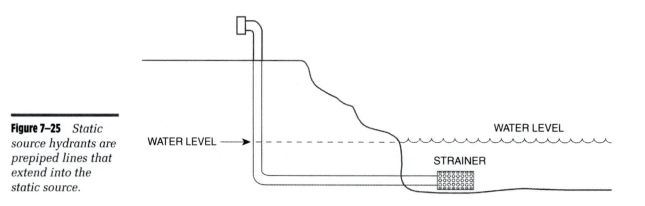

Figure 7–25 *Static source hydrants are prepiped lines that extend into the static source.*

Supply Layout

Setting up a drafting operation takes time, skill, and at least one person to assist the pump operator. The first step is to position the apparatus as close to the source as possible. Next, hard suction hose and the strainer should be coupled, making sure all connections are tight. Then the hard suction is connected to the pump intake and the strainer placed into the water. The hard suction should not be placed over an object that is higher than the pump intake. When this occurs, air pockets are likely to form that may hinder priming and pumping the static source. In addition, the strainer should be placed to ensure unrestricted flow. The last step is to ensure that the pump is airtight. This means that every possible location where air can enter the discharge or intake piping should be checked. Inlet and outlet caps should be tightened and all control valves should be checked to make sure they are completely closed. In addition, the relief valve should be turned off, and bleeder valves should be tightened.

■ **Note**

The hard suction should not be placed over an object that is higher than the pump intake because air pockets are likely to form that may hinder priming and pumping the static source.

RELAY OPERATIONS

relay operation
water supply operations where two or more pumpers are connected in line to move water from a source to a discharge point

Relay operations are those operations in which two or more pumpers are connected in-line to move water from a source to a discharge point. Relay operations are used for several reasons. First, they are used when a water source is distant from the incident and cannot provide appropriate flows and pressures to the scene. Second, they are used when a water source is relatively close to the incident but lacks pressure. Third, relay operations are used to overcome loss of pressure from elevation gains.

The original water supply can be either a static source or a hydrant in a municipal source. Therefore, the considerations for static sources and hydrants are the same for relay operations as they are for drafting operations. This section, then, discusses the unique considerations and setup requirements of relay operations.

■ **Note**

Relay operations require at least two pumps: a supply pump and an attack pump.

Relay operations require at least *two pumps*: a supply pump and an attack pump. The supply pump, obviously is positioned at the supply, while the attack pump is the last pump in the relay. Additional pumps in the relay are called *in-line pumps*. Relay operations tend to take longer to set up and are usually more complicated to operate (Figure 7–26).

Figure 7–26 *Relay operations require at least two pumps, usually take longer to set up, and are more complicated to operate.*

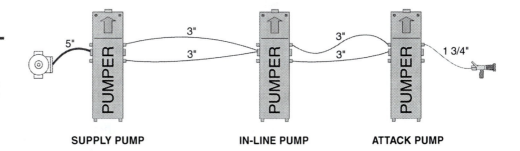

SUPPLY PUMP   IN-LINE PUMP   ATTACK PUMP

A relay operation is similar to the operation of a two-stage pump operating in the series mode. Recall from Chapter 4 that in the series mode, water is discharged from one impeller to the intake of the second impeller. In a relay operation, water is discharged from one pump to the intake of a second pump. However, water flows through supply hose between the two pumps rather than through fixed piping within the pump. In the series mode, the flow available to the second impeller is limited by the flow produced by the first impeller. The same holds true for relays in that the flow available to the second pump is limited by the flow generated by the first pump. For this reason, the largest pump should be placed at the source when possible. In a series mode, pressure is doubled as the second impeller takes the pressure generated by the first impeller and adds the pressure it generates. The same concept holds true for relay operations in that the second pump can take advantage of the pressure provided by the first pump. However, in a relay the pressure generated by the first pump will be reduced by friction in hose lines and elevation when it enters the intake of the second pump.

The major difference between a relay operation and a pump operating in the series mode is the ability to independently control changes in pressure and flow. In a series mode, each impeller rotates at the same speed. Therefore, a change in flow in the first impeller will automatically be pumped by the second. Further, a change in pressure by the first impeller will automatically be doubled as it passes through the second impeller. In each case, the change is predictable and automatic. In a relay operation, the impellers in one pump do not rotate at the same speed as other pumps and are therefore able to generate different flows and pressures. When the pumps are connected and flowing, changes in pressures and flows in one pump affect the entire system.

One of the major difficulties in relay operations is how to control changes in pressures and flows. When additional water is needed for the incident, flow must increase from the supply pump and all other pumps in the relay. If an attack line is shut down, pressure can increase throughout the system. The manner of controlling pressures and flows in a relay depends, in part, on the type of relay used. In general, relay operations can be either open or closed.

Closed Relay Operations

In a **closed relay** system, water enters the relay at the supply and progresses through the system to the attack pump. This system is similar to the pump operating in the series mode in that all the water and pressure is delivered from one pump to the next pump in the relay. Because all the water is contained, controlling pressure and flow requires that each pump be changed. For example, if an attack line is shut down, the attack pump would compensate for reduced flows as would each of the pumps in the relay. Controlling pressure and flow with this system is difficult, at best, in that changes occur quickly and affect the entire relay operation. Pump operators would find it difficult to keep up with needed changes at each pump as well as to coordinate efforts with other pumps in the relay.

■ **Note**

One of the major difficulties in relay operations is how to control changes in pressures and flows.

closed relay

relay operation in which water is contained within the hose and pump from the time it enters the relay until it leaves the relay at the discharge point; excessive pressure and flow is controlled at each pump within the system

Open Relay Operations

open relay

relay operation in which water is not contained within the entire relay system; excessive pressure is controlled by intake relief valves, pressure regulators, and dedicated discharge lines that allow water to exit the relay at various points in the system

In an **open relay** system, flow and pressure are not contained within the total system. In other words, all the flow and pressure from one pump is not always delivered to the next pump. This design allows a continuous flow through the relay, which reduces the need to make constant changes as well as reducing the overall effect of pressure changes within the system. Open systems can be set up according to one of three methods.

One method simply dedicates and maintains an open discharge line from the attack pump to flow excess water. In this case, once flow is initiated, only the attack pumper is required to make changes. If one or more attack lines are shut down, the pump operator simply increases flow through the dedicated line. This also has the benefit of quicker responses to change as well as having additional water readily available. Care must be taken concerning where the excess water is flowing. This system tends to pump more water than is actually needed.

Another method is to have the relay deliver its water directly into a portable tank rather than directly into the intake of the attack pump. The attack pump would simply draft from the portable tank. In this case, the relay operates continuously without needing constant changes. If attack lines are shut down, the attack pump reduces flow and the portable tank simply overfills. This system also provides ready access to additional water if needed.

Finally, the relay can be set up to take advantage of new automatic intake and discharge relief valve requirements. In this case, each pump sets its relief valves(s) to control pressure increases within the relay. When pressures rise above the setting, relief valves automatically open to dump excess pressure. This system has the added benefit of being able to automatically increase flows by simply increasing the relief valve setting. Relief valve requirements are discussed in both NFPA 1901 and NFPA 1962.

Designing Relays

Step One The first step in relay design is to evaluate each of the following factors:

- Amount of water to flow
- Available water at the supply
- The size and number of pumps available
- The size and length of supply hose available
- The distance from the source to the incident

The weakest of these factors will be the limiting factor of the relay design. For example, a relay operation capable of flowing 1,000 gpm is limited when the supply can only provide 500 gpm. In addition, a 1,000-gpm pump equipped with 500 feet of 2½-inch supply hose is not able to pump the quantity or distance as the same pump equipped with 500 feet of 4-inch supply hose.

Two of the more important factors to consider in the design of a relay are the amount of water the relay is expected to flow and the water available at the supply. The basic design of the entire system will be affected by these factors. Another important consideration is the number of pumps available and their rated capacity. Obviously, pumps must be of sufficient capacity to provide the required flow in the relay. Pumps must also be able to generate sufficient pressures to move this flow over a distance. When designing a relay, it is important to keep in mind the relationship between flow and pressure. When pressure increases, flows decrease. Recall, as mentioned earlier in this chapter that pumps at draft are expected to flow 100% capacity at 150 psi, 70% capacity at 200 psi, and 50% capacity at 250 psi. In essence, increased pressures reduce the quantity of water that can be pumped, which is often the case during relay operations.

The size of hose available for use in the relay is another important consideration from two perspectives. First, it is important to keep in mind the relationship of hose size with the amount of water it can flow and the friction loss it develops (see classification of hoses in Chapter 6). In general, larger diameter hose will flow more water and have less friction loss than smaller diameter hose. Second, the highest operating working pressure of the hose must be considered. For medium and small diameter hose attack lines, the highest operating pressure is 275 psi, while the highest operating pressure for LDH supply lines is 185 psi according to NFPA 1961 and 1962. Because of the nature of relay operations, supplying water over distances, pressures can easily exceed the 185 psi operating pressure of LDH supply hose. To safely utilize LDH, NFPA 1962 requires a discharge relief device with a maximum setting no higher than the service test pressure of the hose in use.

There are two additional considerations related to pressure in relay operations. The first relates to the gain or loss in pressure resulting from changes in elevation within the system. In general terms, pumping uphill is harder than pumping downhill; more pressure is needed to pump water uphill and less pressure is needed to pump water downhill. For every one foot of elevation gain, pressure will increase .434 psi. If the average height of a single story is 10 feet, then the pressure gain will be 4.34 psi (.434 psi/ft × 10 ft = 4.34 psi). As a rule of thumb, add or subtract 5 psi for each story (or each 10 feet) in elevation gain or loss. (Additional information on pressure gain and loss is presented in Section 4 of this book.) Another consideration in the design of a relay is the pressure required at the intake of each pump within the relay. When one pump flows water to the next, friction loss in the hose will reduce the discharge pressure. The goal is to provide enough pressure from the first pump to cover the loss in pressure from friction so that at least 20 psi remains when water enters the second pump. The purpose of maintaining a minimum of 20 psi is to ensure that the pump will not cavitate. Recall that cavitation can damage pumps, cause a loss of prime, and reduce pumping efficiency. Cavitation is discussed in greater detail in Chapter 9.

Finally, the distance between the source and the incident is important. In general, the farther the distance, the more resource-intensive the relay will be. When large flows are required, either large pumps with LDH will be spaced quite some distance apart or smaller pumps with medium diameter hose will be spaced rather close together in the relay (Figure 7–27). In reality, a variety of combinations can occur, each depending on the weakest component in the system.

Step Two When all of these factors are considered, the next step is to determine the distance between pumpers. Unless each pump has the same capacity and available hose, the distance between one pump and the next will be different. The goal is to maximize the distance between pumps while providing the required flow. This is accomplished by first determining the pump discharge pressure that can provide the required flow in the relay. For example, a relay operation requiring 500 gpm

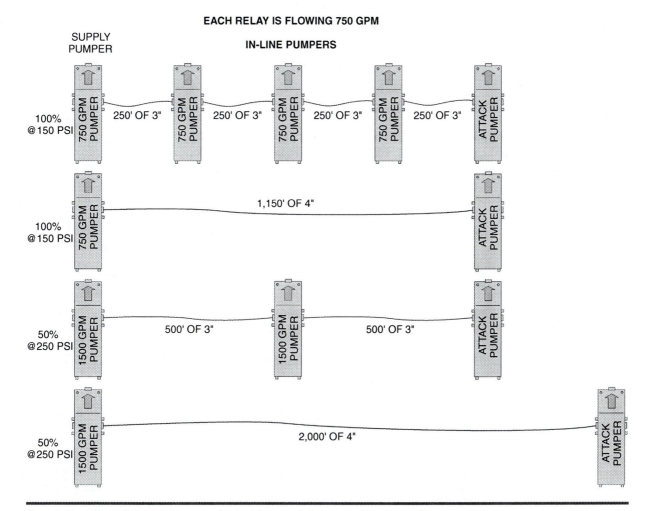

Figure 7–27   *When distances between the incident and the source increase, relay operation resources increase as well.*

can be provided by a 750-gpm pump operating at 200 psi (70% capacity) or a 1,000-gpm pump at 250 psi (50% capacity).

The next step is to determine the distance between pumps. When the water leaves the discharge side of the pump, the pressure will be reduced by friction as water travels to the next pump. The question, then, is how far can the water travel until the pressure is reduced to 20 psi (the minimum intake pressure for pumps in a relay)? The following formula can be used to determine this distance:

$$(PDP - 20) \times 100 / FL \tag{7–3}$$

where   PDP   =   pump discharge pressure
           20   =   reserved intake pressure (psi) at the next pump
        100   =   length (feet) of one section of hose (the most common length of hose is 50 feet; however, friction loss calculations use 100-foot increments/sections)
           FL   =   Friction loss (psi) per 100-foot section of hose

Take, for example, a 750-gpm pump flowing 500 gpm through 4-inch hose at 200 psi pump discharge pressure. The friction loss in 4-inch hose flowing 500 gpm is 5 psi per 100 feet (see chart on friction loss in Appendix F). In this example, the pressure in the hose will be 20 psi after water travels a distance of 3,300 feet [(185 − 20) × 100 / 5 = 3,300]. Note that the pump discharge pressure was reduced to 185 to comply with supply hose maximum operating pressure. What distance will water travel if the 500-gpm relay flow is provided by a 1,000-gpm pump with 3-inch supply line? First, the 1,000-gpm pump can provide 500 gpm at a pump discharge pressure of 250 psi. Next, the friction loss in 3-inch hose flowing 500 gpm is 20 psi per 100 feet. In this example, water will travel 1,150 feet before it reaches 20 psi [(250 − 20) × 100 / 20 = 1,150]. In comparison, if the 3-inch hose is replaced with 2½-inch hose in the last example, water will travel 460 feet when the pressure is reduced to 20 psi [(250 − 20) × 100 / 50 = 460].

Step Three   The last step is to lay the supply lines and position the pumps. The largest pump should be positioned at the water source whenever possible. The next pump in the relay simply lays a line equal to the predetermined distance. When all the lines are in place, the relay operation can begin. The steps for initiating relay flows are discussed in Chapter 8.

TANKER SHUTTLE OPERATIONS

tanker shuttle

water supply operations in which the apparatus is equipped with large tanks to transport water from a source to the scene

Tanker shuttles are those operations where apparatus equipped with large tanks transport water from a source to the scene (Figure 7–28). Tanker shuttle operations are used for two general reasons. First, they are used when pressures and flow from the supply source or pump, hose size, and distance limit the ability to move water from the supply to the incident. Second, they are used when obstacles such as ele-

Figure 7–28

Apparatus equipped with large tanks are used in tanker shuttle operations. Courtesy KME Fire Apparatus.

vation, winding roads, intersections, and railroad crossings limit the use of other supply methods.

The components of a tanker shuttle include multiple tankers, pumps, a fill site where tankers receive their water, and a dump site where tankers unload their water (see Figure 7–29). As with relay operations, the fill site supply can be either a static source or a hydrant. This section, then, discusses the unique considerations and setup requirements for tanker shuttle operations.

Shuttle Equipment

The equipment used in shuttle operations includes pumpers, tankers, portable dump tanks, and jet siphons. At least one pumper is required at the scene and uses the water delivered by the shuttle to supply attack and exposure lines. Typically, the pumper will either draft from a portable tank or will be supplied under pressure by a nurse tanker. Other pumps can be located at the fill site to assist with rapid filling of tankers or at the dump site to assist in the movement of water to the scene. For example, limited access to the scene may require that tankers deliver their water some distance from the scene. In this case, an additional pumper is required to move water from the dump site to the attack pumper.

Tankers used in shuttle operations should be sufficient in capacity to provide required flows. According to NFPA 1901, tankers should have a minimum capacity of 1,000 gallons. In addition, the tank must be able to both fill and unload at a rate of 1,000 gpm. Adequate ventilation is also required to ensure that the tank is not damaged during filling and unloading operations. The weight of the tanker is an important consideration for maintaining control as well as weight limits of roads and bridges. Consider the weight of water in a 1,000-gallon tanker. Since the weight of one gallon of water is 8.35 pounds, the weight of 1,000 gallons is 8,350 pounds (8.35 lbs per gal × 1,000 gal = 8,350 lbs), or 4.175 tons (8,350 lbs ÷ 2,000 lbs per ton = 4.175 tons). In general, smaller tankers are better for short-distance

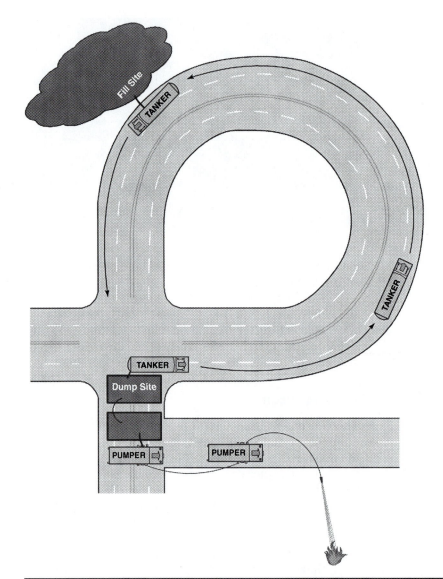

Figure 7–29   *Major components within a tanker shuttle operation.*

portable dump tank
a temporary reservoir
used in tanker shuttle
operations that
provides the means to
unload water from
a tanker for use by a
pump

jet siphon
device that helps
move water quickly
without generating a
lot of pressure and
that is used to move
water from one
portable tank to
another or to assist
with the quick off-
loading of tanker
water

shuttles because they are able to load and unload water at a faster rate than larger tankers, which are better suited for longer distances.

Portable dump tanks and **jet siphons** are valuable pieces of equipment for
shuttle operations. Portable tanks provide the means to unload tanker water for use
by a pump (Figure 7–30). These tanks can be set up in almost any location where
the ground is relatively level. Most portable dump tanks have a capacity of 1,000

Figure 7–30 *Portable tanks are an essential piece of equipment in most shuttle operations. Courtesy Ziamatic Corporation.*

to 3,000 gallons. Some are equipped with special devices to ease filling and transferring from one tank to another. Jet siphons (Figure 7–31) are devices that help move water quickly without generating a lot of pressure (Figure 7–32). They are used to move water from one portable tank to another (Figure 7–33) or to assist in the quick off-loading of tanker water.

Figure 7–31 *Example of two different jet siphons. Courtesy Ziamatic Corporation.*

Figure 7–32 *Jet siphons use very little water and pressure to move large quantities of water. Courtesy Greg Burrows.*

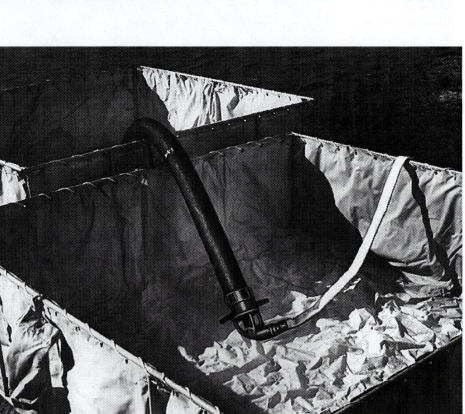

Figure 7–33 *Jet siphons move water from one drop tank to another. Courtesy Ziamatic Corporation.*

Designing a Tanker Shuttle

Although the specific design of a tanker shuttle will vary from one operation to another, three basic components exist in all shuttle operations: the fill site, the dump site, and the shuttle flow capacity. Each of these components must be carefully coordinated in order to efficiently and effectively supply water.

fill site
location where tankers operating in a shuttle receive their water

Fill Site Refilling tankers occurs at the **fill site**. Obviously, the goal is to fill the tankers in a fast and efficient manner. One consideration for a fast and efficient fill operation is the source used to supply the tankers. Both hydrants and static sources can be used to fill tankers. The most efficient method of filling a tank is to connect a hydrant directly to the tank's intakes (Figure 7–34). When hydrant pressures are weak, a pumper may be used to decrease fill times. When hydrant flows are low, a portable dump tank can be used to allow a pumper to draft with greater flows. If a tank does not have an intake valve, the less efficient method of filling the tank from the top must be used.

A second consideration for a quick and efficient fill operation is to provide adequate access to the fill site. In addition, the fill station should be set up to quickly connect and disconnect fill lines. If possible, the fill site should have two complete sets of fill lines available. While one tank is being filled, the second tanker can be connected and standing by. Filling two tankers at the same time may actually increase fill times.

■ **Note**
The goal at the fill site is to fill the tankers in a fast and efficient manner.

Figure 7–34 *The fill site is the location where tankers receive water. Courtesy Greg Burrows.*

Finally, safety should be considered at fill sites. One potential safety concern is that apparatus will be moving in close proximity to personnel and equipment. Another concern is adequate venting when a tank is being filled. Improper venting can be a significant hazard to both equipment and personnel.

Dump Site   The **dump site** is the location where tankers deliver their water. The goal, again, is to quickly unload the water and head back to the fill site. As with the fill site, issues of access and safety must be considered. Several options are available when delivering the tanker's water. Tankers can deliver their water directly to the attack pump, to another tanker (called a *nurse tanker*), or to a portable dump tank (Figure 7–35). Each tanker should carry a portable dump tank so it can quickly unload its water and head to the fill site. In some cases, the dump site will have multiple portable dump tanks allowing several tankers to unload their water at the same time. The dump site should be located where adequate room is available for tankers to maneuver.

Shuttle Flow Capacity   **Shuttle flow capacity** is the volume of water that can be pumped without running out of water. The flow capacity of a shuttle is limited by the volume of water being delivered and the time it takes to complete a shuttle cycle. The volume of water being delivered depends on the size and number of tankers. The **shuttle cycle time** is the total time it takes to dump water and return with another load. The cycle time includes the time it takes to fill the tanker, the time it takes to dump its water, and the travel distance between the fill and dump sta-

dump site
location where tankers operating in a shuttle unload their water

shuttle flow capacity
the volume of water a tanker shuttle operation can provide without running out of water

shuttle cycle time
the total time it takes for a tanker in a shuttle operation to dump water and return with another load; including the time it takes to fill the tanker, to dump the water, and the travel distance between the fill and dump stations

Figure 7–35 *Tanker dumping water directly into a portable dump tank. Courtesy Ziamatic Corporation.*

tions. In addition, the time a tanker must wait, if any, while other tanks are filling or dumping is included in the cycle time.

Individual shuttle tanker flow capabilities can be determined by dividing tank volume by the time it takes to complete a cycle. For example, a 1,500-gallon tanker that can complete a cycle in 10 minutes will have a shuttle flow rating of 150 gpm (1,500 / 10 = 150 gpm). Cycle times may vary because different size tanks can fill and dump their water at different speeds. In addition, several fill sites may be used that will have different travel times. A 1,000-gallon tanker with an 8-minute cycle time will have a shuttle flow capacity of 125 gpm. When the two tankers are utilized in the same shuttle, the combined shuttle flow will be 275 gpm.

Combined shuttle flows can be calculated by adding individual tank flows (as in the previous example) or by adding the volumes of all tankers and dividing the average cycle time for all tankers. In this case the total volume would be 2,500 and the average cycle would be 9 minutes, for a total combined flow of 277 gpm (2,500 / 9 = 278 gpm). The slight difference between the two methods is the result of averaging the cycle times.

❗Safety

● Improper venting while filling a tank can be a significant hazard to both equipment and personnel.

SUMMARY

The first step in fire pump operations is securing a water supply. The basic water supplies available to pump operators are onboard tanks, static sources, and municipal systems. These sources can be used in combination with relay operations and tanker shuttle operations. In reality, any and all of the water supplies can be used in a variety of configurations. The onboard water source is by far the fastest, yet most limiting. Municipal water supplies are by far the most common. A variety of hose layouts are available depending on required flow, pump size, hose size, and appliances available. Drafting operations from static sources tend to take longer to set up and are demanding on both the apparatus and the pump operator. Relay operations and tanker shuttle operations should be designed rather than just pieced together.

The most important concept in securing a water supply is knowing the expected flow. Simply selecting the first available source may lead to inefficient utilization of the source and pump. It is also important to understand the considerations for each of the water sources. Preplanning will make the hectic first minutes of an incident a little less stressful for the pump operator.

Chapter Formulas

7–1   Determine the flow from hydrant pitot readings.

$$Q = 29.84 \times c \times d^2 \sqrt{p}$$

7–2   Calculate the discharge at the specified residual pressure and/or desired pressure drop.

$$Q_R = Q_F \times (h_r^{0.54} / h_f^{0.54})$$

7–3   Determine the distance between pumps.

$$(PDP - 20) \times 100 / FL$$

REVIEW QUESTIONS

Key Terms and Concepts

On a separate sheet of paper, identify and/or define each of the following.

1. Required flow
2. Supply layout
3. Onboard supply
4. Municipal supply
5. Wet barrel hydrant
6. Dry barrel hydrant
7. Reverse lay
8. Forward lay
9. Static source
10. Drafting
11. Static source hydrants
12. Relay operation
13. Tanker shuttle
14. Fill site
15. Shuttle cycle time

Multiple Choice and True/False

Select the most appropriate answer.

1. The estimated flow of water needed for a specific incident is called
   a. available flow.
   c. critical flow.
   b. required flow.
   d. incident flow.

2. In general, larger diameter supply hose will provide more friction loss over longer distances than smaller supply hose.

   True or False?

3. Each of the following factors contribute to the availability of water except

   a. quantity.   **c.** accessibility.

   b. flow.   **d.** pump capacity.

4. Supply reliability is the extent to which the supply will consistently provide water.

   True or False?

5. Which of the following water supplies is usually the easiest to secure yet often the most limited?

   a. municipal systems   **c.** rivers

   b. ponds   **d.** onboard tank

6. How long can a 1,000-gallon onboard water tank sustain two 150-foot 1¾-inch hose lines each flowing 125 gpm?

   a. 2 minutes   **c.** 6 minutes

   b. 4 minutes   **d.** 8 minutes

7. Typically, the same municipal water supply system will provide water for both normal consumption such as household and industrial uses, as well as for emergency use to hydrants and fixed fire-protection systems.

   True or False?

8. The type of hydrant usually operated by individual control valves for each outlet is called a(an)

   a. individual outlet hydrant (IOH).

   b. dry barrel hydrant.

   c. wet barrel hydrant.

   d. None of the above are correct.

9. Which of the following NFPA standards suggests that hydrants be classified and color coded based on their rated capacity?

   a. 291   **c.** 1002

   b. 1500   **d.** 1901

10. A hydrant with a red bonnet would most likely flow _____ gpm according to NFPA's hydrant classification system.

   a. less than 500

   b. between 500 to 750

   c. between 500 and 1,000

   d. greater than 1,000

11. If the pump is located at the hydrant, which of the following has not occurred?

   a. reverse lay

   b. forward lay

   c. first pump in a relay

   d. boosting pressure using a four-way hydrant valve

12. Drafting is the process of moving or drawing water away from a source through hard suction hose to the suction side of a pump using a suction process.

   True or False?

13. If the priming system reduced the pressure to 0 psi, a perfect vacuum, water would be forced to a height of _____ feet, also known as theoretical lift.

   a. 7.35   **c.** 22.5

   b. 14.7   **d.** 33.81

14. Because priming systems come nowhere near creating a perfect theoretical lift, the general rule of thumb is to attempt to pump no more than two-thirds of the theoretical lift, which is

   a. 7.35 feet.   **c.** 22.5 feet.

   b. 14.7 feet.   **d.** 33.81 feet.

15. Which of the following best describes a "static hydrant"?

   a. prepiped lines that extend into a water source

   b. wet or dry barrel hydrant when no water is flowing

   c. hydrant that is identified as being out of service

   d. no such hydrant

16. All of the following are factors that affect static water supply reliability except
    a. environmental and seasonal conditions.
    b. condition of hydrants and water mains.
    c. pump and equipment used to draft.
    d. pump operator's drafting knowledge and skill.

17. One of the major difficulties in relay operations is
    a. walking from one pumper to the next.
    b. having enough pumpers to use in the relay.
    c. ensuring that an adequate water supply is chosen.
    d. how to control changes in pressures and flows within the system.

18. A relay where water enters at the supply and progresses through the system to the attack pump is called a(an)
    a. relay system.
    b. open relay system.
    c. traditional relay.
    d. closed relay system.

19. According to NFPA 1961, supply hose operating pressure should not exceed
    a. 150 psi.　　　　c. 200 psi.
    b. 185 psi.　　　　d. 250 psi.

20. The second pumper, all in-line pumpers, and the attack pumper should maintain at least _____ psi intake pressure.
    a. 5　　　　　　　c. 35
    b. 20　　　　　　d. 50

21. The weight of water in a 1,500-gallon tanker weighs a little over
    a. 2 tons.　　　　c. 6 tons.
    b. 4 tons.　　　　d. 8 tons.

22. A hydrant with an orange bonnet can deliver which of the following flows ranges?
    a. 200 to 500 gpm　　c. 1,000 to 1,500 gpm
    b. 500 to 999 gpm　　d. 1,500 gpm or greater

Short Answer

On a separate sheet of paper, answer/explain the following questions.

1. Explain the importance of preplanning water supplies.

2. When the required flow is unknown, what should govern the selection of a water source?

3. What effect do higher pump pressures have on the ability of a pump to flow capacity?

4. What effect does the size of hose have on the ability to move water?

5. What factors contribute to "water availability?"

6. What affects supply reliability for the water sources?

7. What affects the supply hose layout?

8. Why is it important to fully open or close dry barrel hydrants?

9. List four factors that may reduce the reliability of municipal water supply systems.

10. If the pressure inside a centrifugal pump is reduced to 8.7 psi, how high will atmospheric pressure raise water?

11. Explain the limitations of using LDH as a supply between pumps in a relay.

12. List two reasons for using a tanker shuttle operation.

ACTIVITIES

1. Identify all usable static sources within your response district.

2. Determine two potential relay operations within your response district. If hydrants are the predominate water supply, assume several hydrants are out of service.

3. Develop specific procedures and forms to conduct flow testing using NFPA 291 as a guide.

PRACTICE PROBLEMS

1. Determine how long a 500-gallon tank and a 1,000-gallon tank can sustain two 1½-inch lines flowing 100 gpm each.

2. Design a relay using the following information and provide the information requested:
   - Required relay flow of 750 gpm
   - Distance between pump and incident is 1,500 feet
   - Water supply is a hydrant with an orange bonnet
   - Pumpers and hose available:
     P1 = 750-gpm pump with 500 feet of 2½-inch
     P2 = 750-gpm with 1,000 feet of 4-inch
     P3 = 1,500-gpm with 600 feet of 3-inch
   a. Where will you place each of the pumpers?

   b. What will be the distance between each pump?

3. Determine the individual and combined shuttle flows using the following information, where fill time and dump time includes connecting and disconnecting hose:

T1:	Tank size	1,000 gallons
	Fill time	3 minutes
	Dump time	2 minutes
	Travel time	5 minutes round-trip
T2:	Tank size	1,500 gallons
	Fill time	3 minutes
	Dump time	4 minutes
	Travel time	5 minutes round-trip
T3:	Tank size	2,000 gallons
	Fill time	4 minutes
	Dump time	4 minutes
	Travel time	6 minutes round-trip

BIBLIOGRAPHY

Eckman, William F. *The Fire Department Water Supply Handbook.* Saddle Brook, NJ: Penn-Well Publishing, 1994.
   Provides in-depth information on water supplies, especially relay operations and tanker shuttle operations.

ISO Fire Suppression Rating Schedule. New York: Insurance Service Office, 1980.
   Water supply section provides information on how water supplies affect rating.

Chapter

8

PUMP OPERATIONS

Learning Objectives

Upon completion of this chapter, you should be able to:

■ List and discuss the steps in each of the following basic pump procedures: engaging the pump, priming the pump, operating the transfer valve, operating the throttle control, and setting the pressure regulating device.

■ Explain pump procedures when utilizing the onboard water tank, hydrants, drafting operations, and relay operations.

■ List and discuss the three primary methods used to transfer power from the engine to the pump.

■ Discuss basic considerations for positioning pumping apparatus.

NFPA 1002
Standard for Fire Apparatus Driver/Operator
Professional Qualifications
(2003 Edition)

This chapter addresses parts of the following knowledge elements within sections 5.2.1, 5.2.2, and 5.2.4:

Safe operation of the pump

Ability to position a fire department vehicle and to operate at a fire hydrant and at a static water source

Power transfer from vehicle engine to pump

Draft

Operate pumper pressure control systems

Operate the volume/pressure transfer valve (multistage pumps only)

Operate auxiliary cooling systems

Make the transition between internal and external water sources

INTRODUCTION

After a water supply has been selected and secured, the next activity in fire pump operations is to operate the pump. This activity moves water from the intake to the discharge side of the pump. Although a variety of pump sizes and configurations exists, the same basic steps must be taken to move water from the supply to the discharge point. These basic steps include the following:

Step 1: position apparatus, set parking brake, and let engine return to idle.

Step 2: engage the pump.

Step 3: provide water to the pump (prime if necessary).

Step 4: set transfer valve (if so equipped).

Step 5: open discharge line(s).

Step 6: throttle to desired pressure.

Step 7: set the pressure regulating device.

Step 8: maintain appropriate flows and pressures.

In general, these steps are presented in the order in which they are carried out during pump operations. However, the order may vary slightly depending on department policy, manufacturer recommendations, and type of water supply being used. Regardless of the information provided, pump operators should always follow department standard operating procedures (SOPs) and especially manufacturer's recommendations. Typically, pump manufacturers will provide operating and maintenance manuals as well as other pump training material such as slides, overheads, and videos. Pump operators should be intimately familiar with fire department SOPs, and department training manuals can provide excellent information on operating procedures.

This chapter presents the second task of fire pump operations, operating the pump. Basic pump procedure is discussed first, and then specific pump operations are presented for each of the five water supplies described in Chapter 7.

BASIC PUMP PROCEDURES

The difficulty in discussing pump procedures is that procedures can vary from department to department and from manufacturer to manufacturer. Therefore, terms such as "always" and "never" must be used with extreme care. Several caveats that come close to being universal among departments and manufacturers (or at least should be) for centrifugal pump operations include the following:

- Never operate the pump without water.
- Always keep water moving when operating the pump at high speeds.
- Never open, close, or turn controls abruptly.
- Always maintain awareness of instrumentation during pumping operations.
- Never leave the pump unattended.
- Always maintain constant vigilance to safety.

Beyond these, there are few ironclad rules for carrying out pump procedures. Basic pump procedures, then, are those general procedures commonly utilized during most pumping operations. Understanding these procedures will increase the ease and reduce the complexity of demanding pump operations.

Positioning Pumping Apparatus

One step required in every pump operation is positioning the pumper. In Chapter 7, we discussed a few considerations for positioning the pumping apparatus as it relates to securing a water supply. Here, we discuss some general considerations for positioning the pumping apparatus as it relates to attacking the fire. A few additional considerations are provided later in this chapter when discussing specific pump operating procedures.

In some cases, the positioning decision will be made by a higher level position within the incident command system. In such cases, the pump operator must position the apparatus as directed. However, there are times when the pump operator and/or officer have considerable autonomy when positioning apparatus. Even when the pump operator is directed where to locate the apparatus, the instructions are usually in general terms and the responsibility to position the apparatus efficiently and effectively rests with the pump operator. Regardless, a number of variables and considerations must be evaluated when selecting a location to position fire apparatus.

One of the more important considerations is the initial size-up of the incident. When approaching the scene, a careful assessment of the incident will help determine the most efficient and effective location. For example, if no fire or smoke is visible, the best location to park is near the main entrance of the occupancy. In such cases, the pump operator should plan ahead for a water source in case attack lines are needed. Other considerations include:

- *Follow SOPs.* Following SOPs takes some of the initial stress off the driver. In addition, the SOPs might require specific placement of initial arriving apparatus in order to maintain space for later arriving apparatus.

- *Consider the water supply.* Positioning apparatus for best use of water supplies should always be considered. As indicated in Chapter 7, preplanning can help ensure this is accomplished.

- *Consider tactical priorities.* The tactical priorities of the incident will influence where to position the pumping apparatus. For example, if rescue is the highest priority, the first in pumper might be positioned to help effect rescue either through an aerial device, if so equipped, or through the use of ground ladders. If the fire is small and a quick attack is needed, the apparatus should be positioned so that preconnects can be pulled and the onboard water supply used. If this is the case, the pumper must be positioned close enough for preconnected lines to reach the fire. If the fire is advanced, the driver/operator might be asked to provide large quantities of water to a master stream device. In such cases, a strong water supply must be secured.

- *Consider Surroundings.* There are several things to consider with regard to surroundings. One important consideration is radiant heat from the fire. Whenever possible, apparatus should not be subjected to high levels of radiant heat. In many cases, fire extension is not considered in apparatus placement. The growth of fire could, and has, occurred faster than firefighters could break down hose lines to reposition the pumper. Another important surrounding consideration is the potential for collapse. When possible, the apparatus should be placed outside the collapse zone. Preplanning can help determine the appropriate collapse zone distance. At a minimum, the collapse zone should be at least the same distance as the height of the structure. Other considerations include:

  - Power lines. Don't park under them.
  - Escape route. Consider developing one before one is needed.
  - Wind direction. Try to position the vehicle upwind.
  - Terra ferma. The harder the ground, the better and the higher the apparatus, the easier it is to pump and the less likely hazardous runoff will flow under the apparatus.

Pump Engagement

pump engagement
the process or
method of providing
power to the pump

One step required in most every pump operation is **pump engagement**. Typically this means providing power to the pump from either a separate engine or from the engine that drives the apparatus (often referred to as the *drive engine*). When powered by a separate engine, sometimes called an auxiliary engine, the pump is typically connected directly to it. In this case, the pump can be operated independently of the drive engine. Auxiliary engine sizes range from small gasoline engines, such as used in wildland vehicles, to large diesel engines used in airport rescue and firefighting (ARFF) apparatus. Several benefits of this system include:

- Increased versatility in placing/locating the pump on the apparatus
- Allows pump-and roll-capability—important for wildland apparatus
- Pump pressure is not controlled by the speed of the apparatus

When powered by the drive engine, the pump is typically connected indirectly to the engine. The majority of main pumps receive their power in this manner. The procedure for engaging the main pump will depend on the location of the pump and how power is transferred from the drive engine. The power to main pumps can be transferred from the drive engine by one of three general methods: power takeoff (PTO), front crankshaft, or split-shaft.

PTO method (pump engagement)
method of driving a pump in which power is transferred from just before the transmission to the pump through a PTO; a method of power transfer that allows either stationary or mobile operation of the pump

Power Takeoff (PTO) The **PTO method** transfers power to the pump through a PTO on the transmission (Figure 8–1). In doing so, the pump can operate in either a stationary position or while the vehicle is in motion. Wildland, mobile water supply, and, increasingly, initial attack apparatus use the PTO method to transfer power from the drive engine to the pump. Older vehicles that use the PTO method tend to power smaller pumps up to 500 gpm. Today, pumps up to 1,250 gpm can be powered using the PTO method. When the pump is engaged and the vehicle is in motion, the speed of the pump is controlled by the speed of the apparatus. When the apparatus speeds up, the pump will speed up as well.

The basic steps to engage a pump utilizing a PTO are as follows:

Step 1: bring apparatus to complete stop and let engine return to idle speed.

Step 2: disengage the clutch (push in the clutch pedal).

Figure 8–1 *Pumps driven by a PTO can be operated in a stationary position or while the vehicle is in motion. Courtesy Waterous Company.*

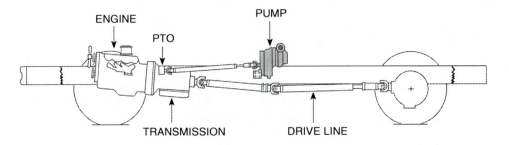

ENGINE PUMP PTO TRANSMISSION DRIVE LINE

Step 3:   place transmission in neutral.

Step 4:   operate the PTO lever.

Step 5a: *for mobile pumping,* place the transmission in the proper gear.

Step 5b: *for stationary pumping,* place transmission in neutral and be sure to set parking brake.

Step 6:   engage the clutch slowly.

front crankshaft method (pump engagement)
method of driving a pump in which power is transferred directly from the crankshaft located at the front of an engine to the pump; this method of power transfer is used when the pump is mounted on the front of the apparatus and allows for either stationary or mobile operation

To disengage the pump, simply return the engine to idle, disengage the clutch, and operate the PTO lever.

Front Crankshaft   Another way to transfer power to the pump is the **front crankshaft method**. In this method, the pump is connected directly to the crankshaft located at the front of an engine. This method is used when the pump is mounted on the front of the apparatus (Figure 8–2). As with the PTO method, the pump can be operated while in motion or in a stationary position. Pump capacities for older vehicles are typically limited to pumps up to 750 gpm while newer pumps up to 1,250 gpm can be powered in this manner.

The basic steps to engage a pump connected to the front crankshaft are as follows:

Step 1:   bring apparatus to complete stop, apply parking brake, and let engine return to idle speed.

Step 2:   put transmission in neutral.

Step 3:   operate the pump control (may be in the cab or on the front of the pump). If the pump control is located on the pump, a "pump engaged" warning light is usually provided in the cab.

Step 4:   *for mobile pumping,* place transmission in the proper gear.

To disengage the pump, return the engine to idle and operate the pump control.

split-shaft method (pump engagement)
method of driving a pump in which a sliding clutch gear transfers power to either the road transmission or to the pump transmission; this method of power transfer is used for stationary pumping only

Split-Shaft   Finally, a **split-shaft method**, or midship transfer, can be used to transfer power to the pump. This system is used for large pumps mounted in the middle or toward the back of the apparatus. It allows power to be delivered to either

Figure 8–2 *Front-mounted pumps receive power from the front of the engine crankshaft. Courtesy Waterous Company.*

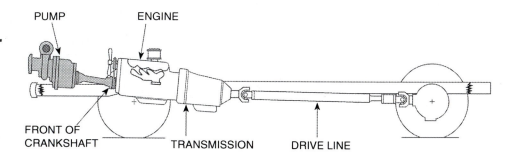

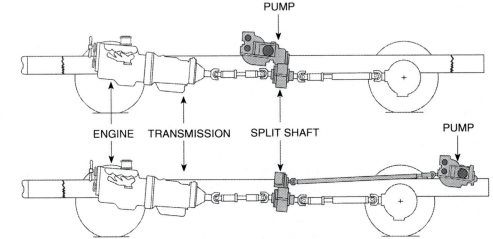

PUMP

ENGINE   TRANSMISSION   SPLIT SHAFT   PUMP

Figure 8–3   *The most common and versatile manner to power the main pump is via a split-shaft transmission. Courtesy Waterous Company.*

the road transmission or to the pump transmission through a sliding clutch gear (Figure 8–3). The major concern with transferring power in this manner is to ensure that the road transmission is properly disconnected when operating the pump. If the road transmission is not properly disconnected, the apparatus may attempt to move when the pump throttle is increased. Often, indicating lights are provided to let the pump operator know when the road transmission is properly disconnected and when the pump is properly engaged.

The basic steps to engage a pump utilizing a split-shaft arrangement are as follows:

Step 1: bring apparatus to a complete stop and let the engine return to idle speed.

Step 2: put transmission in neutral.

Step 3: apply parking brake.

Step 4: move the pump shift control from the "ROAD" position to the "PUMP" position (see Figure 8–4).
Note: the shift-indicating light should come on, often a green "PUMP ENGAGED" light, which indicates the shift is complete (if so equipped).

Step 5: shift transmission into pumping gear (usually highest gear).
Notes: an indicating light should come on, often a green "OK TO PUMP" light (indicates road transmission properly disengaged). Also, the speedometer should increase (5–15 mph) and truck should not attempt to move if the engine speed is increased.

To disengage the pump, complete the following:

Step 1: return engine to idle.

Step 2: put transmission in neutral.
Note: you may need to wait 5 to 10 seconds to allow drive shaft to stop turning.

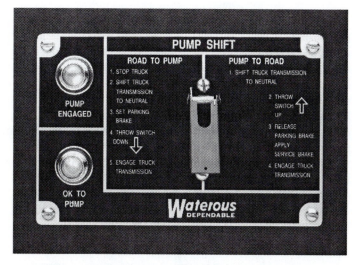

Figure 8–4
Indicating lights on pump shift control devices let the pump operator know when it is safe to pump. Courtesy Waterous Company

Step 3: move pump shift lever from "PUMP" to "ROAD" position
Note: shift warning light should turn off.

Chocking the Vehicle

As mentioned in Chapter 3, a good safety habit to get into is to ensure the apparatus is chocked whenever the pump operator leaves the driver's seat. This is especially important during pumping operations in that improper pump engagement may not disengage the drive transmission. In addition, setting the parking brake may be accidentally overlooked or the parking brakes may fail, allowing the apparatus to move.

❗Safety
● Whenever the pump operator leaves the driver's seat, he or she should be sure the apparatus is chocked.

Priming the Pump

Recall that priming is simply the act of getting air out and water into a pump. When water is routinely carried in a pump, priming may not be necessary. When using the onboard tank or a hydrant as a water source, priming can occur as the supply pressure forces air out. Typically, this will only work for small air pockets in the pump and piping. The steps for priming the pump using the supply are as follows:

Step 1: open water supply to the intake side of the pump.

a. Onboard source: open the tank to pump.

b. Hydrant: open intake after bleeding air from supply line.

Step 2: open discharge gate (to allow supply pressure to force air out).

Step 3: engage the pump if not already engaged.

Step 4: slightly increase engine speed.
Note: a corresponding increase in the master discharge gauge means the pump is primed.

If these simple steps fail to get all the air out of the pump, the positive displacement priming pump must be used. This is usually the case if the pump is maintained without water. If the pump is being primed in conjunction with a drafting operation, the priming system must be used. The priming system is usually the fastest and most effective means to prime the pump. Some pump manufacturers suggest that priming occur before the pump is engaged to prevent damage caused by overheating. When the pump is engaged during the priming process, some experienced pump operators suggest slightly increasing engine rpms to decrease the priming time. However, the engine speed should not be increased over 1,200 rpms. Keep in mind that operating the pump dry even for short periods of time may quickly damage it. According to NFPA 1901, *Automotive Fire Apparatus*, a dry pump must be able to achieve a prime within 30 seconds for pumps rated at less than 1,500 gpm, and within 45 seconds for pumps rated at 1,500 gpm or larger. The steps to prime the main pump using the priming pump are as follows:

■ **Note**

Operating the pump dry for even short periods of time may quickly damage it.

Step 1: engage the pump.
    Note: some pump manufacturers recommend priming the pump before it is engaged.

Step 2: secure a water source (connect supply lines).

Step 3: close all discharge control valves.

Step 4: open intake for water supply (tank-to-pump, intake valve).

Step 5: engage primer.
    Note: the primer should not be operated more than 30 seconds, for pumps rated at less than 1,500 gpm, 45 seconds for 1,500-gpm or larger pumps.

Step 6: ensure pump is primed. If the pump is not engaged, you can tell that it's primed when the discharge from the priming pump changes from air/oil mixture to water/oil mixture. If the pump is engaged, when the master discharge gauge reads a positive pressure, sometimes a change in engine sound can be heard as it is placed under load when water enters the pump, indicating that it's primed.

Step 7: engage the pump if not already engaged.

Transfer Valve

transfer valve

control valve used to switch between the pressure and volume modes on two-stage centrifugal pumps

As mentioned in Chapter 4, two-stage centrifugal pumps typically provide the ability to operate in either a pressure or volume mode. The **transfer valve** is used to switch between the two modes. Recall that when pumping in the volume mode, each impeller is providing half of the flow while generating the same pressure. When pumping in the pressure mode, one impeller pumps to the next impeller where the same flow is pumped but the pressure is doubled.

In essence, when a large volume of water is expected to be pumped, the pump should be placed in the volume mode. When higher pressures are needed, the pump should be placed in the pressure mode. The general rule of thumb is to pump in volume when expected flows will be greater than 50% of the pump's rated capacity or when pressures will be less than 150 psi. When the flows are expected to be less than 50% of the rated capacity or when pressures will be greater than 150 psi, the pump should be operated in the pressure mode. For example, a 1,000-gpm pumper expecting to flow 750 gpm at 140 psi should have the transfer valve in the volume position. If the same pumper expects to flow 400 gpm at 220 psi, the transfer valve should be in the pressure mode position.

The transfer valve should be operated before pump pressure is increased. If the transfer valve is manually operated, higher pressures will make switching modes nearly impossible. When power-assisted transfer valves are provided, switching modes at higher pressures is possible. However, switching from the volume mode to the pressure mode may significantly increase pressures. In addition, switching modes may cause a loss in prime during drafting operations. As a general rule, slow the pump speed to idle and then change the mode.

Control Valve Operations

Recall from Chapter 6 that control valves start, stop, or direct water and should always be operated slowly. They are designed to be operated by either pulling and pushing, or by rotating up and down or sideways. Newer pump panel designs incorporate electronic control valve operation. The benefits of these electronic control valve actuators include slow activation, resistance to valve movement after setting, and ease of operation with higher pressures. Control valves are provided with a locking mechanism to ensure that they do not move after being set. Control valves can typically be locked by twisting the handle. Some control valves automatically lock, as is the case with most electronic actuators. Care should be taken to unlock the handle prior to changing the valve position. Most important, though, control valves should be operated in a slow and smooth manner. Abrupt operation of control valves can cause extensive damage and excessive wear and tear.

Throttle Control

throttle control
device used to control the engine speed, which in turn controls the speed of the pump, when engaged, from the pump panel

The purpose of the **throttle control** is to control engine speed, which in turn controls the speed of the pump. When the throttle is increased, the engine rpms increase and subsequently the speed of the pump impellers increases. The end result is an increase in water discharge pressure. The total discharge pressure of a pump is often referred to as *engine pressure*, *engine discharge pressure*, or *pump discharge pressure*. The term *engine discharge pressure* describes the relationship between the engine rpms and the discharge pressure. In general, an increase or decrease in the throttle produces a corresponding increase or decrease in the discharge pressure. This increase or decrease in engine pressure will be indicated on the master discharge gauge, located on the pump panel.

The throttle is typically found on the pump panel (Figure 8–5). To increase pressure, the throttle is usually turned clockwise, and to decrease pressure it is usually turned counterclockwise. The throttle should be increased and decreased

Figure 8–5 *The throttle, found on the pump panel, controls engine speed which in turn controls the speed of the pump. Courtesy W. S. Darley & Company.*

slowly. A red emergency button is often found on the throttle. Depressing the button allows the throttle to be moved in or out without turning. The intent of the button is to allow the pump operator to quickly reduce engine pressure in cases of emergency. Normal pump operations do not require the use of this button. When the button is used to increase pressure, the chances of high-pressure surges and damage to equipment may occur. When the pump is dry or when no water is flowing, the throttle should not be increased to high speed or damage may result. The most common throttle device is depicted in Figure 8–5. However, electronic throttle devices are increasingly being installed on newer pump panels (see Figure 8–6).

When throttling up, pump operators should listen for increased engine rpms. In addition, the master intake and discharge gauges should be constantly scanned (Figure 8–7). The master discharge gauge should show a corresponding increase as the throttle is increased. If a corresponding increase is not observed *when first initiating flows*, most likely the pump is not in gear or not primed. Check the pump and initiate pump engagement procedures if it is not properly engaged. If the pump is properly engaged, the pump must be primed. If a corresponding increase in pressure is not observed *during an operation*, the pump is flowing all the water from the supply. In this case, the throttle should be slowly decreased until the pressure begins to drop. If the throttle is increased past the point of a corresponding increase in discharge pressure, the pumping operation may fail and result in:

- Pump cavitation
- Loss of prime
- Intake lines collapsing
- Damage to municipal water mains

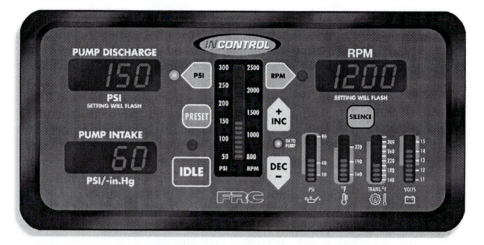

Figure 8–6
Electronic throttles are common on newer pump panels. Courtesy Fire Research Corporation.

Figure 8–7 *When increasing the throttle, pump operators should listen to the engine rpms as well as scan discharge gauges.*

When pumping from a hydrant, the master intake gauge will show a positive pressure. As the throttle is increased and more water is being pumped, the pressure reading will drop. When the pressure reading nears zero, the pump is again flowing all the water that is available from the supply. When using hydrants as a supply source, the intake pressure should not be reduced below 20 psi according

■ **Note**

If the engine speed begins to increase without a corresponding increase in pressure, the pump may be cavitating.

to NFPA 291, *Recommended Practice for Fire Flow Testing and Marking of Hydrants*. Maintaining 20-psi intake pressure will help guard against the pump cavitating or collapsing the intake lines. Finally, if the engine speed begins to increase without a corresponding increase in pressure, the pump may be cavitating. If this occurs, the engine speed should be reduced until the discharge pressure begins to drop. Cavitation is discussed in detail in Chapter 9.

When the throttle is increased and a corresponding increase in pressure is not observed, or when the master intake gauge is reduced to 10 psi, the pump is flowing all the water available from the supply. Pump operators can compensate for this occurrence by increasing the supply flow. Increasing the supply flow can be accomplished by adding another supply line or increasing the size of the supply line. In addition, one or more discharge lines can be either closed or replaced by smaller lines. Finally, nozzle flows can be reduced by using smaller flow nozzles, changing the setting on nozzles to lower flows, or simply decreasing pressure for automatic nozzles, which will adjust for reduced flows.

Pressure Regulator Operation

Pressure regulating devices provide a degree of safety to both personnel and equipment. Recall from Chapter 5 that two types of devices are used on fire department pumps, pressure governors and pressure relief valves. Pressure governors regulate pressure by controlling engine speed. Pressure relief valves regulate pressure by opening a valve to release excessive pressure to the intake side of the pump, to the discharge side of the pump, or to the atmosphere. The general steps for their use are provided as follows:

Pressure Governor

Step 1: turn on pressure governor.
Step 2: put the governor control into the rpm mode.
Step 3: set discharge pressure.
Step 4: put the governor control into the pressure mode.

Main Discharge Pressure Relief Valve

Step 1: increase throttle to desired operating pressure.
Step 2: turn on pressure relief valve.
    Note: the relief valve should be at the highest setting (relief valve hand-turned fully clockwise).
Step 3: turn relief-valve handle counterclockwise slowly until the valve opens and the light comes on. The discharge pressure will decrease and you may hear a change in engine sound.

Step 4: turn relief-valve handle clockwise until light goes off.

Step 5: relief valve is now set slightly above normal operating pressure. *Note:* when the relief valve directs excessive pressure to the intake side of the pump, the ability to operate properly may be reduced when excessive intake pressure occurs.

intake pressure relief valve

pressure regulating system that protects against excessive pressure build-up on the intake side of the pump

Intake Pressure Relief Valve NFPA 1901 now requires the use of **intake pressure relief valves** (Figure 8–8). The intent is to dump excessive pressure before it reaches the pump. According to NFPA 1901, intake relief valves must be adjustable. However, some relief valves are *field adjustable* during an operation while others are not. Field-adjustable intake relief valves (Figure 8–9) can be controlled by turning a handle (or knob) to one of several preset pressures closest to the maximum intake pressure desired. Nonfield-adjustable intake relief valves must be set using an external source and gauge. These intake relief valves are set in the same basic

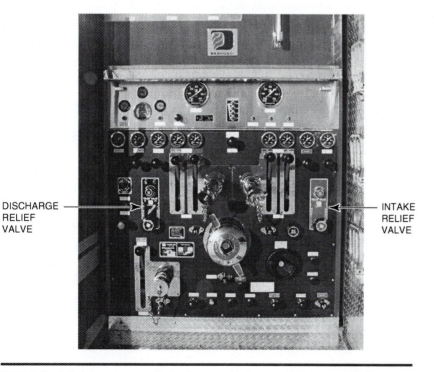

DISCHARGE RELIEF VALVE

INTAKE RELIEF VALVE

Figure 8–8 *Intake pressure relief valves control excessive pressure before entering the intake side of the pump. Courtesy W. S. Darley & Company.*

Figure 8–9 *Example of a field adjustable intake relief valve. Courtesy Waterous Company.*

manner as the main discharge relief device. The general steps for setting a nonfield-adjustable intake relief valve are as follows:

Step 1: connect a line from the discharge of the assisting pump to the intake containing the relief valve.

Step 2: ensure that all discharge and intake valves are closed.

Step 3: increase the discharge pressure of the assisting pump to the desired intake relief valve setting.

Step 4: set the intake relief valve.

    a. If the intake relief valve is already open, turn the control handle clockwise until the valve closes

    b. If the intake relief valve is not open, turn the control handle counter-clockwise until the valve opens.

ONBOARD WATER PROCEDURES

■ Note

Attack crews must be notified in advance of tank water running out.

Operating the pump with the onboard tank as a water supply is perhaps the easiest of all the pumping procedures. Pump operators should know in advance how long the supply will last. Care must be taken to ensure that the pump does not operate without water, or extensive damage will likely occur. In addition, attack crews must be notified in advance of tank water running out.

The general steps for pumping the onboard tank water are as follows:

Step 1: position apparatus, set the parking brake, and let engine return to idle.

Step 2: engage the pump.

Step 3: upon exiting the vehicle, chock the vehicle, and ensure that the pump panel "okay to pump" light is on.

Step 4: set the transfer valve. In most cases, this should be in series mode based on the limited supply of water in the tank.

Step 5: open "tank-to-pump" control valve.

Step 6: check to see that the master discharge gauge shows pressure, and if it does not, prime the pump (may have air leaks). You will have to prime the pump if it was stored dry.

Step 7: connect the discharge lines.

Step 8: open the discharge control valves.

Step 9: increase throttle.

Step 10: set the pressure regulating device.

Step 11: plan for more water!

With an additional water supply secured, the following steps should be taken:

Step 10: ensure that the pressure regulator is properly set.

Step 11: open the supply intake valve.
Note: intake pressure will increase, causing the governor to reduce engine speed or the relief valve to open, depending on which one is used.

Step 12: close the tank-to-pump valve.

Step 13: if equipped with a relief valve, reduce engine speed and reset the pressure regulator.

Step 14: slightly open the pump-to-tank valve to refill the onboard tank.

HYDRANT PROCEDURES

Hydrant operations range from simple to complex. The most time-consuming aspect is connecting the supply lines. When the apparatus is positioned next to the hydrant large diameter hose (LDH) is often used. Care should be taken to position the vehicle close enough to the hydrant, yet not too close. Positioning too close to the hydrant may cause kinking in the supply hose and cramped working conditions for the pump operator. When connecting supply hose to the side intake, the vehicle should be positioned just before or after the hydrant in relation to the intake to help reduce kinks in the supply hose. In addition, placing a couple of counterclockwise twists in the hose will not only help prevent kinks from forming while opening the hydrant, it has the effect of causing the hose to tighten on the intake. Clockwise rotations prior to charging can cause the hose to loosen on the intake and potentially coming off, causing serious injury. When connecting to the front intake, the vehicle must be positioned a few feet before the hydrant to reduce kinks in the hose. Positioning the vehicle with wheels turned at an angle allows minor distance adjustments to be made if necessary. After supply lines are connected, the basic steps for operating the pump with a hydrant as a supply source are as follows:

Step 1: position the apparatus, set the parking brake, and let the engine return to idle.

Step 2: engage the pump.

Step 3: upon exiting the cab, chock the wheels and ensure that the pump panel "okay-to-pump" light is on.

Step 4: connect hose from the hydrant to the intake side of the pump.

Step 5: slowly open the hydrant.

Step 6: bleed off air in the supply line.

Step 7: slowly open the intake valve.

Step 8: set the transfer valve to volume or pressure.

Step 9:   slowly open discharge(s).

Step 10: throttle up to desire pressure.
Note: ensure that the master intake gauge does not fall below 10 psi.

Step 11: set the pressure-regulating device.

TRANSITION FROM TANK TO HYDRANT

Often, pump operators must transition from onboard tank water to an external supply of water such as a hydrant. The pump operator must be able to make the transition to an external water supply without significant increase or decrease in pressure. Inadvertent flow interruptions could be extremely dangerous to internal attack crews. External water supplies should be connected as soon as possible to ensure tank water is not exhausted. The basic steps for transitioning from tank water to hydrant water include the following:

Step 1:   connect supply hose to intake as soon as possible after initiating flow from the tank.

Step 2:   slowly charge the hydrant.

Step 3:   bleed off the air within the supply hose.

Step 4:   make sure the pressure-regulating device is set.

Step 5:   slowly open the intake valve and close the tank-to-pump control valve while constantly monitoring discharge pressures.

Step 6:   when the tank-to-pump is fully closed, continue to open the intake control valve while also maintaining appropriate discharge pressure (either reduce pump speed or close individual discharge valves).

Step 7:   adjust pressure-regulating device as necessary.

Step 8:   crack the pump-to-tank (sometimes labeled "Tank Fill"), or tank-to-pump if no check valve is installed, to refill the tank.

DRAFTING PROCEDURES

Drafting operations tend to be complex and time-consuming. The procedure for pumping from draft is as follows:

Step 1:   position the apparatus (as close to the static water source as safety permits) and let the engine return to idle.

Step 2:   set the parking brake.

Step 3:   engage the pump.
Note: some manufacturers recommend that the pump be engaged after priming.

Step 4:   connect hard suction from the intake to the static water source.
Note: Some departments require the hard suction be connected before final positioning of the apparatus. After the hard suction is connected, the apparatus slowly moves into position. A rope can be tied to the end of the hard suction hose to help maneuver it into the water and help keep the strainer from resting on the bottom. Another method to keep the strainer off the bottom is to use a float or ground ladder.

Step 5:   ensure that all discharges, caps, and drains are closed.

Step 6:   prime the pump (engage priming device).
Note: if the pump is engaged, do not increase engine speed over 1,200 rpms.

Step 7:   the intake master gauge should read a negative pressure.
Notes: the primer should not be operated more than 30 seconds for pumps rated at less than 1,500 gpm, 45 seconds for 1,500-gpm or larger pumps. If the intake does not read negative pressure or if oil/water is not discharging on the ground within the above time limits, disengage the priming system and look for leaks.

Step 8:   verify that the pump is primed.

    a. If the pump is engaged, the discharge pressure gauge will increase.

    b. If the pump is not engaged, you may hear or see an oil/water mixture discharging on the ground.

Step 9:   engage the pump if not already engaged.

Step 10: set the transfer control valve.

Step 11: gradually open the discharge outlet.

Step 12: slowly increase pump speed.
Note: monitor hard suction to ensure that drafting does not cause a whirlpool or that debris does not block the screen.

Step 13: set pressure regulating device.

RELAY OPERATIONS

Relay operations are required when more than one pump is needed to move water from the supply to the incident. Once the relay is designed and the lines laid, flow can be initiated. When initiating flow, during the operation, and when shutting down the operation, communications between each of the pump operators within the relay is vital. Pressure can quickly rise within the system, requiring constant evaluation of gauges and communications between the pump operators. The following basic steps may be followed to initiate flow in a relay operation:

Step 1: supply pump secures the water source (can be either hydrant or static source).

Step 2: each pump in the relay, except the supply pump, opens one discharge outlet and sets an intake relief valve to 20 psi (if adjustable).

Step 3: supply pump initiates flow by opening a discharge.

Step 4: supply pump increases discharge pressure to desired setting (keep in mind that the next pump in relay should have a minimum of 20 psi intake pressure).

Step 5: when water reaches the second pump, close the discharge gate.

Step 6: engage the pump if not already engaged.

Step 7: second pump increases discharge pressure to desired setting.
Note: the second pump must ensure that intake does not drop below 20 psi.

Step 8: set the pressure regulating device.

Step 9: repeat steps 5 and 8 for each pump in the relay.

The system should now be flowing water. The use of pressure-regulating devices will help reduce constant changes in relay pressures; however, the pump operator must continuously monitor instrumentation.

DUAL AND TANDEM PUMPING OPERATIONS

Two special pumping configurations are dual and tandem pumping. Dual pumping operations are similar to the volume mode in a multistage pump whereas tandem pumping is similar to the pressure mode.

dual pumping
a hydrant that directly supplies two pumps through intakes

Dual pumping is when one hydrant supplies two pumps. This process is used when the hydrant is strong in terms of pressure and volume and when large quantities of water are required. In the basic setup, one pumper connects to a hydrant and then connects another section of hose from its unused intake to the intake of the second pumper (see Figure 8–10). In this process, the first pumper is not pumping water to the second pumper. Rather, the excess water not used by the first pumper is diverted to the second pumper. Note the similarity to the volume mode: Both pumps receive water and then discharge the water independently of each other. Add the total discharge flow from each pumper to determine the total flow from the hydrant. The basic procedures for setting up a dual pump operation are as follows:

Step 1: connect supply hose from the hydrant to the first pumper (this first pumper can begin pumping).

Step 2: position the second pump close to the first pumper and connect supply hose from the unused intake of the first pumper to the intake of the second pumper.
Note: If the first pumper has a gated valve, the valve should be opened. If the first pumper does not have a gated valve on the unused

intake, the hydrant must be slowly closed until the intake pressure on the first pump reads only a slight pressure. At this time, the cap on the unused intake can be removed to facilitate connection of the supply hose. If the intake pressure is not reduced, excessive water will discharge from the intake, causing problems with connecting the supply hose. After the supply hose is connected between the two intakes, the hydrant can be fully opened.

Step 3: the second pumper can then begin pumping water as described earlier in this chapter.

tandem pumping

a hydrant that directly supplies one pumper and then discharges to the second pumper's intake

Tandem pumping is similar to dual pumping in that one hydrant typically supplies two pumps. The difference is that in tandem pumping the first pumper pumps all its water to the second pumper as in a relay operation (see Figure 8–11).

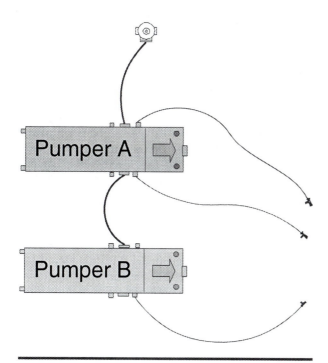

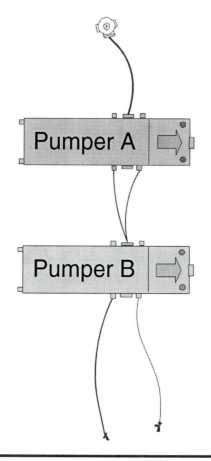

Figure 8–10 *Example of a dual pumping operation. Excess water from pumper A is transferred to pumper B.*

Figure 8–11 *Example of a tandem pumping operation. The pressure generated from pumper A is added to the pressure generated by pumper B.*

Note the similarity with operating in the pressure mode of a multistage pump. The pressure provided by the first pumper is added to the pressure generated by the second pump. Tandem pumping is used when higher pressures are required than can be provided by a single pump. Because of this, caution must be exercised to ensure hose line pressures do not exceed their designed operating pressures. Higher pressures are sometimes required when supplying high-rise sprinkler systems, standpipes, or long hose lays. The basic procedures for setting up a tandem pump operation are as follows:

Step 1: connect supply hose from the hydrant to the first pumper.

Step 2: connect supply hose from the discharge gate(s) of the first pumper to the intake of the second pumper.

Step 3: initiate flow from the first pumper and bleed air from the supply hose at the second pumper.

Step 4: the second pumper can then begin pumping water as described earlier in this chapter.

PUMP TEST PROCEDURES

Pump tests are conducted to determine the performance of the pump and related components. In general, there are five types of tests performed on pumps: manufacturer's tests, certification tests, delivery tests, annual service tests, and weekly/monthly tests. The manufacturers, certification, and acceptance tests are typically conducted on new apparatus. The basic intent of these tests is to verify apparatus and pump performance prior to acceptance by a fire department. NFPA 1901 *Standard for Automotive Fire Apparatus* specifies the requirement for each of the tests. The annual service test and weekly/monthly tests focus on all in-service pumps. The basic intent of these tests is to ensure the pump continues to perform at an acceptable level. NFPA 1911 *Standard for Service Tests of Fire Pumps on Fire Apparatus* contains the requirement for annual service tests while manufacturer's recommendations typically determine monthly/weekly tests.

Manufacturer, Certification, and Acceptance Tests

The requirements of these tests are clearly delineated in NFPA 1901. Manufacturer's tests are typically conducted by the manufacturers at the manufacturer's facilities. Certification tests are conducted by an independent testing organization, usually Underwriters Laboratory (UL). The acceptance test is normally conducted upon delivery of a new apparatus to a fire department to verify and document stated performance levels of the apparatus, pump, and related components. In addition, the test provides a benchmark for comparison of future pump tests. Finally, the test allows pump operators to become familiar with the new apparatus. In essence, the acceptance test is simply a repeat of some of the manufacturer and certification tests.

Annual Service Test

According to NFPA 1911 all pumps rated at 250 gpm or higher must be tested annually and when any major repair or modification has taken place to ensure the pump maintains appropriate performance levels. Chapter 4 of NFPA 1911 focuses on test site requirements while Chapter 5 presents specific tests.

Site Requirements NFPA 1911 suggests that the test be conducted at draft to help determine the true performance of the pump. When conducting the test at draft, the water supply should be clear, at least four feet deep, not more than ten feet below the pump, and the hard suction hose must be at least two feet below the surface of the water. The environmental conditions established by NFPA 1911 include:

Temperature must be between 0°F and 110°F

Water temperature must be between 35°F and 90°F

Barometric pressure must be a minimum of 29 in. Hg corrected to sea level

Discharge hose must be provided that will flow the rated capacity through the nozzle(s) or other flow measuring device without exceeding 35 ft/sec flow velocity. The flow measuring device can be either a pitot tube or flow meter with an accuracy of ±5%. When nozzles are used, they must be smooth bore connected to a monitor, or otherwise properly secured, and the specific nozzle size and hose configuration must be calculated or looked up in a chart to ensure the specific capacity ratings are achieved during the test. In most cases, this may mean shutting down the pump to change the discharge nozzle and hose configuration. When flow meters are used, the device directly measures flow and no calculations are required. In most cases, this means the test is performed more quickly in that the same discharge configuration is used for each capacity test and shutting down the pump is not necessary. All test gauges must be calibrated within 60 days of the annual pump test.

The following tests are included in the annual service test.

Engine Speed Check The no-load governed engine speed shall be checked and compared to the results taken when the apparatus was new. Any variance not within ± 50 rpm must be investigated and corrected prior to starting any test.

Vacuum Test This test ensures the interior of the pump can maintain a vacuum. The test begins by operating the primer, in accordance with manufacturer's recommendations, to obtain a minimum vacuum of 22 in. Hg. The vacuum must be maintained for five minutes with no more than a 10 in. Hg drop in vacuum. After the test begins, the primer cannot be operated.

The basic steps for conducting the vacuum test are as follows:

Step 1: position apparatus on a level service, leave the vehicle running, and chock the wheels.

Step 2: drain all water from the pump.

Step 3: inspect and connect 20 feet of hard suction hose to the intake and place a cap on the open end.

Step 4: ensure intake valve is open, all connections are tight, all discharge control valves are closed, and discharge control valve outlet caps are removed.

Step 5: ensure priming pump oil reservoir is full.

Step 6: connect the vacuum test gauge to the test gauge connection.

Step 7: engage the priming device until the test gauge reads 22 in. Hg. Note the time and compare the test gauge reading with the master intake gauge.

Step 8: turn off the engine and listen for air leaks.

Step 9: after five minutes, note the test gauge reading.

Priming Device Test This test is conducted to ensure the priming device is able to develop a sufficient vacuum to draft. This is a timed test and consists of priming the pump and discharging water. The timed test begins when the primer is started and ends when water is discharging. The time must not exceed 30 seconds for pumps rated at 1,250 gpm or less and not exceed 45 seconds for pumps rated at 1,500 gpm or greater. For pump systems with 4-inch or larger intake pipe, an additional 15 seconds is allowed.

The steps for conducting the priming test are as follows:

Step 1: set up intake hose for a drafting operation and at least one discharge line.

Step 2: open the intake valve, close all discharge valves, and remove discharge caps.

Step 3: engage the priming device as previously discussed in the chapter and note the time.

Step 4: open a discharge line to increase the pressure to the correct pressure/ flow.

Step 5: note the time when the discharge pressure/flow is achieved.

Step 6: note the time when the discharge pressure/flow is achieved.

Pumping Test The annual pumping test helps ensure the pump is capable of flowing its rated capacity. The overall operating condition of the pump is established and can be compared to previous year's tests to determine performance trends. The annual pump test is a 40-minute test consisting of the following:

20 minutes: 100% rated capacity at 150 psi

10 minutes: 70% rated capacity at 200 psi

10 minutes: 50% rated capacity at 250 psi

Pumps with a rated capacity of 750 gpm must also undergo an overload test consisting of flowing the rated capacity for 5 minutes at 165 psi. The test is usually conducted immediately following the 20-minute 100% rated capacity test. For two-stage pumps, the transfer valve shall be in volume mode for the 100% rated capacity test, either volume or pressure for the 70% rated capacity test, and in the pressure mode for the 50% rated capacity test. During the 20-minute test, a complete set of readings shall be taken and recorded a minimum of five times (approximately every four minutes). During the 10-minute tests, readings shall be taken and recorded a minimum of three times. Any five % or greater variance in pressure or flow readings shall be determined and corrected and the test continued or repeated. The only time engine pressure should be reduced is to change hose and nozzle configuration between capacity tests. During the pump test, the operator should scan all instrumentation to ensure the engine and pump are within normal operating ranges.

The pump test can begin immediately following the priming device test previously discussed. The basic steps for pump test are as follows:

Step 1: verify hose and nozzle configure that will enable 100% rated capacity flow (hose and nozzle configures can be calculated as discussed in section 4 or looked up on charts and table).

Step 2: open intake valve and gradually increase pump speed while slowly opening discharge valves. Continue until the discharge valve is fully open and 150 psi pump discharge pressure is obtained. (For two-stage pumps the transfer valve should be in volume mode.)

Step 3: verify that 100% capacity is flowing by either using a pitot tube gauge to verify pre-established pressure is obtained or a flow meter to verify discharge rate. Adjust engine pressure speed and discharge control valves to obtain 100% rated capacity flow at 150 psi discharge pressure.

Step 4: the test begins upon verification of pressure and flow.

Step 5: during the test, record a complete set of readings about every 4 minutes.

Step 6: after 20 minutes, increase the discharge pressure to 165 psi to conduct the overload test for 5 minutes (only required for pump rated capacities of 750 gpm and higher).

Step 7: increase pump pressure to 200 psi and verify a flow of 70% rated capacity. For two-stage pumps, either pressure or volume mode can be used. In some cases, hose and nozzle configures may need to be changed. Record a complete set of readings at least three times during the 10-minute test.

Step 8: increase the pump pressure 250 psi and verify a flow of 50% rated capacity. As in Step 7, the hose and nozzle configuration may need to be changed and a minimum of three readings should be recorded. Two-stage pumps should be in the pressure mode.

Step 9: after 10 minutes at 250 psi, the test is complete and the engine pressure should be decreased and all discharge and intake valves closed.

Pressure Control Test This test is conducted to ensure the pressure control device is capable of adequately maintaining safe discharge pressures. The pressure control device is tested at the following pressures:

150 psi flowing 100% of rated capacity

90 psi (this is achieved by flowing rated capacity with the discharge pressure reduced from 150 psi to 90 psi by using the throttle only)

250 psi flowing 50% of rated capacity

While operating at each of these pressures, the pressure control device is set according to manufacturer's recommendations. Next, the discharge valves are closed no faster than in 3 seconds and no more slowly than in 10 seconds. To pass the test, the discharge pressure shall increase no more than 30 psi.

The basic steps are as follows:

Step 1: establish the rated capacity flow at 150 psi net pump pressure.

Step 2: set the pressure control device per manufacturer's recommendations.

Step 3: slowly close all discharge control valves and note the rise in discharge pressure; the rise should not exceed 30 psi.

Step 4: open all discharge control valves and re-establish 100% capacity flow at 150 psi.

Step 5: using the engine throttle only, reduce the pump discharge pressure to 90 psi (do not change the discharge valve setting, hose, or nozzles).

Step 6: set the pressure control device per manufacturer's recommendations.

Step 7: slowly close all discharge control valves and note the rise in discharge pressure; the rise should not exceed 30 psi.

Step 8: open all discharge control valves and establish 50% rated capacity flow at 250 psi.

Step 9: set the pressure control device per manufacturer's recommendations.

Step 10: slowly close all discharge control valves and note the rise in discharge pressure; the rise should not exceed 30 psi.

Step 11: test is complete, slowly open discharge control valves and reduce engine speed to idle.

Tank to Pump Piping Flow Tests This test is conducted to ensure the piping between the onboard water supply and the pump is capable of flowing the manufacturer's designed flow rate.

The steps for conducting this test are as follows:

Step 1: fill water tank until it overflows.

Step 2: close all intakes, fill line, and bypass cooling line.

Step 3: connect discharge lines (hose and smooth bore nozzle) capable of flowing specific flow rating of the tank to pump piping.

Step 4: open discharge control valves and increase engine speed until the maximum consistent pressure reading is obtained.

Step 5: without changing the engine speed, close discharge valves and refill water tank. If necessary the bypass valve can be temporarily operated to maintain acceptable water temperature range.

Step 6: open discharge control valves fully and take pitot tube or flow meter readings. The engine speed can be adjusted to maintain specified flow rates.

Step 7: after the flow rate is recorded, the test is finished.

Gauge and Flow Meter Test This test is conducted to ensure all pressure gauges and flow meters provide an appropriate level of accuracy. Any pressure gauge reading off by more than 10 psi and flow meter reading off by more than 10% should be recalibrated, repaired, or replaced.

The steps for testing pressure gauges is as follows:

Step 1: cap all discharge outlets and open the discharge control valves.

Step 2: increase the pressure until the calibrated test gauge reads 150 psi.

Step 3: verify that each discharge pressure reading is within ± 10 psi.

Step 4: repeat step three for a pressure of 200 psi and again at 250 psi.

The steps for testing flow meters are as follow:

Step 1: connect a hose and smooth bore nozzle to each discharge outlet; it is not necessary to test all flow meters at the same time.

Step 2: established flow at each discharge and verified via pitot tube reading at the nozzle. NFPA 1911 suggests specific test flows for various pipe sizes. Several examples include: 1-inch pipes at 128 gpm, $2\frac{1}{2}$-inch pipes at 300 gpm, and 3-inch pipes at 700 gpm.

Other Tests Depending on the type of operation, environmental conditions, or specific needs, other tests can be included in the annual service tests. For example, if the need for foam operations is significant, foam system tests can be included in the annual service test. In addition, intake relief devices, when equipped, should be tested to ensure they meet manufacturer's operations and specifications.

Weekly/Monthly Tests Weekly, monthly, and periodic tests are conducted for two primary reasons. First, they are conducted to ensure components are in proper working order. Second, they are conducted in order to keep them working properly. For example, testing the priming pump ensures that it is operating properly while also lubricating its close fitting parts. Several examples of weekly and monthly tests typically required by manufacturers include the following:

testing pressure regulating device

operating priming system

conducting a dry vacuum test

operating the transfer valve

Safety

As with any activity carried out by pump operators, safety should be considered when conducting tests. One way to help ensure safety is to not rush through the tests. Hurrying to finish can increase the risk of an accident, increase the chance of inaccurate results, and increase the chance that a safety problem is overlooked. Another important safety consideration is to ensure the work area is free from hazards. This can be accomplished by walking around the apparatus, looking under and above, for slippery surfaces and loose equipment. Finally, increased safety can be achieved by always keeping it in mind. Several common test safety considerations includes:

> wear protective gloves, helmet, and hearing protection.
>
> slowly open and close all control valves to prevent dangerous water surges
>
> allow only essential personnel within the testing area
>
> do not straddle hose under pressure
>
> ensure hose and nozzles are properly secured

SUMMARY

Operating the pump, the second activity of fire pump operations, moves water from the intake side of the pump to the discharge side of the pump. In general, the same basic steps must be taken to move water from the intake to the discharge regardless of the size and configuration of the pump. However, pump procedures may vary based on different pump manufacturers and different fire departments.

REVIEW QUESTIONS

Key Terms and Concepts

On a separate sheet of paper, identify and/or define each of the following.

1. Pump engagement
2. PTO method (pump engagement)
3. Front crankshaft method (pump engagement)
4. Split-shaft method (pump engagement)
5. Transfer valve
6. Throttle control
7. Intake pressure relief valve

Multiple Choice and True/False

Select the most appropriate answer.

1. Although a variety of pump sizes and configurations exists, the same basic steps must be taken to move water from the supply side to the discharge point.

   True or False?

2. The power to main pumps can be transferred from the drive engine by each of the following methods except
   a. split-shaft.
   b. power take-off (PTO).
   c. front crankshaft.
   d. midship transfer (MST).

3. When the pump is powered by the drive engine through a PTO, the transmission is placed in _____ for stationary pumping.

   a. low gear   c. neutral
   b. high gear   d. either 4th or 5th

4. One unique feature of a pump connected to the front crankshaft is that the engine speed does not need to be at idle to be engaged.

   True or False?

5. A pump connected to the front crankshaft of the drive engine is usually mounted
   a. on the front of the apparatus.
   b. toward the back of the apparatus.
   c. in the middle of the apparatus.
   d. on top of the drive engine.

6. A pump connected to a split-shaft transmission is usually mounted
   a. on the front of the apparatus.
   b. toward the back of the apparatus.
   c. in the middle of the apparatus.
   d. either toward the back or the middle of the apparatus.

7. For both the PTO and the front crankshaft method of powering a pump, the transmission is placed in drive prior to operating the PTO level or the pump control lever.

   True or False?

8. The split-shaft method of powering a pump requires shifting the transmission into pumping gear, usually the highest gear, after the pump shift control has been moved to the pump position.

   True or False?

9. Which NFPA standard focuses on the annual service test of pumps?
   a. 1901   **c.** 1921
   b. 1911   **d.** 1500

Short Answer

On a separate sheet of paper, answer/explain each
of the following questions:

1. List the four ways a pump can receive its
   power.
2. Can a pump be primed without using the
   priming system? If not, why? If so, how?
3. How can you tell if the pump is primed?
4. How do you determine in which mode (pressure or volume) to operate a two-stage centrifugal pump?
5. Why is it important to set the transfer valve
   before pressure is increased?
6. How can you tell if the pump is flowing all
   the water it is receiving from the supply?
7. Explain how to set a pressure relief valve.
8. When using a hydrant as a supply source,
   the intake should not be reduced below what
   pressure?

9. When conducting a drafting operation,
   should the pump be engaged before or after
   the pump is primed?
10. Before increasing the throttle on the pump
    panel, what important safety item should the
    pump operator look for?
11. How long can the priming pump be safely
    activated?
12. How can you tell if the pump is properly
    engaged when power is transferred using the
    split-shaft method?
13. What is the purpose of the throttle control on
    a pump panel?
14. What should occur when the throttle is
    increased?
15. If the throttle is increased past the point of a
    corresponding increase in discharge pres-
    sure, what might be occurring?
16. List three tests required by NFPA 1901.
17. List the various tests associated with the
    annual service testing of pumps.

ACTIVITY

1. Obtain a pump manufacturer's operating
   manual and develop a step-by-step procedure for pumping the onboard tank, a
   hydrant, and a drafting operation.

PRACTICE PROBLEM

1. Which position should the transfer valve be in
   for the following pumping operations (each
   pumper is drafting from a static water supply):
   a. 1,000-gpm pumper flowing the following
   lines:
   - Two 1¾-inch preconnect lines; each line
     is 150 feet, flowing 120 gpm @ 133 psi
   - one 2½-inch line, 400 feet, flowing 250
     gpm @ 150 psi

   - one 3-inch line, 500 feet, flowing 300
     gpm @ 136 psi
   b. 750-gpm pumper flowing the following
   lines:
   - two 1¾-inch preconnect lines; each line
     is 200 feet, flowing 125 gpm @148 psi
   - one 2½-inch line, 1,200 feet, flowing
     275 gpm @ 182 psi.

Chapter 9

DISCHARGE MAINTENANCE
AND TROUBLESHOOTING

Learning Objectives

Upon completion of this chapter, you should be able to:

- Discuss the importance of continuously monitoring instrumentation.
- Explain the various procedures for initiating, adding, and removing discharge lines.
- List and explain the procedures for maintaining pump operations.
- Explain the unique concerns with hot and cold weather operations.
- Define water hammer and explain how to prevent its occurrence.
- Define cavitation and explain how it can be detected and controlled.
- Explain the basic process of troubleshooting pumping problems.
- Explain the basic types, components, and operation of sprinkler and standpipe systems.
- Discuss considerations when supplying water to sprinkler and standpipe systems.

NFPA 1002
Standard for Fire Apparatus Driver/Operator Professional Qualifications
(2003 Edition)

This chapter addresses parts of the following knowledge elements within section 5.1.1:

Recognize system problems, and correct any deficiency

The following knowledge elements within sections 5.2.1, 5.2.2, and 5.2.4 are also addressed:

Rated flow of the nozzle is achieved and maintained

Apparatus is continuously monitored for potential problems

Safe operation of the pump

Operate pumper pressure control systems

Make the transition between internal and external water sources

The following knowledge elements in section 5.2.4 are addressed as well:

Supply water to fire sprinkler and standpipe systems

Operating principles of sprinkler systems as defined in NFPA 13, 13D, and 13R

Fire department operations in sprinklered properties as defined in NFPA 13E

Operating principles of standpipe systems as defined in NFPA 14

INTRODUCTION

discharge maintenance
process of ensuring that pressures and flows on the discharge side of the pump are properly initiated and maintained

When the water supply is secured and pump procedures initiated, the next task of fire pump operations is discharge maintenance. In essence, **discharge maintenance** is simply ensuring that pressures and flows on the discharge side of the pump are properly initiated and maintained. To accomplish this task, pump operators must understand how to determine appropriate discharge settings, set and adjust these settings, temporarily pause operations, and maintain the pump engine within safe operating parameters. In addition, pump operators must understand, recognize, and minimize conditions such as cavitation and environmental factors that affect pump operations. Finally, pump operators must be able to quickly and efficiently troubleshoot pump problems. Effectively carrying out the task of discharge maintenance requires pump operators to both continuously monitor instrumentation and plan ahead for potential changes.

Monitoring pump instrumentation is important for ensuring that discharge pressure and flow settings are maintained. Gradual changes in pressures and flows can occur on both the intake and discharge sides of the pump. Abrupt changes to pressures and flows can also occur on both the intake or discharge sides. For example, supply or discharge lines can break, seriously affecting pumping operations. Constant monitoring of instrumentation is essential for detecting and compensating for these gradual and abrupt changes. Finally, constant evaluation of engine instrumentation may detect potential problems. For example, a gradual increase in engine temperature could cause serious problems if not detected early enough.

Pump operators should plan ahead for potential changes in pump operation. Planning ahead will ensure that changes to pump operations are completed in an efficient and effective manner. Common tasks that may require planning include:

- Pausing operations
- Extending pump operations
- Securing an alternate supply source should the current supply fail
- Determining what to do should a supply line or discharge line break
- Judging the ability to add lines and the best manner to supply them

This chapter presents the task of discharge maintenance by first discussing procedures for maintaining flows and pressures, then pump operation considerations, sprinkler and standpipe operations, and, finally, pump troubleshooting. The process for determining discharge settings is presented in Section 4, Chapter 12.

INITIATING AND CHANGING DISCHARGE FLOWS

Discharge flows are controlled by *pump speed* and *discharge control valves*. Pump speed is increased and decreased to control total flow from the pump. As discussed in Chapter 5 of this text, the primary means to control pump speed is the throttle located on the pump panel. Discharge control valves are opened and closed to provide individual control over flows from discharge outlets. Used in combination, a variety of flows and pressure settings can be achieved. Recall that the master discharge gauge indicates total discharge pressure from the pump, while individual discharge gauges indicate pressure readings for their specific discharge.

Multiple lines of different pressures and flows require a technique called feathering of discharge control valves. **Feathering** is simply the process of partially opening or closing control valves to regulate pressure and flow for individual lines (Figure 9–1). The term *gate* is also used to describe this process. Feathering, or gating a valve is most often used when initiating, changing, or shutting down a discharge line while other discharge lines are flowing.

■ **Note**

Monitoring pump instrumentation is important for ensuring that discharge pressure and flow settings are maintained.

feathering
the process of partially opening or closing control valves to regulate pressure and flow for individual lines

Figure 9–1
Discharge lines of different pressure and volume require feathering of discharge control lines.

Initiating Discharge Flows

Initiating flow on the discharge side of the pump entails opening the discharge control valve, increasing engine speed to the desired pressure, and setting the pressure regulator. When a discharge line is first charged, water enters the hose and pressure increases in relation to the discharge pressure of the pump. Pump operators should check the hose for sharp bends and kinks that may restrict flow. Kinks and sharp bends that restrict flow will cause higher pressure readings. When the kinks and bends are removed, less restriction will reduce pressure. In addition, when nozzles are momentarily opened and closed, pressure readings on the pump panel will rise and fall, creating difficulty in setting appropriate pressures. The best and easiest time to set pressures occurs when nozzles are open and water is flowing. This is especially true when multiple lines of varying pressures are used.

Pump operators must recognize and compensate for these difficulties in setting discharge lines. A good pump operator will not have a noticeable increase or decrease in pressure (5 to 10 psi) when adding or removing lines. The discharge lines in Figure 9–2 will be used to explain procedures for initiating and changing discharge flows. Pressure settings for these lines were determined using the friction loss formula $C\,Q^2\,L$, which is discussed in the section on friction loss calculations in Chapter 11.

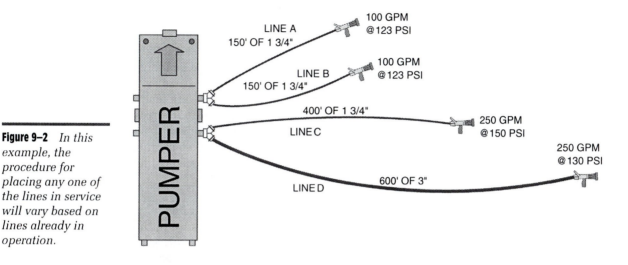

Figure 9–2 *In this example, the procedure for placing any one of the lines in service will vary based on lines already in operation.*

Initiating One Line When one line is placed in service, the pump operator first fully opens the appropriate discharge control valve. Remember, to ensure the safety of personnel and equipment, all discharge control valves should be operated in a slow and smooth manner. Next, the throttle is slowly increased until the desired pressure setting is reached. For example, to initiate line A in Figure 9–2, the pump operator would simply open the discharge control valve for line A and then increase the throttle to 123 psi.

Initiating Two Lines *Multiple lines of the same configuration* (same hose size, length, and nozzle setting) require the same pressure and flow setting. Therefore, the lines can be opened and the discharge pressure set at the same time. For example, to place both lines A and B into service at the same time, the pump operator would simply open the control valves for lines A and B and increase the throttle until the pressure is increased to 123 psi.

Multiple lines of different configurations may require different pressure and flow settings. In this case, the line with the lower pressure must be feathered. For example, to place lines B and C into service at the same time, the pump operator first opens both control valves for lines B and C. Then, the pressure is increased to 123 psi in both lines. The next step is to continue increasing the pressure to 150 psi for line C while maintaining 123 psi for line B. Maintaining 123 psi on line B is accomplished by slowly closing (feathering) the discharge control valve for line B.

Changing Discharge Flows

More often than not, discharge lines are added and removed during pumping operations. When this occurs, discharge flow settings must be changed at the pump panel. The procedure for changing discharge flow settings depends on the pressure and flow of the line being added or removed.

Adding a Line of Less Pressure When adding a line of less pressure than those currently in operation, the pump operator simply feathers the control valve open for the new line until the proper setting is reached. Because the total gallons-per-minute flow has increased, the pump speed may need to be increased to maintain the discharge pressure of existing line(s). This occurs because pressure is a function of restriction. When an additional line is opened, less restriction occurs on the discharge side of the pump, resulting in a reduction in pressure.

Take for example, placing line D in Figure 9–2 in service when lines A and C are currently in operation. To accomplish this task, the control valve for line D is slowly feathered opened to 130 psi while increasing the throttle to maintain 123 psi on line A and 150 psi on line C.

Adding a Line of Equal Pressure When adding a line of equal pressure, the new line is simply feathered opened until the desired pressure setting is achieved, while the throttle is increased to maintain the original pressure setting. For example, to initiate flow to line B when line A is currently in operation, the pump operator simply feathers open line B to 123 psi while maintaining 123 psi on line A by increasing the throttle.

Adding a Line of Greater Pressure When adding a line of greater pressure than those currently in operation, the pump operator must feather existing lines. For example, to place line C in service while line D is currently in operation, the pump operator slowly feathers open the control valve for line C while maintaining 130 psi on line D by simultaneously increasing the throttle. When line C reaches 130 psi, line D must be feathered closed as the pressure in line C is increased to 150 psi. Adding a line of higher pressure requires the pressure regulating system to be reset.

Removing a Line Removing lines from service requires the pump operator to slowly close the control valve of the line being removed. If the line being removed is the highest of those flowing, the throttle should be reduced to the new highest operating line. If the line being removed is a lower-pressure line, no change in engine pressure is required. However, when the line is removed, the pressure may increase as a result of greater restriction. If the rise in pressure is great enough to cause the pressure regulator to activate, the throttle should be slowly reduced to the appropriate pressure setting. In doing so, the pressure regulator should reset itself.

Changing Pressure Regulating Setting

Pressure regulating systems should not be used to change the discharge pressure setting. Rather, the control valves and the pump panel throttle should be used. Using the pressure regulator to control discharge pressures can result in excessive pump speeds, which increase pump wear and tear as well as increase safety concerns should the regulating device fail or be accidentally turned off.

If the pump discharge pressure is increased, the pressure regulating system must first be reset above the new discharge pressure setting. The pressure regulating system should be turned off before pressure is increased. Turning the pressure regulator off when it is activated will cause an increase in pressure on the discharge side of the pump, which may cause a water hammer intense enough to damage equipment and injure personnel. Water hammer is further discussed later in this chapter.

After the discharge pressure is increased, the pressure regulating system is then set to the new pressure setting. When the pump discharge pressure is decreased, the pressure regulating setting must be reduced as well. The need to change pressure regulating settings can arise from:

- Adding a line of higher discharge pressure
- Increasing the pressure of an existing line
- Removing the highest pressure discharge line
- Reducing the pressure of an existing line

MAINTAINING PUMP OPERATIONS

The pump operator does not relax after pump operations are initiated. Maintaining pump operations is an important activity that requires constant attention and planning. Common activities for maintaining pump operations include temporarily pausing operation, planning for extended operations, and maintaining the engine and pump within operating parameters.

Pausing Operations

If discharge lines will not be flowing water for a short period of time, the pumping operation should be paused to reduce wear and tear on the pump and to avoid damage from overheating and cavitation. When pausing pump operations, slowly reduce engine speed and slowly close discharge control valves. If drafting from a static source, a slight discharge pressure should be maintained. This pressure can be accomplished by opening an unused discharge outlet with a hose directing water back into the static source. This procedure is also a good practice with reduced flows such as during overhaul. If the operation will be paused for an extended period, the pump should be disengaged.

Extended Operations

Special consideration should be given to pump operations that are expected to last for 4 or more hours. Long pumping operations tend to reduce attention to instrumentation, especially when the operation does not require constant changes in flow. Special attention should be given to the engine and pump to ensure that overheating does not occur, and attention should be given to oil pressure and fuel levels. Modern electronic instrumentation often provides visual and audible warning signals as systems and components move outside normal operating parameters. Arrangements should be made well in advance if refueling is required.

Maintaining Engine and Pump Operating Parameters

Even during short pump operations, engine and pump parameters can change quickly. Pump operators must be diligent in monitoring both pump and engine instrumentation. Of particular concern during pump operations is the detrimental effect of excessive heat.

■ **Note**
Pump operators must be diligent in monitoring both pump and engine instrumentation.

Engine Cooling Systems As discussed in Chapter 5, engine cooling systems are designed to maintain normal operating temperatures while the vehicle is in motion or while the vehicle is stationary at normal idle speed. During stationary operations under load, as is the case during pumping operations, the engine cooling system is inadequate to remove excessive heat buildup, so pumping apparatus are equipped with an auxiliary cooling system (Figure 9–3). Water from the pump is diverted through piping to the engine, absorbs heat, and is then returned to the pump. This system should be activated when the engine temperature first begins to rise. Under most situations, turning on the auxiliary cooling system is all that is required to maintain proper engine temperature. When the cooling system is not effective in maintaining engine temperatures within normal operating parameters, other measures such as opening the engine compartment to increase ventilation must be taken. If measures taken to reduce engine temperatures within safe levels fail, the operation should be shut down to protect the engine from extensive damage. Conversely, care should also be taken to avoid cooling the engine below its normal operating range because excessive cooling can reduce efficiency, cause excessive wear and tear, and potentially damage the engine.

Figure 9–3 *Auxiliary cooling systems are required on pumping apparatus to maintain the drive engine within normal operating temperatures.*

Pump Cooling Systems When the pump is engaged and water is flowing, the temperature inside the pump will normally be maintained within the normal temperature operating range. However, when discharge lines are closed or flows are reduced to a minimum, the temperature of water in the pump can quickly rise. Increased water temperature inside the pump can cause serious injury to personnel through burns and can cause significant damage to the pump. To compensate, simply maintain water movement within the pump. Either flow water from an unused discharge or open the tank-to-pump and the pump-to-tank valves to recirculate water.

SHUTTING DOWN AND POSTOPERATIONS

Pumping operations are not completed until the pump is shut down and properly readied for the next operation. The desire to return to the station can tempt pump operators to quickly wrap up the operation. Shutting down pumping operations requires the same care and diligence to safety required when initiating them.

Shutting Down Operations

The steps to shutting down operations are similar to those used to initiate the operation but in the reverse order. The basic steps to shut down a pumping operation include:

Step 1: slow engine to idle speed.

Step 2: close all discharge control valves.

Step 3: check water tank level (refill if necessary).

Step 4: disengage the pump drive.

Step 5: bleed off any remaining pressure from the discharge line.

Step 6: turn off pressure regulating devices and engine cooling systems (return to their normal position).

Step 7: close the intake supply.

Step 8: disconnect intake and discharge hose.

Step 9: remove wheel chocks before moving the apparatus.

Although the sequence of these steps may change depending on department policy and pump manufacturer's recommendations, care should taken regardless of the sequence. Carelessness in the shutdown sequence can cause damage and even injury to personnel resulting from:

- Excessively high pressure
- Generation of steam or near boiling water temperatures
- Operating the pump dry
- Water hammer
- Cavitation

Postoperations

Postoperation activities ensure that the pump is returned to a ready status. Pump operators should follow pump manufacturers' recommendations for postoperations. Postoperation activities include the following:

Flushing the Pump The pump should be thoroughly flushed if foam, saltwater, dirty water, or contaminated water was used. When flushing the pump, use a clean water supply. *Backflushing* may be required and is simply reversing the flow so that water enters the discharge side and leaves the intake side of the pump.

Inspect and Fill Liquid Levels All liquid levels should be inspected and replenished as necessary. For example, the tank water level and the priming oil tank reservoir should be checked and filled if necessary. In addition, engine coolant, oil, and fuel levels should be checked and filled if necessary.

Intake and Regulator Screens If water used during pump operations is suspected of containing dirt, sand, or other debris, intake screens, as well as regulator screens, should be inspected and cleaned if necessary. Backflushing may also be required.

Control Valves Pump operators should ensure that all discharge and intake control valves, drain valves, transfer valves, and other components are returned to their normal positions. If the primer was used during pump operations, it is a good idea to briefly operate the priming device. In doing so, the close-fitting parts of the priming device are lubricated, helping to ensure the system remains in good operating order.

PUMP OPERATION CONSIDERATIONS

Several conditions may be encountered that affect the ability to safely and efficiently conduct pump operations. Pump operators must be familiar with these conditions in order to both detect them and to compensate for them. The most common conditions include water hammer, cavitation, and the effects of environmental factors.

Water Hammer

water hammer
a surge in pressure created by the sudden increase or decrease of water during a pumping operation

Water hammer is a surge in pressure created by the sudden increase or decrease of water. This surge is created by the velocity of the water as well as its weight within a hose. The potential for damage from water hammer is significant given the velocity and weight of water during pumping operations. The velocity of water on the discharge side of the pump is often high, with normal pump pressure in excess of 100 psi. In addition, the weight of water in the hose during normal pumping operations is also high. For example, a 400-foot section of 3-inch hose filled with water weighs 1,253 lbs. Imagine the effect of a little more than half a ton of water at 100 psi coming to an immediate stop!

Obviously, this surge in pressure can be high enough to damage the pump and hose as well as cause personnel to lose control of hose. Pump operators can reduce the likelihood and extent of water hammer in several ways. The best protection, perhaps, is simply knowing that it can occur. All personnel, not just the pump operator, should understand the effects and consequences of water hammer because hydrants, nozzles, and appliances are not always operated by just the pump operator. The best way to reduce water hammer is to start and stop water flows slowly by slowly opening and closing control valves for pump intakes and discharges, as well as hydrants, nozzles, wyes, hose clamps, and other similar appliances. The use of pressure regulating devices can also help reduce the effects of water hammer.

Cavitation

cavitation

the process that explains the formation and collapse of vapor pockets when certain conditions exist during pumping operations

The damage caused by **cavitation** can be significant, resulting in reduced pump efficiency and, when severe enough, extensive pump repairs. Often referred to as "the pump running away from the water supply," cavitation occurs when a pump attempts to flow more water than the supply can provide. Most often, cavitation is associated with drafting operations but can just as easily occur when using a hydrant as a supply source. The primary cause of cavitation is the effect of pressure zones in the pump as they act on water.

Pressure Zones   When more water is being pumped than the supply can deliver, two pressure zones are produced within the pump (Figure 9–4). One is a low-pressure zone developed at the eye (center) of the impeller. This low-pressure zone occurs when the supply is unable to deliver the amount of water being discharged. The other is a high-pressure zone that is the result of increased velocity (pressure) as water moves from the center of the impeller to the outer edge of the impeller. The outer edge of the impeller has greater velocity than the center of the impeller, thus creating more velocity (pressure) toward the outer edge.

Effects of Pressure on Water   At normal atmospheric pressures (14.7 psi at sea level), water boils at 212 degrees Fahrenheit. As shown in Table 9–1, the boiling point of water is affected by pressure. When pressure increases, the boiling point of water increases, meaning that higher temperatures are required to boil water. Conversely, when pressure decreases, the boiling point of water is reduced. When this occurs, lower temperatures are required to boil water. The rate at which water evaporates is closely associated with its boiling point at a specific temperature. The closer water comes to boiling, the greater the rate of evaporation. Increased pressure reduces the ability of a liquid to vaporize, while decreased pressure increases the ability to vaporize. In other words, the more pressure there is pushing down on water, the less it can evaporate and, consequently, higher temperatures are required for it to boil. When less pressure is pushing down, it is easier for water to evaporate and, consequently, lower temperatures are required for it to boil.

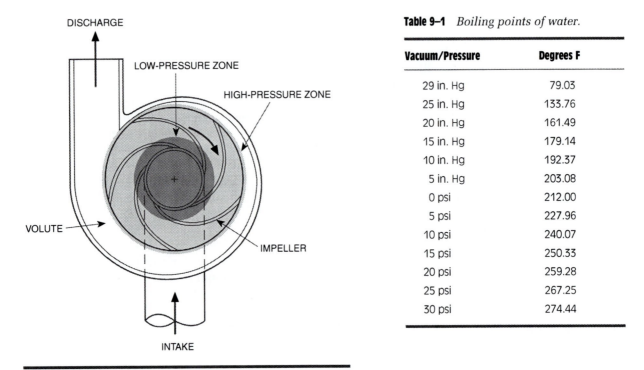

DISCHARGE

LOW-PRESSURE ZONE

HIGH-PRESSURE ZONE

VOLUTE

IMPELLER

INTAKE

Table 9–1   *Boiling points of water.*

Vacuum/Pressure	Degrees F
29 in. Hg	79.03
25 in. Hg	133.76
20 in. Hg	161.49
15 in. Hg	179.14
10 in. Hg	192.37
5 in. Hg	203.08
0 psi	212.00
5 psi	227.96
10 psi	240.07
15 psi	250.33
20 psi	259.28
25 psi	267.25
30 psi	274.44

Figure 9–4   *Two pressure zones are created when the supply cannot provide enough water to the discharge. The low-pressure zone occurs near the center of the impeller, while the high-pressure zone occurs near the outer edge of the impeller.*

■ Note

The return of vapor pockets to a liquid state is instantaneous and forceful and is typically described as an implosion.

Effects of Cavitation When water passes through the low-pressure zone, its boiling point is reduced and its ability to vaporize increases. The low-pressure zone allows water to readily vaporize and form vapor pockets. When these vapor pockets pass through the high-pressure zone, the boiling point of water increases and the ability of water to vaporize decreases. The high-pressure zone, then, forces the vapor pockets back into the original liquid state. Often the return of vapor pockets to a liquid state is instantaneous and forceful and is typically described as an *implosion.*

Thousands of small vapor pockets commonly form in the low-pressure region. Consequently, thousands of implosions occur in the high-pressure region of the pump. These small implosions can cause pitting on the smooth surface of the impeller, resulting in increased friction loss, and can cause the impeller to become unbalanced. In either case, the efficiency of the pump is diminished.

Signs of Cavitation In some cases the only warning sign of cavitation is the knowledge that pumping more water than the supply can handle will cause cavitation. That is to say, cavitation may occur with no outward warning signs. In other cases, the signs that cavitation is occurring are obvious. Several such signs include:

- No corresponding increase in pressure as engine speed is increased
- Engine speed automatically increases during a pumping operation
- Fluctuating discharge pressure readings
- Rattling sounds, like sand or gravel going through the pump
- Excessive pump vibrations
- Sudden pressure or capacity loss

Stopping Cavitation The quickest way to stop cavitation is to simply reduce the pump speed. This results in the immediate reduction or elimination of the two pressure zones. Cavitation can also be stopped by increasing the flow of the supply to adequately meet the demands of discharge flows. Finally, cavitation can be stopped by reducing the amount of water being discharged. Examples for reducing discharge flows include:

- Removing one or more discharge lines
- Reducing the size of discharge lines
- Reducing nozzle gallons-per-minute settings

Environmental Considerations Normal changes in weather conditions typically do not adversely affect pumping operations. Cold weather can actually help engine performance. Hot weather can usually be controlled by the auxiliary cooler and by recirculating pump water. However, the extremes can be menacing to pump operators.

Cold Weather Operations Extremely cold weather can cause major problems with pumps. When the pump is susceptible to freezing, all water must be removed from the pump and associated piping such as discharge lines, auxiliary cooling system, pressure regulating line, and primer lines. During cold weather operations, intake and discharge lines should be laid out so that water can flow as soon as the pump is primed. The movement of water in the pump should not be stopped until the operation is complete and the pump and piping can be drained. Whenever possible, leaks should be stopped to reduce the accumulation of ice. Following are the steps that should be taken to drain the pump:

Step 1: open all intake and discharge control valves and their caps.

Step 2: open the main pump drain valve (some pumps may require discharge drain valves to be opened as well).

Step 3: for multistage pumps, operate the transfer valve several times.

Step 4: drain the pressure regulating devices if required by the manufacturer.

Step 5: after the pump is drained, close all discharge and intake control valves as well as drain valves.

Hot Weather Operations Extremely hot weather can also cause problems during pump operations. As discussed earlier in this chapter, both the engine and the pump must be maintained within normal operating temperatures. When operating in hot weather, pump operators must take into consideration the temperature of the supply water, especially if the supply is a static source. The major concern is that warmer water is more suspectable to cavitation because less energy is required to bring the water to its boiling point.

SUPPORTING SPRINKLER AND STANDPIPE SYSTEMS

Pump operators may be called on to support sprinkler and standpipe systems. Sprinkler and standpipe systems are fixed systems within buildings and structures used to help protect lives and property by providing immediate detection and/or extinguishing capabilities. To properly support these fixed systems, the pump operator must be familiar with the basic types of systems, their operations, and support considerations. NFPA 13E, *Recommended Practice for Fire Department Operations in Properties Protected by Sprinkler and Standpipe Systems*, suggests that inspecting and preplanning sprinkler and standpipe systems is critical to the development of appropriate support procedures. Selecting a water source, locating where to connect to the system, and determining what pressure to pump should be accomplished during preplanning and should be incorporated into standard operating procedures.

Sprinkler Systems

According to NFPA 13E, when sprinkler systems are properly designed, installed, and maintained, they can provide water to the fire in a more effective manner than using manual fire suppression methods. Sprinkler systems are designed to move water from a source, through piping, to one or more discharge points (sprinkler heads). The water supply for most sprinkler systems is designed to supply only a small number of sprinkler heads. For larger fires, when multiple sprinkler heads are activated or when pipes break, the pump operator must provide additional water to the system. Sprinkler systems can be installed for both interior and exterior use. Interior systems are designed to keep fire growth contained or to extinguish the fire. Exterior sprinkler systems are designed primarily to protect exposed properties from the spread or extension of fire.

There are several common parts or components to most sprinkler systems. One common component is a water supply source. Sprinkler systems are connected to a main water supply, typically either a municipal or private water supply source. In addition, sprinkler systems have a fire department connection that provides a means for a secondary water supply source. NFPA 13, *Standard for the Installation of Sprinkler Systems*, requires that fire department connections have NH standard thread with caps that can easily be removed by fire department personnel. Most fire department connections are siamese with two 2½-inch female connections. One-way check valves are most commonly installed on a sprinkler system especially if connected to a municipal water system. These valves help keep water from within the sprinkler system from entering the connected water supply. Each type of sprinkler system has a valve that is used to initiate the flow of water from the connected water supply to the system. The basic types corresponding to the types of systems include: main operating valve (wet pipe system), dry pipe valve, preaction valve, and deluge valve.

Piping or tubing move the water from the supply to one or more sprinkler heads. First, the riser moves water from the supply source to the feeder mains then to the cross mains and finally the branch mains onto which sprinkler heads are connected. The water supply control valve is usually located on the riser. This valve is most often an outside stem and yoke (OS&Y) indicating valve and allows for quick visual determination of the valve location (open or shut). A main drain control valve allows the system to be drained for maintenance or repair. Finally, an inspector's test valve, located at the farthest end of the system, is used to simulate the single head flow to test system response.

Sprinkler systems are also equipped with waterflow alarms that activate when the system is in operation. According to NFPA 13, waterflow alarms must activate when the flow of water is equal to or greater than that from a single sprinkler head with the smallest orifice. The alarm must initiate within 5 minutes after flow begins and continue until the flow stops. A retard chamber is usually provided to help prevent the system alarm activation from sudden water supply pressure surges. This is accomplished by collecting water within the chamber prior to activating the alarm. Water supply surges will normally not provide the sustained flow into the chamber that is required to activate the system. A small opening at the base of the chamber allows the water to drain from the retard chamber.

Finally, all sprinkler systems have sprinkler heads. Most sprinkler heads, except those installed on deluge systems, are closed heads equipped with some sort of heat-sensitive device or operating mechanism. This sensing device is usually a fusible link or a chemical pellet that melts at a fixed temperature or a liquid-filled tube that bursts at a fixed temperature. Sprinkler heads come in a variety of temperatures ratings. The NFPA 13 required sprinkler head temperature rating and classification are provide in Table 9–2. The three basic sprinkler head designs are upright, pendant, and sidewall (see Figure 9–5). Finally, sprinkler heads operate at a fixed discharge rate and pattern.

There are four basic types of sprinkler systems: wet pipe systems, dry pipe systems, deluge systems, and preaction systems.

Wet Pipe Systems As the term implies, wet pipe systems contain water within the piping from the water source to the sprinkler head. Because of this, these systems are mainly used within heated structures or where freezing is not a concern. The water within the system is under pressure and the sprinkler heads are closed. Before activation, the pressure within the system is higher than the pressure from the connected water supply. Pressure gauges are usually located on either side of the main check valve, providing direct reading of both pressures.

Heat from the fire will fuse the fusible element on the sprinkler head which opens a discharge orifice. When a sprinkler head is activated, water immediately discharges from the system. As the pressure drops, several actions occur within the system. First, the main check valve opens, allowing water from the supply source

Table 9–2   *NFPA 13 required sprinkler head temperature rating, classification, and color coding.*

Maximum Ceiling Temperature		Temperature Rating		Temperature Classification	Color Code	Glass Bulb Colors
°F	°C	°F	°C			
100	38	135–170	57–77	Ordinary	Uncolored or black	Orange or red
150	66	175–225	79–107	Intermediate	White	Yellow or green
225	107	250–300	121–149	High	Blue	Blue
300	149	325–375	163–191	Extra high	Red	Purple
375	191	400–475	204–246	Very extra high	Green	Black
475	246	500–575	260–302	Ultra high	Orange	Black
625	329	650	343	Ultra high	Orange	Black

to flow into the system. The check valve prevents backflow from the system to the water supply when the system operates and in some cases operates a flow alarm. Second, water flows into the retard chamber and activates the alarm system. The alarm system can be configured for a local alarm, such as a water gong, and/or send an electronic signal to a monitored or supervised system or to notify a fire department. Water continues to flow through the fused sprinkler heads until the water supply control valve is closed.

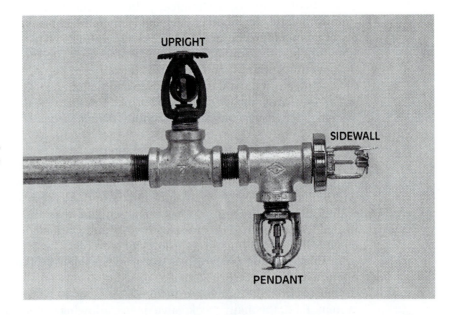

Figure 9–5   *The basic sprinkler head designs. Note that the sensing device for the upright is a fusible link, the sidewall is a glass bulb, and the pendant is a chemical pellet.*

Dry Pipe Systems Dry pipe systems contain compressed air as opposed to water within the system. These systems are used where freezing temperatures occur. The compressed air in the system is of sufficient pressure to cause the water supply control valve to remain closed. This is usually accomplished by a differential valve, which allows a smaller amount of air pressure within the system to hold back a higher pressure of the connected water supply. When a sprinkler head is fused, air initially escapes from the head. The pressure in the system is reduced enough to cause the dry pipe control valve (water supply valve) to open, allowing water to enter the sprinkler system. Because air is initially contained in the system, water does not immediately discharge from the sprinkler head as it does with wet pipe systems. Some systems use an exhauster or accelerator to quickly remove air from the system to reduce the time it takes to start discharging water. According to NFPA 13, dry pipe systems must be capable of discharging water to the most remote sprinkler in not more than 60 seconds.

Preaction Systems A preaction system is similar to both a dry pipe system and a wet pipe system. First, as with a dry pipe system, the piping is void of water and the sprinkler heads are closed. Air within the system may or may not be compressed. The main difference with this system is that a preaction valve is used to open the connected water supply. This preaction valve is activated via a connected detection system. When heat or smoke is detected, the preaction valve opens, allowing water to flow into the sprinkler system. At this point, the system operates similarly to a wet pipe system. When a sprinkler head is fused, water immediately discharges.

Deluge Systems The difference between a preaction system and a deluge system is that all the sprinkler heads in a deluge system are open. A separate detection system is used to open the water supply valve. Water entering the system flows through all the sprinkler heads at the same time. These systems are used when rapid fire spread is a concern. Because of the large water flow requirements of these systems, larger piping is typically installed and separate pumps may be required to ensure appropriate flows and pressures are provided to the system.

Support Considerations As discussed earlier, sprinkler system support should be preplanned and developed into standard operating procedures. Several preplanning considerations include the following:

- Identification of the primary water supply for the system.
- Location of control valves.
- Secondary water supply that will not degrade the system's primary water source. Open water supplies should not be used when the sprinkler system is directly connected to a municipal or potable water supply unless appropriate backflow devices are installed.
- Location of fire department connections and possible hose layout.
- Appropriate discharge pressures and flow for system operations.

According to NFPA 13E, first-arrive companies should take prompt action to supply sprinkler systems. Specifically, at least one hose line should be connected to the fire department connection and more lines added as required by fire conditions. Prior to connecting to the fire department connection, check to ensure no debris has accumulated within the siamese (see Figure 9–6). The siamese should have a clapper valve that allows one 2½-inch hose line to be charged at a time. If not, then connect both lines prior to charging the lines or place a gated valve on the second 2½-inch inlet. Unless otherwise indicated on the system, the supply line should be pumped at 150 psi. If hose streams are used, pump operators should secure a water supply that will not reduce water flow to the sprinkler system. NFPA 13 suggests securing a water source from one of the following when hose lines will be used:

- Large mains that can support both hose line operations and the sprinkler system
- Water mains not needed for sprinkler supply
- Drafting from static sources.

Other considerations for sprinkler system support include the following:

- When fire and smoke is visible, start pumping to the sprinkler system immediately.
- If the sprinkler system has been activated but no fire or smoke is evident, wait until interior crews verify a fire is present before pumping to the system. The system could have malfunctioned or extinguished the fire.
- Do not shut down a sprinkler system for improved visibility.
- When the fire is extinguished, turn off the sprinkler system and stop flowing water to the fire department connection. Some fire department connection piping bypass the sprinkler control valve.
- Place the transfer valve into the volume position.
- Look to see if the discharge pressure is on the fire department connection.

Figure 9–6   *Fire department connections should be inspected for debris prior to connection to the supply hose.*

- As a general rule of thumb, pump the fire department connection at 150 psi unless otherwise noted.

Standpipe Systems

Standpipe systems are installed to provide a prepiped water supply to hose lines within large structures such as warehouses, high-rise, and industrial buildings. In much the same manner as sprinkler systems, standpipe systems move water from a source, through fixed piping, to hose connections. Standpipe systems can be supplied from one of two different water sources. First, the system can be permanently connected to a water supply source in the same fashion as automatic sprinkler systems. Second, the system can be connected to a fire department connection. In this case, all the water required in the system must be provided by a pumper connected to the system. With such systems, the fire department connection must be marked as "STANDPIPE" and the required pressure must be indicated. The system is comprised of control valves, check valves, risers, and cross mains similar to sprinkler systems. The discharge point for standpipe systems are hose connections as opposed to sprinkler heads.

According to NFPA 14, *Standard for the Installation of Standpipe and Hose Systems*, there are three classes of standpipe systems that are based on their intended users. Class I standpipe systems provide 2½-inch hose connections and are intended to be used by firefighters or fire brigade members. Class II standpipe systems provide 1½-inch hose stations and are intended primarily for trained personnel during initial attack efforts. Some class II systems may have 1-inch hose for areas of light hazard occupancies. Class III standpipe systems provide 1½-inch and 2½-inch hose connections for use by firefighters and fire brigade members. Some class III systems may also have 1-inch hose for light hazard occupancies. Class I and III systems have a minimum flow rate of 500 gpm and each additional standpipe shall flow 250 gpm not to exceed 1,250 gpm. Class II standpipe systems have a minimum flow rate of 100 gpm. Adequate water supply must be available to sustain the standpipe for a minimum of 30 minutes.

NFPA 14 also lists the five basic types of standpipe systems as follows:

Automatic Dry. This dry pipe system is filled with air under pressure. When a hose valve is opened, the dry pipe valve opens, allowing water to flow into the system. This water supply must be capable of supplying the required flow rate.

Automatic Wet. An automatic wet system is filled with water. The water supply must be able to automatically supply the required flow for the system.

Semiautomatic Dry. This system is a dry pipe system with a deluge valve activated by a remote device at each hose connection. The water supply must also be capable of supplying the system demand.

Manual Dry. This is a dry pipe system without a permanently connected water supply. These systems require a fire department to supply the standpipe through a fire department connection.

Manual Wet. This is a wet pipe system without a permanently connected water supply that can fully provide the required fire flow. As with a manual dry pipe system, manual wet pipe systems require a fire department to supply the systems through a fire department connection.

Support Considerations As with sprinkler system support, preplanning is critical to properly supporting standpipe systems. The same basic water supply considerations for sprinkler systems should be assessed when selecting a water supply source to support standpipe operations. Determining the correct pressure and flow for the standpipe system is of critical importance. NFPA 13E suggests pump operators should consider the following when calculating pump discharge pressure:

- Friction loss in the hose line connected to the fire department connection
- Friction loss in the standpipe itself
- Pressure loss due to elevation of nozzle
- Number and size of attack lines operating from the standpipe
- Desired nozzle pressure

The best time to calculate pump discharge pressure is before an incident occurs. Friction loss calculations are discussed in detail in Section 4 of this text.

TROUBLESHOOTING

Problems encountered during pumping operations can be reduced by becoming familiar and experienced with pumps and their systems, including knowledge of pump construction, operating principles, operating procedures, and proper preventive maintenance.

■ Note
Pump troubleshooting can be grouped into two categories: procedural problems and mechanical problems.

In general, pump troubleshooting can be grouped into two categories: procedural problems and mechanical problems. If proper procedures were followed, it is more than likely that a mechanical problem exists. In either case, the best method for troubleshooting a pump problem is to follow the flow of water from the intake to the discharge while attempting to determine the problem. In most cases, problems are readily identifiable and correctable using this method.

Pump manufacturers typically provide troubleshooting guides specific to their pumps (see Appendix D).

SUMMARY

The task of discharge maintenance deals with those activities related to initiating and maintaining discharge lines. The procedures for initiating discharge lines vary depending on pressures and flows of both the new line(s) and those already in operation. Maintaining discharge lines focuses on those procedures that keep the engine and pump within normal operating parameters. Discharge maintenance also includes procedures for operating pumps over extended periods, pausing operations, shutting down operations, and postoperations. Two common conditions that can occur in almost any pump operation are water hammer and cavitation. Understanding why these conditions occur and how to prevent them is vital for the safety of both equipment and personnel. Preplanning is perhaps the most important activity directed toward ensuring that sprinkler and standpipe systems are properly supported. Troubles encountered during pump operations are most likely the result of equipment malfunction or procedural error.

REVIEW QUESTIONS

Key Terms and Concepts

On a separate sheet of paper, identify and/or define each of the following:

1. Discharge maintenance
2. Feathering/Gating
3. Water hammer
4. Cavitation

Multiple Choice and True/False

Select the most appropriate answer.

1. To ensure safety, the pressure regulator should be used to reduce discharge line pressures.

   True or False?

2. When the pump attempts to deliver more water than is provided by the supply, a _____ pressure zone is developed at the eye of the impeller, while a _____ pressure zone occurs at the outer edge the impeller.

   a. high, low   c. vapor, cavitation

   b. low, high   d. cavitation, vapor

3. At normal atmospheric pressures (14.7 psi at sea level) water boils at _____ degrees Fahrenheit.

   a. 100   c. 212

   b. 185   d. 250

4. When pressure increases, the boiling point of water

   a. increases.

   b. decreases.

   c. stays the same.

   d. None of the above are correct.

5. Increased pressure will _____ the ability of a liquid to vaporize, while decreased pressure will _____ the ability of a liquid to vaporize.

   a. increase, reduce

   b. reduce, increase

   c. initiate, stop

   d. None of the above are correct.

6. Which of the following is not a method of stopping cavitation?

   a. reducing pump speed

   b. increasing the flow of the supply

   c. removing one or more discharge lines

   d. All of the above are correct.

Short Answer

On a separate sheet of paper, answer/explain the following questions.

1. List five instrumentations that should be continuously monitored during pump operations.
2. Explain how discharge flows are controlled.
3. Discuss the general procedures for placing two discharge lines into service at the same time when one line requires a higher pressure setting than the other.
4. Explain how to place a discharge line into service when several lines are already in operation.
5. When should a pumping operation be paused? Explain the steps for doing so.
6. Explain how the drive engine can be maintained within normal operating procedures.
7. Briefly explain how the pump water temperature can be prevented from rising.
8. What should be considered when the pumping operation is over?
9. What is water hammer? Explain how to prevent it from occurring.
10. Define cavitation, providing a brief explanation of the phenomenon.
11. List the warning signs of cavitation.
12. Explain how to control cavitation.
13. If problems occur during pumping operations, what are the most likely causes?
14. List the basic types of sprinkler and standpipe systems and discuss their components and operation.
15. Discuss basic support considerations for sprinkler and standpipe systems.

ACTIVITIES

1. Develop a lesson plan for teaching firefighters about water hammer. Include the following information in your lesson plan:
   a. Definition of water hammer
   b. Explanation of how water hammer occurs
   c. Procedures for ensuring that water hammer does not occur
2. Develop a standard operating procedure for both a sprinkler system and a standpipe system installed in a local building or structure.

PRACTICE PROBLEMS

1. During a drafting operation Engineer Smith could not secure a prime. He checked to ensure that no air leaks were present. In addition, the hard suction was checked for blockage and proper deployment. What really baffled Engineer Smith was that the priming pump had recently been replaced. In addition, the priming system had functioned properly during numerous training sessions within the last two weeks. What is Engineer Smith overlooking?
2. Develop a similar scenario presented in the preceding problem. Provide enough detail to narrow the problem down to one or two possible causes. Be sure to provide enough detail to determine the cause(s).

Refer to Figure 9–7 for the remaining questions.

Use a separate sheet of paper for your explanations.

3. When no lines are flowing, explain how to initiate flow for lines B and C at the same time.

4. When line D is flowing, explain how to initiate flow in line C.

5. Explain how to initiate flow in lines A and B when line C is already in operation.

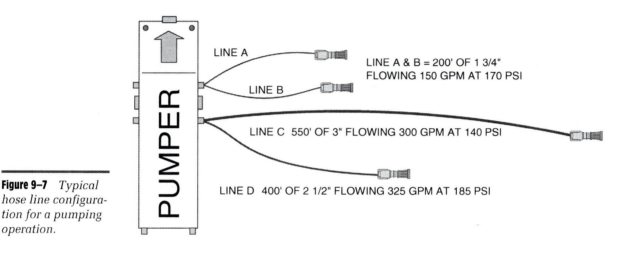

Figure 9–7 *Typical hose line configuration for a pumping operation.*

LINE A

LINE B

LINE A & B = 200' OF 1 3/4"
FLOWING 150 GPM AT 170 PSI

LINE C   550' OF 3" FLOWING 300 GPM AT 140 PSI

LINE D   400' OF 2 1/2" FLOWING 325 GPM AT 185 PSI

PUMPER

Section

4

WATER FLOW CALCULATIONS

The first three sections of this book discuss pump operators and their three main duties: preventive maintenance, driving apparatus, and pump operations. In addition, how and why pumps work as well as the peripherals and equipment used during pump operations are discussed. Section 4 now introduces a new element focusing on information related to water flow calculations. In general, concepts related to determining proper flows and pressures for fire pump operations are presented. Chapter 10 focuses on definitions and concepts basic to the science of hydraulics; Chapter 11 discusses methods for estimating the amount of water to deliver to the scene; and Chapter 12 covers calculations specific to friction loss and engine pressure. Appendix G summarizes the information presented in this section as it pertains to hydraulic calculations.

Chapter

10

INTRODUCTION TO HYDRAULIC THEORY AND PRINCIPLES

Learning Objectives

Upon completion of this chapter, you should be able to:

■ List and discuss the basic properties of water.

■ Explain the difference between density, weight, and pressure.

■ Calculate volume and weight of water.

■ Define specific heat and latent heat of vaporization.

■ Discuss the five principles of pressure.

■ List and discuss the different types of pressures.

■ Discuss the four principles of friction loss.

■ Explain the concept of nozzle reaction.

NFPA 1002
Standard for Fire Apparatus Driver/Operator Professional Qualifications
(2003 Edition)

This chapter addresses parts of the following knowledge element within sections 5.2.1 and 5.2.2:

Hydraulic calculations for friction loss and flow using both written formulas and estimation methods

The following knowledge element within section 5.2.4 is also addressed:

Calculation of pump discharge pressure

INTRODUCTION

hydraulics
the branch of science dealing with the principles and laws of fluids at rest or in motion

hydrodynamics
the branch of hydraulics that deals with the principles and laws of fluids in motion

hydrostatics
the branch of hydraulics that deals with the principles and laws of fluids at rest and the pressures they exert or transmit

Hydraulics is a branch of science dealing with the principles of fluids at rest or in motion. **Hydrodynamics** is the term given to the study of fluids in motion, while **hydrostatics** is the term given to the study of fluids at rest. The principles and laws associated with both hydrodynamics and hydrostatics are applicable to fire pump operations. A good grasp of hydraulic concepts will enable a better understanding of pump operations as well as of water flow calculations.

Hydraulic concepts and calculations are often perceived as being difficult and confusing. The reasons are many and varied. For some, the difficulty may stem from fear or lack of math skills. Others may try to learn rules of thumb without comprehending the basic principles that govern them. Admittedly, many of the equations are complicated and confusing to learn. Mastering this material requires thought, reflection, and dutiful study.

Three suggestions are offered to help reduce the difficulty and confusion of learning hydraulic concepts and calculations. First, *learn the basics*, including basic math skills. Lack of basic math skills will make learning hydraulic calculations, as well as a whole array of other subjects, difficult. Consider taking at least a college-level math course. The author of this textbook has also written *Practical Problems in Mathematics for Emergency Services*, as a study guide and resource to help review basic math concepts. In addition, learn the basic concepts of hydraulics. Rules of thumb on the fireground are certainly important, but their usefulness will be increased if the basic concepts behind them are known and understood.

Second, *keep the units straight*. Hydraulic calculations involve units associated with the numeric value. For example, 1 gallon of water weighs 8.34 pounds, which can be expressed as 8.34 lb/gal (pounds per gallons). Often, these units can be abbreviated in several ways. Consider the following examples:

gallon per minute = gpm or gal/min (flow rate)

pounds per square inch = psi or lb/in² (pressure)

pounds per cubic foot = lb/cu ft or lb/ft³ (density)

Learning to approach these calculations methodically and keeping track of units will ease a lot of the frustration and confusion. Consider the following equation used to determine the weight of 40 gallons of water:

$$40 \text{ gal} \times 8.34 \text{ lb/gal} = 333.6 \text{ lb}$$

Note that the volume units (gallons) cancel each other and yield the described units of mass (lb). To avoid error and confusion, try to consistently assign units within all equations.

Third, *practice . . . practice . . . practice*. The more you practice, the easier it will become. This concept is certainly important while learning hydraulic calculations. It is equally important, if not more important, to maintain and increase the knowledge and skills developed.

This chapter first presents the basic physical characteristics of water, and then the closely related concepts of density, weight, and pressure are discussed. Finally, important friction loss concepts and nozzle reaction calculations are presented.

PHYSICAL CHARACTERISTICS OF WATER

Fire pump operations is all about water—moving it from a source to a discharge point. Pump operators should therefore have a solid grasp of water's physical characteristics. This is also essential for understanding hydraulic principles and concepts. Water is a chemical compound comprised of two parts hydrogen and one part oxygen (H_2O).

Pure water is a colorless, odorless, and tasteless liquid that readily dissolves many substances. The water used for fire suppression is rarely in a pure state, so values for freshwater (water prepared for domestic use) are typically used for fire protection hydraulic calculations. Water is a noncombustible liquid and is considered to be virtually incompressible.

Water exists in a solid state at temperatures below 32 degrees Fahrenheit (°F) and, unlike nearly all other compounds, expands when it freezes. Between 32°F and 212°F, water tends to exist predominately in a liquid state. At temperatures above 212°F, it exists almost exclusively in a vapor (gas) state. The temperatures at which these phase transitions occur are affected by atmospheric pressure and will vary relative to changes in pressure. Boiling points and freezing points are also affected by the purity of the liquid.

evaporation
the physical change of state from a liquid to a vapor

vapor pressure
the pressure exerted on the atmosphere by molecules as they evaporate from the surface of the liquid

boiling point
the temperature at which the vapor pressure of a liquid equals the surrounding pressure

The physical change of state from a liquid to a vapor is known as **evaporation**. A very small amount of liquid water can produce a large amount of water vapor. This large increase in the volume occupied by water molecules can have the effect of increasing pressure when confined in a container. The closer the temperature of a liquid gets to its boiling point, the faster it tends to evaporate. It should be noted that evaporation takes place even with cold liquids. The pressure exerted on the atmosphere by molecules as they evaporate from the surface of the liquid (Figure 10–1) is known as **vapor pressure**. The higher the vapor pressure, the faster the rate of evaporation of the liquid. Conversely, the higher the atmospheric pressure, the slower the rate of evaporation.

Temperature affects the rate of evaporation and consequently vapor pressure. When heated, the vapor pressure of water increases, causing the water to evaporate at a faster rate. The temperature at which the vapor pressure of a liquid equals the surrounding pressure is known as its **boiling point**.

Atmospheric pressure can also affect the boiling point of water. Higher atmospheric pressure on a liquid reduces the rate of evaporation. For example, water in an open container boils at a lower temperature than in a closed container. Remember that boiling occurs when the vapor pressure of a liquid equals atmospheric pressure. Consider a pressure cooker. Water in the container boils at a temperature much higher than 212°F because of the increase in pressure within the cooker. It stands to reason then, that higher temperatures will be required to increase vapor pressure to equal that of the surrounding pressure (boiling point) within a closed container. It should also be pointed out that lowering the pressure will increase the rate of evaporation.

Recognition of how pressure and temperature affect vapor pressure and boiling point can be important considerations for pump operations, especially for

Figure 10–1 *The pressure exerted on the atmosphere by molecules as they evaporate from the surface of a liquid is known as vapor pressure.*

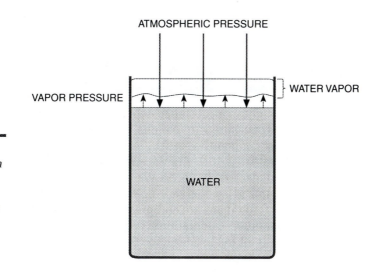

latent heat of fusion
the amount of heat that is absorbed by a substance when changing from a solid to a liquid state

specific heat
The amount of heat required to raise the temperature of a substance by 1°F. The specific heat of water is 1 Btu/lb °F

latent heat of vaporization
the amount of heat absorbed when changing from a liquid to a vapor state

■ Note
Recognition of how pressure and temperature affect vapor pressure and boiling point can be important considerations for pump operations.

understanding the phenomenon of cavitation. In addition, drafting operations can be affected by pressure and temperature. For example, when atmospheric pressure is reduced and the temperature of water increased, the height to which water can be drafted will decrease and it is easier for cavitation to occur.

When water changes physical states (solid, liquid, and gas) it absorbs or releases heat. The amount of heat that is absorbed or released is called **latent heat of fusion** and is measured in British thermal units (Btu). The Btu is defined as the amount of heat required to raise one pound of water one degree Fahrenheit (F). The equivalent metric measurement is calorie and is defined as the amount of heat required to raise 1 gram of water 1 degree Celsius (C). One Btu is approximately equivalent to 252 calories. The term **specific heat** refers to the amount of heat required to raise the temperature of a substance by 1°F. The specific heat of water is 1 Btu. For example, it takes 152 Btu to raise one pound of water from 60°F to 212°F (ambient temperature to boiling point). If we use 8.34 pounds as the approximate weight of one gallon of water, then one gallon of water absorbs 1,268 Btu (152 Btu/lb × 8.34 lb/gal = 1,267.65 Btu/gal). Another way of saying this is that it takes 1,268 Btu to raise the temperature of one gallon of water from 60°F to 212°F. At sea level, 14.7 psi, water is ready to change from a liquid to a gas (or vapor). The term **latent heat of vaporization** refers to the amount of heat absorbed or released when changing from a liquid to a vapor state. When one pound of water changes from a liquid state to a vapor state it absorbs 970.3 Btu. Another way of saying this is that it takes 970.3 Btu to convert one pound of water to vapor or, in fire suppression terms, steam. Again, if we use 8.34 pounds as the weight of one gallon of water, then the amount of heat required to convert one gallon of water from a liquid state to steam, sea level at 212°F, is 8,092.3 Btu (970.3 Btu/lb × 8.34 lb/gal = 8,092.3 Btu/gal). Note that the amount of heat absorbed by one gallon of water as it changes to a vapor state is more than 6 times as much as the amount of heat absorbed by the same amount of water as the temperature changes from 60°F to 212°F. This high heat absorption coupled with the high expansion (approximately 1,700 to 1) makes water an excellent extinguishing agent.

DENSITY OF WATER

density
the weight of a substance expressed in units of mass per volume

Density is the weight of a substance expressed in units of mass per volume. Typically, density is measured in pounds per cubic foot and can be expressed as "lb/cu ft" or, more commonly, "lb/ft³." Freshwater has a density of 62.4 lb/ft³ when the water temperature is 50°F and the atmospheric pressure is 14.7 psi (sea level). Saltwater has a density of 64 lb/ft³. Water used in fire suppression is not always a constant 50°F and is used at varying elevations. Although the density of water varies slightly with temperature and pressure, these differences are not significant for fire pump operation calculations. With this in mind, calculations in this text assume the density of freshwater to be 62.4 lb/ft³.

WEIGHT AND VOLUME OF WATER

weight
the downward force exerted on an object by the Earth's gravity, typically expressed in pounds (lb)

The most common U.S. unit of **weight** is the pound (lb), and it represents the downward force exerted on the object by the Earth's gravity. Consider a container measuring 1-foot wide by 1-foot long by 1-foot high (Figure 10–2). When 1 cubic foot (ft³) of water is added to this container, the weight (downward force) of water is 62.4 lb. Now consider a second container of equal measure placed beside the first for a total of 2 ft³. The density of water is a constant and remains the same at 62.4 lb/ft³. Therefore, the total weight of water increases to 124.8 lb for 2 ft³ (62.4 lb/ft³ × 2 ft³ = 124.8 lb).

volume
three-dimensional space occupied by an object

Volume is the amount of space occupied by an object and is considered a three-dimensional measurement. The most common U.S. unit of volume used in the fire service is gallon (gal) followed by cubic feet (ft³) and cubic inches (in³).

Calculating Weight and Volume

Determining the weight of small volumes of water can usually be done by simply placing the water on a balance. However, this is impractical for larger quantities of water. To calculate the weight of water, the following formula can be used:

$$W = D \times V \tag{10–1}$$

where W = weight in pounds (lb), D = density (pounds per unit volume) (lb/ft³), and V = volume. When the density of water is known, all that is needed to calculate the weight of water is the volume. It is important to make sure that the units of volume are the same as those given in the constant used for density.

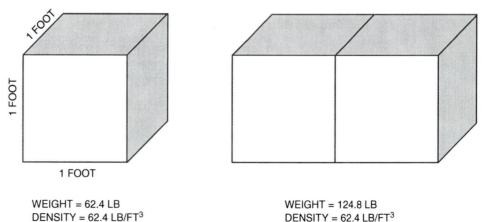

Figure 10–2   *Weight is expressed in pounds; density is expressed in pounds per unit of volume.*

1 FOOT

1 FOOT

1 FOOT

WEIGHT = 62.4 LB
DENSITY = 62.4 LB/FT³

WEIGHT = 124.8 LB
DENSITY = 62.4 LB/FT³

Weight Calculations Using Cubic Feet If the volume in cubic feet is known, the calculations are straightforward. For example, what is the weight of water in a vessel containing 50 ft^3 of water? Recall that the density of water in each cubic foot of water is 62.4. The calculation, then, is as follows:

$$W = D \times V$$
$$= 62.4 \text{ lb/ft}^3 \times 50 \text{ ft}^3$$
$$= 3{,}120 \text{ lb}$$

By using the appropriate value for the density of water, the cubic feet cancel and yield an answer with the units of pounds (lb), the desired units. The weight of 50 ft^3 of water is 3,210 lbs.

Volume Calculations Using Cubic Feet In many cases, the volume of a vessel is not known and will need to be calculated. A good example of this is a rectangular tank of unknown volume. The formula for calculating the volume of a square or rectangular vessel is as follows:

$$V = L \times W \times H \tag{10–2}$$

To determine the volume of a square or rectangular tank in cubic feet, the units will be as follows:

V = volume in cubic feet

L = length in feet

W = width in feet

H = height in feet

To determine the weight of water in this tank, simply plug the volume into the formula used in the last example. What is the weight of water in an onboard tank measuring 4 feet long, 5 feet wide, and 3 feet high? The first step is to find the volume (number of cubic feet of water) of the tank.

$$V = L \text{ (ft)} \times W \text{ (ft)} \times H \text{ (ft)}$$
$$= 4 \text{ ft} \times 5 \text{ ft} \times 3 \text{ ft}$$
$$= 60 \text{ ft}^3$$

In this case the volume is 60 ft^3. Notice that the correct answer will naturally include the described units (ft^3). To find the weight of water, multiply the density of water (62.4 lb/ft^3) by the volume (60 ft^3):

$$W = D \times V$$
$$= 62.4 \text{ lb/ft}^3 \times 60 \text{ ft}^3$$
$$= 3{,}744 \text{ lb}$$

By using lb/ft^3 for the density of water and ft^3 for the volume, the answer (3,744) includes the appropriate units of weight (lb).

To find the volume in cubic feet for a cylinder, such as municipal water tank or section of hose, use the following formula:

$$V = .7854 \times d^2 \times H \tag{10-3}$$

where V = volume in cubic feet
.7854 = constant for calculating the area of a circle based on its diameter
d = diameter of the cylinder
H = height (or length) of the cylinder

What is the volume of water in a tank 50 feet wide by 75 feet tall?

$$\begin{aligned} V &= .7854 \times d^2 \times H \\ &= .7854 \times (50 \text{ ft})^2 \times 75 \text{ ft} \\ &= .7854 \times 2{,}500 \text{ ft}^2 \times 75 \text{ ft} \\ &= .7854 \times 187{,}500 \text{ ft}^3 \\ &= 147{,}262.5 \end{aligned}$$

Weight Calculations Using Gallons During fire pump operations, gallons rather than cubic feet are normally used to express volume. To calculate the weight of a volume expressed in gallons, a different value for the density of water must be used. The same formula is used except the density is expressed in gallons as opposed to cubic feet. In this case, D is expressed in pounds per gallon rather than pounds per cubic foot, and V is expressed in gallons. The weight of a gallon of water can be calculated by dividing the weight of one cubic foot of water (62.4 lb/ft³) by the number of gallons in a cubic foot (7.48 gal/ft³):

$$\begin{aligned} 1 \text{ gallon of water} &= \frac{62.4 \text{ lb/ft}^3}{7.48 \text{ gal/ft}^3} \\ &= 8.34 \text{ lb/gal} \end{aligned}$$

Note that the lb/ft³ cancel yielding a new volume unit of lb/gal. One gallon of water, then, weighs 8.34 lb/gal.

A variety of situations may occur when the volume, in gallons, is known. When this is the case, simply multiply the density of water (lb/gal) by the total number of gallons. Typically, the number of gallons in an onboard tank or tanker will be known, so the weight of water can easily be calculated. Consider a 500-gallon onboard tank. Calculate the weight of water in this tank as follows:

$$\begin{aligned} W &= D \times V \\ &= 8.34 \text{ lb/gal} \times 500 \text{ gal} \\ &= 4{,}170 \text{ lb} \end{aligned}$$

The units of gallon cancel leaving the appropriate unit of weight in the answer, 4,170 lb.

Another example in which the volume in gallons is typically known is that of fire streams. Consider, for example, a fire stream flowing 250 gpm into a structure.

If the stream is maintained for 5 minutes, the fire stream will deliver 1,250 gallons of water into the structure (250 gpm × 5 minutes = 1,250 gallons). Note that the units of minutes cancel, leaving the units of gallons. The weight of water from a fire stream flowing 250 gpm for 5 minutes can now be calculated as:

$$W = D \times V$$
$$= 8.34 \text{ lb/gal} \times 1,250 \text{ gal}$$
$$= 10,425 \text{ lb}$$

Recall that 1 ton is equal to 2,000 lb. In this example, slightly more than 5 tons of water are flowing into the structure every 5 minutes.

Volume Calculations Using Gallons If the volume of a container is unknown, the first step is to calculate the volume in gallons. For square or rectangular containers, simply find the volume in cubic feet and multiply by 7.48 (7.48 = number of gallons in 1 cubic foot). Next, calculate the weight ($W = D \times V$) using the appropriate units. For example, what is the weight of water in a container measuring 5 feet by 6 feet by 4.5 feet? The first step is to select the density and volume in appropriate units.

$$D = 8.34 \text{ lb/gal}$$

$$V = (5 \text{ ft} \times 6 \text{ ft} \times 4.5 \text{ ft}) \times 7.48 \text{ gal/ft}^3$$
$$= 135 \text{ ft}^3 \times 7.48 \text{ gal/ft}^3$$
$$= 1,009.8 \text{ gal}$$

The next step is to calculate the weight using the familiar formula:

$$W = D \times V$$
$$= 8.34 \text{ lb/gal} \times 1,009.8 \text{ gal}$$
$$= 8,421.7 \text{ lb}$$

The weight of water in a container measuring 5 feet by 6 feet by 4.5 feet is 8,421.7 lb.

To find the number of gallons in a cylinder, such as a length of hose, use the formula:

$$V = 6 \times d^2 \times h \tag{10–4}$$

where V = volume in gallons

   6 = constant in gallons per cubic foot (gal/ft^3) derived by multiplying the constant .7854 (from the volume formula) by 7.48 (the number of gallons in one cubic foot) which equals 5.87 and is rounded up to 6 for ease of use on the emergency scene

   d = diameter of the cylinder in feet (in. ÷ 12 in/ft)

   h = height or length of the cylinder in feet

Consider a 150-foot length of 1½-inch hose filled with water. The calculation for the volume in gallons of water contained in this section of hose is given as follows:

$$V = 6 \times d^2 \times h$$

$$= 6 \text{ gal} / \text{ft}^3 \times \left(\frac{\text{in}}{12 \text{ in} / \text{ft}} \right)^2 \times h \text{ (ft)}$$

$$= 6 \text{ gal} / \text{ft}^3 \times \left(\frac{1.5 \text{ in}}{12 \text{ in} / \text{ft}} \right)^2 \times 150 \text{ (ft)}$$

$$= 6 \text{ gal/ft}^3 \times (0.125 \text{ft})^2 \times 150 \text{ ft}$$
$$= 6 \text{ gal/ft}^3 \times 0.0156 \text{ ft}^2 \times 150 \text{ ft}$$
$$= 6 \text{ gal/ft}^3 \times 2.34 \text{ ft}^3$$
$$= 14.04 \text{ gal, or } 14 \text{ gal}$$

Notice how the units cancel and yield an answer of 14 gallons.

Calculating the weight of water in the hose can now be done by:

$$W = D \times V$$
$$= 8.34 \text{ lb/gal} \times 14 \text{ gal}$$
$$= 116.76 \text{ lb}$$

In this example, the 150-foot length of 1½-inch hose contains 14 gallons of water, which weighs 116.76 lb.

PRESSURE

Fire pump operations require pressure to move water from one location to another. The internal pressure developed by fire pumps to accomplish this task is expressed in pounds per unit area (typically pounds per square inch, psi, or lb/in²). Often the terms pressure and weight are mistakenly used interchangeably. The concept of weight refers to the total force of attraction between an object and the Earth (gravity). Recall that the units of weight are pounds but do not involve a unit of surface area. On the other hand, pressure combines the units of weight and area. **Pressure**, then, is the force exerted by a substance in units of weight per unit area, typically pounds per square inch (psi). The formula for calculating pressure is as follows:

pressure
the force exerted by a substance in units of weight per area; the amount of force generated by a pump or the resistance encountered on the discharge side of a pump; typically expressed in pounds per square inch (psi)

$$P = \frac{F}{A} \tag{10–5}$$

where P = pressure in force per unit of area (typically lb/in²)
   F = force (weight, typically in pounds)
   A = surface area (typically in square inches)

Consider a vessel 1 foot by 1 foot by 1 foot containing 1 cubic foot of water (Figure 10–3). Note that the downward force (weight) of the water is spread out over a 1-square-foot area of surface area. Since there are 12 inches in a foot, the area in square inches can be found by multiplying 12 inches by 12 inches. It follows

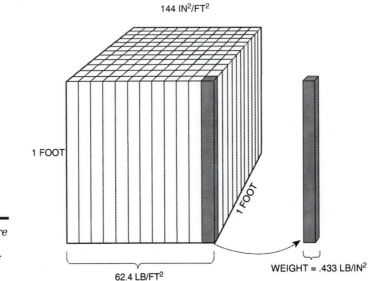

144 IN²/FT²

1 FOOT

1 FOOT

62.4 LB/FT²

WEIGHT = .433 LB/IN²

Figure 10–3 *Pressure is expressed in pounds per unit of area (psi).*

that the 1-square-foot (ft²) area contains 144 square inches (or 144 in²/ft²). Recall that 1 cubic foot of water weighs 62.4 lbs. Note that the weight is exerted over a square foot area. The pressure exerted by this container can now be calculated:

$$P = \frac{F}{A}$$

$$= \frac{62.4 \text{ lb}/\text{ft}^2}{144 \text{ in}^2/\text{ft}}$$

$$= 4.33 \text{ lb}/\text{in}^2$$

force

pushing or pulling action on an object

In essence, weight is the total **force** and can be localized at one point, while pressure is the force averaged over a given surface area of the material. The pressure of water in a 1 ft³ container is .433 psi.

The formula for force is as follows:

$$F = P \times A \tag{10–6}$$

where F = force (weight, typically in pounds)
P = pressure in force per unit of area (typically psi)
A = surface area (typically in square inches)

Using the previous example, what is the force of water in a 1 ft³ container?

$F = P \times A$
   $= .433 \text{ lb/in}^2 \times 144 \text{ in}^2/\text{ft}^2$
   $= 62.35 \text{ lb/ft}^2$ (Note: this figure is rounded to 62.4 in this text.)

Pressure Principles

The manner in which a liquid behaves while under pressure follows several basic principles. Some of these principles are straightforward and logical, others may seem confusing and perhaps illogical. In either case, these principles are important and add to the basic knowledge needed to fully grasp both water flow calculations and fire pump operations.

Principle 1 *The pressure at any point in a liquid at rest is equal in every direction.* One way to view this principle is that the pressure in water is exerted in every direction, downward and outward as well as upward (Figure 10–4). Because liquids have weight, the most obvious direction of force in a liquid is downward. Because we know that liquids exert pressure on the walls of their containers, the lateral forces are also obvious. This outward force can be noted by filling a container, for example a milk container or plastic bag, with water. Note that the sides of the container tend to push outward. The upward pressure in a liquid is, perhaps, not as obvious. This upward pressure can be felt by pushing an empty can into a container of water. The resistance felt is the upward thrust of the water on the bottom of the empty can. This concept is the basis for *buoyancy*. Since each force within the body of a liquid at rest has an equal force in the opposite direction, the sum of all forces is zero.

Principle 2 *The pressure of a fluid acting on a surface is perpendicular to that surface.* Although the sum of all forces acting on a molecule within the body of a liquid is zero, the force created by pressure acting on a surface area does have direction. Specifically, this force is perpendicular to any surface it acts upon (Figure 10–5). Consider a section of 1¾-inch attack line. Before it is charged, the hose is flat. When the line is charged to its operating pressure, the hose becomes round because the pressure is perpendicular to the internal surface of the hose.

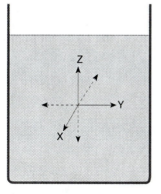

Figure 10–4 *At any point in a liquid at rest, pressure is equal in every direction.*

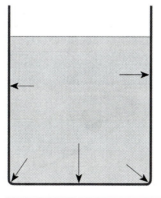
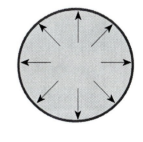

Figure 10–5 *The pressure of water acting on a surface is perpendicular to that surface.*

■ Note

External pressure applied to a confined liquid (fluid) is transmitted equally throughout the liquid.

Principle 3 *External pressure applied to a confined liquid (fluid) is transmitted equally throughout the liquid.* This principle provides the basis for understanding the transmission of pressure through a network of fire hose. In essence, any change in pressure is transmitted throughout the entire hose line. This is true even if the line extends over a great distance. Consider, for example, a pump supplying a discharge layout of 500 feet of $2\frac{1}{2}$-inch hose wyed to two 150-foot sections of $1\frac{1}{2}$-inch line each flowing 95 gpm. The combined flow rate for the two nozzles will be 190 gpm. If one of the $1\frac{1}{2}$-inch lines were closed rapidly, the pump would attempt to deliver its 190 gpm through the other $1\frac{1}{2}$-inch line, resulting in a rapid increase in pressure. Because of principle 3, it should be fairly clear that the resultant pressure surge (water hammer) is transmitted throughout the entire hose lay. This means that both $1\frac{1}{2}$-inch lines as well as the pump will experience a change in pressure. Principle 3 is one of the primary reasons for the use of pressure-regulating devices on pumps.

This principle is also valid when no water is flowing. For the same example, suppose pressure gauges were evenly distributed along the entire lay, as shown in Figure 10–6. Suppose the pump continues to provide pressure and the nozzles of both $1\frac{1}{2}$-inch lines are shut down. Principle 3 states that pressure will be transmitted throughout the network, therefore each of the pressure gauges would have the same reading (as long as elevation is not a factor). To restate the principle, when no water is flowing, pressure is transmitted equally and is undiminished throughout the hose lay.

This principle also provides the basis for calculating pressure in hose line configurations that split from a supply as in the example shown in Figure 10–6. Since pressure is transmitted equally in all directions, the pressure within the $2\frac{1}{2}$-inch hose will be transmitted equally to both $1\frac{1}{2}$-inch lines.

■ Note

The pressure at any point beneath the surface of a liquid in an open container is directly proportional to its depth.

Principle 4 *The pressure at any point beneath the surface of a liquid in an open container is directly proportional to its depth.* An understanding that the magnitude of the downward force created by a column of water is directly proportional to its depth is important in a number of hydraulic calculations.

This principle can be illustrated by examining the downward force acting on the bottom of a vessel measuring 1 foot wide by 1 foot long by 1 foot high. Recall

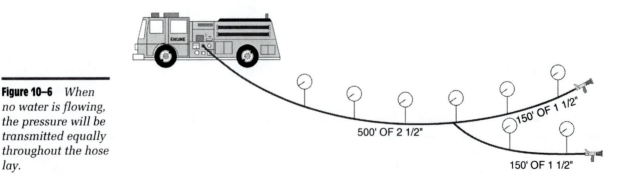

Figure 10–6 *When no water is flowing, the pressure will be transmitted equally throughout the hose lay.*

500' OF 2 1/2"

150' OF 1 1/2"

150' OF 1 1/2"

from earlier in the chapter that pressure is defined as a force per unit of area. As per the previous calculation at equation (10–5), the pressure that 1 cubic foot of water exerts on the bottom of its container is .433 lb/in². Suppose we now place a second cubic foot of water on top of the first (Figure 10–7) and repeat the calculation for pressure. When the second cubic foot of water is added, the weight exerted at the bottom of the vessel doubles to 124.8 lb/ft². Note that the area of the bottom of the vessel has not changed. The pressure at the bottom of the vessel, then, can be calculated as follows:

$$P = \frac{F}{A}$$

$$= \frac{124.8 \ \text{lb/ft}^2}{144 \ \text{in}^2/\text{ft}}$$

$$= .866 \ \text{lb/in}^2$$

This example illustrates that when the height of the water increases, the pressure at the bottom of the vessel also increases proportionally.

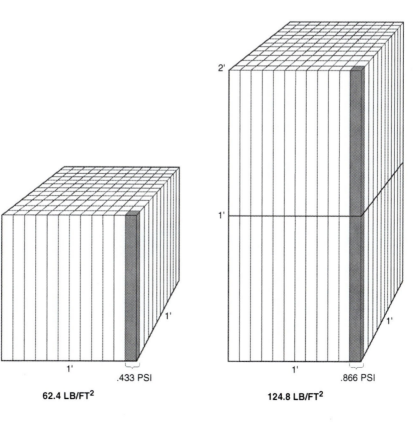

Figure 10–7   *The pressure at a vessel's base is proportional to the height of water.*

62.4 LB/FT²   .433 PSI

124.8 LB/FT²   .866 PSI

■ **Note**

The pressure exerted at
the bottom of a
container is
independent of the
shape or volume of
the container.

Principle 5 *The pressure exerted at the bottom of a container is independent of the shape or volume of the container.* This principle is perhaps the most confusing. The key to understanding this principle is to remember that it discusses pressure rather than weight. Recall that weight is the total force of a substance over a surface area, while pressure is the weight over a specific area, typically a square inch. Note that the principle states the *pressure* is proportional to depth as opposed to *weight* being proportional to depth. For example, the containers in Figure 10–8 each have different shapes and volumes. Obviously, the total weight of water in each container will vary. However, because the level of water in each of the containers is the same, the pressure (psi) at the base will also be the same.

Imagine a 1-square-inch column extending from the base of the vessel to the level of the water. In the preceding principle it was determined that a 1-square-inch column 1 foot high has a weight or exerts a pressure of .433 lb/in² on the base of the container. Consider a 1-square-inch column extending from the bottom to the top in each of the vessels. Because pressure on the bottom is proportional to the height or the column it supports, the pressure will increase by .433 lb/in² for each foot of head. This pressure can also be expressed as .433 psi per foot (.433 lb/in²/ft).

The pressure at the bottom of the vessel can be determined by the following formula:

$$P = .433 \text{ lb/in}^2/\text{ft} \times h \text{ (ft)} \qquad\qquad (10\text{–}7)$$

where P = pressure in psi
        .433 = a constant that represents the pressure exerted by a
                column of water 1 inch by 1 inch by 1 foot
        h = the height (depth) of the water in feet

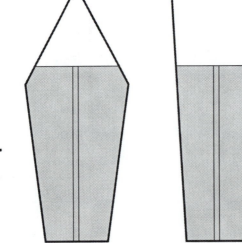

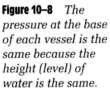

Figure 10–8   *The pressure at the base of each vessel is the same because the height (level) of water is the same.*

In each of the vessels shown in Figure 10–8, the pressure at the bottom can be calculated as follows:

$$P = .433 \text{ lb/in}^2/\text{ft} \times 100 \text{ ft}$$
$$= 43.4 \text{ lb/in}^2$$

Types of Pressures

There are many ways to measure the pressure of a system, especially for flowing systems. Consequently, there are many terms to describe specifically how pressure is measured and exactly what is being measured.

Atmospheric Pressure Surrounding the Earth is a body of air called the atmosphere. Air has weight. Because of this weight, the atmosphere (body of air) exerts pressure on the Earth that is known as **atmospheric pressure**. Air is also easily compressible. Therefore, the weight of air in the upper atmosphere compresses the air in its lower layers. The result is that the higher you go in the atmosphere, the less pressure you encounter. Conversely, the lower you go in the atmosphere, the more pressure you encounter. At sea level, the atmospheric pressure is 14.7 psi. In higher elevations, the atmospheric pressure is reduced, while in lower elevations, the atmospheric pressure increases.

Gauge Pressure (psig) An idle pressure gauge on a pumper at sea level will read 0 psi. This reading is known as gauge pressure (psig). Recall that the pressure of air at sea level is 14.7 psi. **Gauge pressure**, then, is simply a pressure reading less atmospheric pressure. For example, when the gauge reads 200 psi (200 psig), the actual pressure is 214.7 psig (200 psig + 14.7 psi). Pressure gauges used on pumping apparatus typically measure psig and are often identified as such.

Absolute Pressure (psia) The measurement that includes atmospheric pressure is known as **absolute pressure (psia)**. A gauge that measures psia would have a reading of 14.7 psia at sea level. Another way of relating gauge pressure to absolute pressure is to say 14.7 psia is equal to 0 psig. The use of psia is typically limited to measurements within pressurized vessels.

Vacuum (Negative Pressure) The three previous types of pressures are considered positive pressure. Vacuum is considered a negative pressure. Positive pressures are measured in psi while negative pressures are measured in inches of mercury. Pressures less than atmospheric pressure are called **vacuum** and are expressed in units of inches of mercury (in. Hg). Typically at least one gauge on a pump panel measures vacuum, the master intake gauge, which is usually a compound gauge.

Head Pressure **Head pressure** is the vertical height of a column of liquid expressed in feet. Consider a column of water 2.31 feet in height (Figure 10–9). The pressure exerted on the bottom of vessel is calculated as follows:

atmospheric pressure
the pressure exerted by the atmosphere (body of air) on the Earth

gauge pressure
measurement of pressure that does not include atmospheric pressure, typically expressed as psig

absolute pressure (psia)
measurement of pressure that includes atmospheric pressure, typically expressed as psia

vacuum
measurement of pressure that is less than atmospheric pressure, typically expressed in inches of mercury (in. Hg)

head pressure
the pressure exerted by the vertical height of a column of liquid expressed in feet

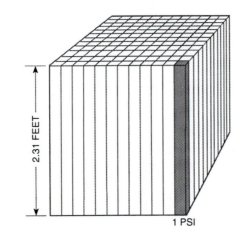

2.31 FEET

1 PSI

Figure 10–9 *One psi will raise a 1 square-inch column of water 2.31 feet.*

$$P = F \times A$$
$$= 2.31 \text{ ft} \times .433 \text{ lb/in}^2/\text{ft}$$
$$= 1 \text{ lb/in}^2 \text{ or (1 psi)}$$

Note that the units of feet cancel. The results of this calculation can also be stated as 1 psi of pressure will raise water 2.31 feet. This information provides the basis for the formula used to calculate head and is stated as follows:

$$h = 2.31 \text{ ft/psi} \times p \qquad (10\text{–}8)$$

where h = the height of water in feet
2.31 = height in feet 1 psi will raise water
p = pressure in psi

Consider a vessel of water with a pressure of 43.3 psi at its base. Using equation (10–8), the height (head) of water can be calculated as follows:

$$h = 2.31 \text{ ft/psi} \times p$$
$$= 2.31 \text{ ft/psi} \times 43.3 \text{ psi}$$
$$= 100 \text{ ft}$$

Note that the units of psi cancel leaving the answer, 100, in units of feet.

static pressure
the pressure in a system when no water is flowing

Static Pressure The term *static* indicates a lack of motion or movement. **Static pressure,** then, is the pressure in a system when no water is flowing. For example, the pressure at a hydrant before it is opened to flow water is called static pressure. Static pressure can be measured by placing a pressure gauge (mounted on an outlet cap) on one of the outlets and turning on the hydrant. Because the gauge is mounted on an outlet cap, no water will be flowing, but the gauge will read static pressure.

Residual Pressure If a second outlet on the hydrant is opened, the pressure reading on the gauge at the first outlet will drop due to friction loss in the system. The new

residual pressure
the pressure remaining in the system after water has been flowing through it

pressure drop
the difference between the static pressure and the residual pressure when measured at the same location

pressure reading is called **residual pressure**. In other words, residual pressure is the pressure remaining in the system after water is flowing.

Pressure Drop   The difference between the static pressure and the residual pressure when measured at the same location is called **pressure drop**. For example, if the static pressure measure is 50 psi and the residual pressure measure is 35 psi, the pressure drop would be 15 psi. The reduction in pressure accounts for loss of pressure caused by friction within the system.

During pump operations, the drop in hydrant pressure from static to residual can be used to estimate the additional flow the hydrant is capable of providing. This is accomplished by first noting the pressure on the master intake gauge on the pump panel after the hydrant is opened but before any discharge lines are opened. Next, initiate and obtain proper flow through one discharge. Again, note the pressure on the master intake gauge. Finally, determine the percentage of the drop in the two readings. Based on the percent drop in pressure, additional flows may be available from the hydrant as follow:

0–10% drop	3 times the original flow
11–15% drop	2 times the original flow
16–25% drop	1 time the original flow

For example, a static reading of 50 psi was noted when the hydrant was opened and a residual reading of 40 psi was noted after a 1½-inch line flowing 100 gpm was initiated. The drop in pressure is 10 psi (50 psi − 40 psi = 10 psi). The percent drop in pressure is 20 (10 psi/50 psi = 20%). In this case, only an additional 100 gpm is available from the hydrant. As a note, some areas use a more conservative percent drop in pressure of 5%, 10%, and 20%.

normal pressure
the water flow pressure found in a system during normal consumption demands

Normal Pressure   Municipal water distributions systems are often designed to meet both consumer needs and fire protection needs. During high consumer demand times, such as in the morning, the pressure within the system will drop. **Normal pressure** is the water flow pressure found in a system during normal consumption demands.

velocity pressure
the forward pressure of water as it leaves an opening

Velocity Pressure   Water pressure within a hose is converted to velocity pressure as it leaves a discharge opening, typically a nozzle. The forward pressure of water as it leaves an opening is called **velocity pressure**. Typically measured with a pitot gauge, velocity pressure can be used to calculate flow.

pressure gain and loss
the increase or decrease in pressure as a result of an increase or decrease in elevation

Pressure Gain and Loss   Previously it was stated that for every 1-foot increase in a 1-square-inch column of water, pressure will increase by .433 psi. In turn, it can be said that to increase the height of water by 1 foot, an increase of .433 psi is required. **Pressure gain and loss** is the increase and decrease in pressure as a result of an increase or decrease in elevation. This is an important concept in fire pump operations when hose lays are advanced either above or below the pump. In essence, for every foot above the pump, the pressure in a hose will decrease by .433 psi. In turn, to compensate for this loss, pressure must be increased by .433 psi for each foot of elevation. For fireground calculations, this pressure can be rounded up to .5 psi.

Consider a hose line taken to the second story of a structure. If the height of an average story is considered to be 10 feet, a pressure reduction of 5 psi will occur (.5 lb/in^2/ft × 10 ft = 5 lb/in^2). For every elevation above the first floor, pressure must be increased by 5 psi. If the hose line is advanced to the fourth floor, the pressure must be increased by 15 psi. Keep in mind that the first floor is typically at ground level and no elevation loss is encountered. The same concept applies when hose lines are advanced below the pump except that pressure will increase. Consider a line taken to the basement of a structure. In this case, pressure will increase and a reduction of 5 psi is required.

When lines are advanced up or down grades rather than floor levels of a structure, .5 psi can be added or subtracted for every increase or decrease of 1 foot. For example, a hose line located 25 feet above the pump would have a 12.5 psi loss in pressure. If the hose is 25 feet below the pump, a 12.5 psi increase will occur.

Nozzle Pressure Nozzles are designed and constructed to provide a specific flow, or range of flows, at a specific pressure. The designed operating pressure of a nozzle is called **nozzle pressure**. When the correct pressure exists at the nozzle, the nozzle will be provided with its designed flow. The main purpose of fireground hydraulic calculations is to calculate the pump discharge pressure required to provide the correct nozzle pressure. When flow meters are used, no calculations are necessary in that when the nozzle is provided its designed flow, the correct operating pressure exists at the nozzle. In either case, the operating pressure is called nozzle pressure. The majority of nozzles used in the fire service are designed to operate at one of the following nozzle pressures:

50 psi	Smooth-bore nozzles used on handline
75 psi	Low-pressure nozzles
80 psi	Smooth-bore nozzles used on master stream device
100 psi	Combination nozzles, including automatic nozzles

It should be noted that some newer combination nozzles have the ability to convert to a low-pressure nozzle while in the field.

FLOW CONCEPTS AND CONSIDERATIONS

There are several concepts basic to the discussion of fluid flow. Perhaps the most fundamental is the conservation of energy law, which states that energy is conserved in that it is neither created nor destroyed. This is not the same as saying the energy will remain the same; rather, the energy can change forms and/or can be

■ **Note**
The main purpose of fireground hydraulic calculations is to calculate the pump discharge pressure required to provide the correct nozzle pressure.

nozzle pressure
the designed operating pressure for a particular nozzle

transferred but the sum total will remain constant. Bernoulli's theorem applies this concept to fluids in motion, specifically incompressible fluids such as water. Basically, the total pressure (head) within a system will be the same anywhere in the system. The equation can be stated as the sum of velocity head, pressure head, and elevation head is constant within a system. The conservation of mass law states that mass cannot be created or destroyed. As it relates to fire pump operations, the same amount of water entering a hose must exit the hose.

Friction Loss

According to Webster's *Ninth New Collegiate Dictionary*, friction is defined as the rubbing of one body against another and the resistance to relative motion between two bodies in contact. This rubbing and resistance to motion causes a reduction in energy called **friction loss**. In reality, anything water comes in contact with, including itself, can cause friction loss. Factors that affect friction loss in fire hoses include the following:

- Rough interior lining (typical of older hose)
- Couplings, adapters, and appliances
- Bends and kinks in the hose
- Length and size of hose
- Flow, pressure, and velocity

Friction loss is commonly expressed in pounds per square inch (psi) and measures the reduction of pressure between two points in a system. In other words, the difference in pressure between two points in a system is the result of friction.

Laminar/Turbulent Flows

A drop in pressure will occur when water flows through a hose. When the velocity of water is low and the hose interior smooth, turbulence in the water will be minimal. This condition is known as **laminar flow** in which thin parallel layers of water develop and move in the same direction together (Figure 10–10). The outer layer moves along the interior lining of the hose, while other layers move alongside one another; therefore, varying velocities can occur among the layers. Because only the very outer layer touches the interior of the hose, friction loss is typically limited.

When the velocity is high and the hose interior rough, turbulence in the water occurs. This condition is known as **turbulent flow** in which water moves in an erratic and unpredictable pattern. This random movement mixes the layers in the water to create a uniform velocity within the hose. Increased pressure loss occurs because more water is subjected to the interior lining of the hose. In comparison, during laminar flow only a thin layer of water touches the interior of the hose. When turbulent flow occurs, water no longer travels in a smooth straight line; rather, movement is erratic, causing increased friction. The pressures in hose are typically high enough to cause turbulent flow.

friction loss

the reduction in energy (pressure) resulting from the rubbing of one body against another, and the resistance of relative motion between the two bodies in contact; typically expressed in pounds per square inch (psi); measures the reduction of pressure between two points in a system

laminar flow

flow of water in which thin parallel layers of water develop and move in the same direction

turbulent flow

flow of water in an erratic and unpredictable pattern creating a uniform velocity within the hose that increases pressure loss because more water is subjected to the interior lining of the hose

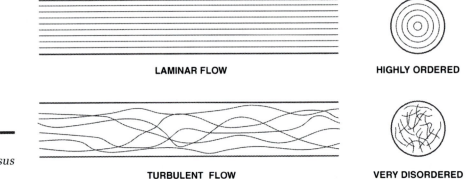

LAMINAR FLOW   HIGHLY ORDERED

TURBULENT FLOW   VERY DISORDERED

Figure 10–10
Laminar flow versus turbulent flow

Fundamental Friction Loss Principles

Although friction loss varies with the age of hose, number of kinks or bends, and numerous other factors, several principles of friction loss remain constant.

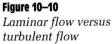

■ Note
Friction loss varies directly with hose length if all other variables are held constant.

Friction Loss Principle 1 *Friction loss varies directly with hose length if all other variables are held constant.* In other words, when the length of the hose doubles, friction loss doubles as well. For example, 500 feet of 3-inch hose flowing 300 gpm has a friction loss of 36 psi (Figure 10–11). If the length is doubled to 1,000 feet, the friction loss will also double to 72 psi. The principle explains, in part, why friction loss formulas typically calculate friction loss per 100-foot sections of hose. (See the friction loss chart in Appendix F.) Because friction loss varies directly with hose length, the friction loss for 100 feet of hose can be multiplied by the number of 100-foot sections to determine total friction loss in the hose.

■ Note
With all other variables held constant, friction loss varies approximately with the square of the flow.

Friction Loss Principle 2 *With all other variables held constant, friction loss varies approximately with the square of the flow.* This rule points out that the rate of increase in friction loss is significantly greater than the increase in flow. In other words, if the flow doubles (2 times as much), the friction loss will increase two times as much squared (2^2, or $2 \times 2 = 4$ times as much). If the flow triples (3 times as much), the friction loss will increase nine times ($3^2 = 9$). Consider a 100-foot section of $2\frac{1}{2}$-inch hose flowing 100 gpm (Figure 10–12). The friction loss would be 2 psi while flowing 100 gpm. If the flow is doubled to 200 gpm, the friction loss increases 4 times (2^2) to 8 psi ($4 \times 2 = 8$). If the original flow is increased to 400 gpm, the friction loss increases 16 times (4^2) to 32 psi ($16 \times 2 = 32$).

As a note, these same friction loss values can also be obtained using the formula $FL=cq^2L$, which is discussed further in Chapters 11 and 12.

■ Note
When the flow remains constant, friction loss varies inversely with hose diameter.

Friction Loss Principle 3 *When the flow remains constant, friction loss varies inversely with hose diameter.* In other words, friction loss will decrease when hose diameter is increased, and friction loss will increase when hose diameter is decreased. Consider 250 gpm flowing through 200 feet of different size hose (Figure 10–13).

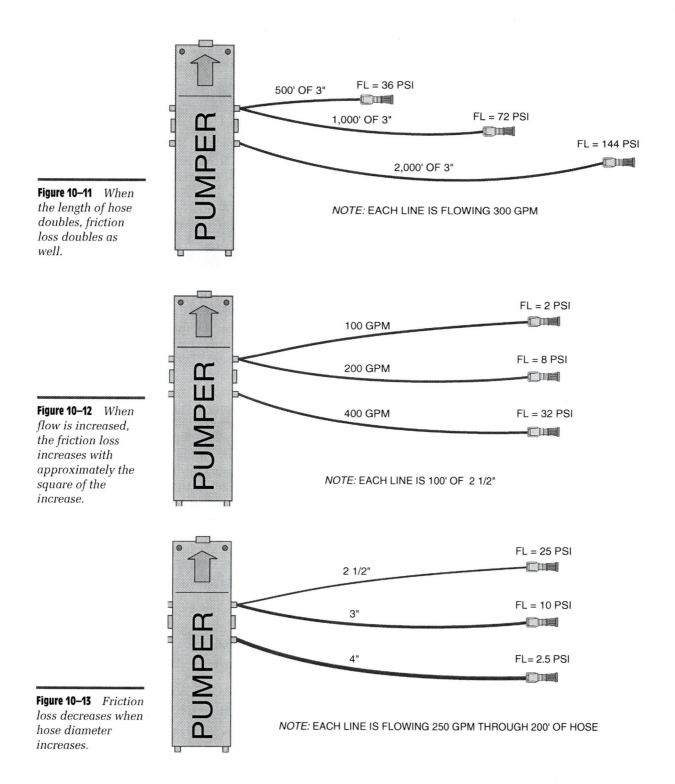

Figure 10–11 *When the length of hose doubles, friction loss doubles as well.*

500' OF 3"   FL = 36 PSI

1,000' OF 3"   FL = 72 PSI

FL = 144 PSI

2,000' OF 3"

NOTE: EACH LINE IS FLOWING 300 GPM

Figure 10–12 *When flow is increased, the friction loss increases with approximately the square of the increase.*

FL = 2 PSI

100 GPM

200 GPM   FL = 8 PSI

400 GPM   FL = 32 PSI

NOTE: EACH LINE IS 100' OF 2 1/2"

Figure 10–13 *Friction loss decreases when hose diameter increases.*

FL = 25 PSI

2 1/2"

FL = 10 PSI

3"

4"   FL= 2.5 PSI

NOTE: EACH LINE IS FLOWING 250 GPM THROUGH 200' OF HOSE

The friction loss in 2½-inch, 3-inch, and 4-inch hose is 25 psi, 10 psi, and 2.5 psi respectively. This principle is the reason for the introduction and use of large diameter hose.

Friction Loss Principle 4 *For any given velocity, the friction loss will be about the same regardless of water pressure.* In other words, the speed (velocity) of water traveling through the hose governs friction loss rather than the quantity of water. For example, if the water within a hose is traveling 20 feet per second, the friction loss will remain approximately the same whether the pressure is 50 psi or 150 psi. Another way of saying this principle is that friction loss is independent of pressure at a given velocity.

NOZZLE REACTION

The basic principle behind nozzle reaction is Newton's third law of motion, which states that every action is accompanied by an equal and opposite reaction. Relating this principle to modern suppression nozzles, the forward discharge of water is accompanied by a recoil of the nozzle in the opposite direction. This tendency of nozzles to move in the opposite direction of water flow is called *nozzle reaction*.

Of concern to pump operators, and perhaps even more for those operating a nozzle, is the fact that the nozzle's opposite reaction is proportional to the amount and velocity of water being discharged Thus, greater discharge flows and pressures will increase nozzle reaction. Recall from Chapter 6 that nozzles are designed to operate at specific nozzle pressures. In most cases, providing proper nozzle pressure will ensure that the nozzle reaction is manageable. When designed nozzle pressures are exceeded, nozzle reaction increases rapidly. Even though modern nozzles are capable of providing variable flows while maintaining proper nozzle pressure(s), nozzle reaction can still increase. An increase in flow through an automatic nozzle will maintain proper nozzle pressure while increasing nozzle reaction because more water is being discharged creating additional force. Often, the focus of hydraulic calculations is on friction loss, flow, and nozzle pressure, with little consideration given to nozzle reaction. Nozzle operators being lifted off the ground or losing control of the nozzle are not safe indicators of excessive nozzle reaction.

Nozzle reaction is measured in pounds (force) and can be calculated for both smooth-bore nozzles and combination nozzles. (See, for example, the nozzle reaction chart in Appendix F.)

Smooth-Bore Nozzles

The formula for calculating nozzle reaction in smooth-bore nozzles is as follows:

$$NR = 1.57 \times d^2 \times NP \qquad\qquad (10\text{--}9)$$

where NR = nozzle reaction in pounds
1.57 = constant
d = diameter of nozzle orifice in inches
NP = operating nozzle pressure in psi

Consider a 1-inch smooth-bore nozzle discharging water with a nozzle pressure of 50 psi. The calculation for nozzle reaction is as follows:

$$NR = 1.57 \times d^2 \times NP$$
$$= 1.57 \times (1 \text{ in.})^2 \times 50 \text{ lb/in}^2$$
$$= 1.57 \times 1 \text{ in.}^2 \times 50 \text{ lb/in}^2$$
$$= 78.5 \text{ lb}$$

The nozzle reaction for this example is 78.5 pounds.

Combination Nozzles

The formula for calculating nozzle reaction in combination nozzles is as follows:

$$NR = \text{gpm} \times \sqrt{NP} \times .0505 \qquad (10\text{–}10)$$

where NR = nozzle reaction in pounds
gpm = gallons per minute
NP = nozzle pressure in psi
.0505 = constant with in^2/gpm as units

The nozzle reaction for a combination nozzle flowing 250 gpm with a nozzle pressure of 100 psi is calculated as follows:

$$NR = \text{gpm} \times (\sqrt{NP} \times 0.0505 \text{ in}^2/\text{gpm})$$
$$= 250 \text{ gpm} \times (\sqrt{100 \text{ lb}/\text{in}^2} \times .0505 \text{ in}^2/\text{gpm})$$
$$= 250 \text{ gpm} \times (10 \text{ lb/in}^2 \times 0.0505 \text{ in}^2/\text{gpm})$$
$$= 250 \text{ gpm} \times .505 \text{ lb/gpm}$$
$$= 126.25 \text{ lb}$$

Note that the square root of 100 is 10. Further note that 10 multiplied by .0505 is .505. Therefore, when combination nozzles are operated at 100 psi nozzle pressure, which they typically are, the nozzle reaction formula can be changed to

$$NR = \text{gpm} \times .505 \text{ (.5 for fireground use)} \qquad (10\text{–}11)$$

Using this condensed formula for the preceding example, the nozzle reaction force is calculated as follows:

$$NR = \text{gpm} \times .505 \text{ lb/gpm}$$
$$= 250 \times 0.505 \text{ lb/gpm}$$
$$= 126.25 \text{ lb}$$

Note that the same results are obtained.

SUMMARY

Hydraulics is a branch of science that deals with principles of water at rest (hydrostatics) and water in motion (hydrodynamics). Understanding the basic principles of hydraulics is fundamental to water flow calculations. Complicated formulas and calculations are typically not used on the fireground. Those formulas and calculations that are used on the fireground are approximations of the more complicated (scientific) formulas that describe hydraulic behavior.

A summary of water characteristics is provided followed by a listing of chapter formulas.

Basic Characteristics of Water

Density of water	62.4 lb/ft^3
Weight of 1 gallon	8.34 lb
Freezes at	32°
Boils at	212°
Number of gallons in 1 cubic foot	7.38 gal

Chapter Formulas

10–1 Determine the weight of water.

$W = D$ (density) $\times V$ (volume)

10–2 Determine the volume of a rectangular vessel.

$V = L$ (length) $\times W$ (width) $\times H$ (height

10–3 Find the volume in cubic feet for a cylinder.

$V = .7854$ (constant) $\times d^2$ (diameter) $\times H$ (height or length)

10–4 Find the number of gallons in a cylinder.

$V = 6$ (constant) $\times d^2$ (diameter) $\times h$ (height or length)

10–5 Determine pressure

$P = F$ (force) $/ A$ (surface area)

10–6 Determine force

$F = P$ (pressure) $\times A$ (surface area)

10–7 Calculate pressure at the bottom of a vessel

$P = .433$ (constant in lb/in^2/ft) $\times h$ (height in feet)

10–8 Calculate head

$h = 2.31$ (height in ft 1 psi will raise water) $\times p$ (pressure in psi)

10–9 Determine nozzle reaction for smooth-bore nozzles

$NR = 1.57$ (constant) $\times d^2$ (diameter in inches) $\times NP$ (nozzle pressure in psi)

10–10 Determine nozzle reaction for combination nozzles

$NR = \text{gpm} \times \sqrt{NP}$ (nozzle pressure in psi) $\times .0505$ (constant with in^2/gpm as units)

10–11 Condensed nozzle reaction for combination nozzles

$NR = \text{gpm} \times .505$ (constant with lb/gpm as units)

REVIEW QUESTIONS

Key Terms and Concepts

On a separate sheet of paper, identify and/or define each of the following.

1. Hydraulics
2. Hydrodynamics
3. Hydrostatics
4. Evaporation
5. Vapor pressure
6. Boiling point
7. Density
8. Specific heat
9. Latent heat of vaporization
10. Weight
11. Pressure
12. Atmospheric pressure
13. Gauge pressure
14. Absolute pressure
15. Vacuum
16. Head pressure
17. Static pressure
18. Residual pressure
19. Pressure drop
20. Normal pressure
21. Velocity pressure
22. Pressure gain and loss
23. Nozzle pressure
24. Friction loss
25. Laminar flow
26. Turbulent flow

Multiple Choice and True/False

Select the most appropriate answer.

1. The temperature at which the vapor pressure equals the surrounding pressure is known as
   a. vapor pressure.　　c. flash point.
   b. boiling point.　　d. vapor density.

2. The temperature of boiling water is 212 degrees Fahrenheit regardless of whether it is in an open pot or in a pressure cooker.

   True or False?

3. The physical change of state from a liquid to a vapor is known as
   a. boiling point.　　c. vapor density.
   b. flash point.　　d. evaporation.

4. The density of fresh water is approximately
   a. 62.4 lbs.　　c. 8.34 lbs.
   b. 62.4 lb/ft³.　　d. 7.48 lbs.

5. What is the weight of water in a vessel containing 67 cubic feet of water?
   a. 559.45 lbs
   b. 501.16 lbs
   c. 4,180.8 lbs
   d. None of the answers are correct.

6. There are _____ gallons in a cubic foot.
   a. 7.48
   b. 8.35
   c. 62.4
   d. None of the answers are correct.

7. What is the weight of water in a tank containing 500 gallons of water?
   a. 3,740 lbs
   b. 4,170 lbs
   c. 31,200 lbs
   d. None of the answers are correct.

8. What is the weight of water from a fire stream flowing 125 gpm for 10 minutes?
   a. 1,043.75 lbs
   b. 7,800 lbs
   c. 104,375 lbs
   d. None of the answers are correct.

9. A 200-foot section of 3-inch hose contains how many gallons when completely filled, with no water flowing?

   a. 36 gallons
   b. 75 gallons
   c. 162 gallons
   d. 3,600 gallons

10. The pressure at any point in a liquid in motion is equal in every direction.

    True or False?

11. The pressure of a fluid acting on a surface is perpendicular to that surface.

    True or False?

12. A gauge reading at sea level of 137 psia is equivalent to a gauge reading of

    a. 122.3 psig.
    b. 151.7 psig.
    c. 100 psig.
    d. None of the answers are correct.

13. Positive pressure is measured in psi, while negative pressure is measured in

    a. −psi.
    b. psia.
    c. psig.
    d. inches of mercury.

14. A pressure drop of 13% was calculated after one line was placed in service. How many additional like lines will the hydrant support?

    a. 1 like line
    b. 2 like lines
    c. 3 like lines
    d. None of the answers are correct.

15. For every one-foot increase in a one-square-inch column of water, pressure will increase by

    a. 433 psi.
    b. 1 psi.
    c. 8.35 lbs.
    d. 62.4 lb/ft³.

16. Friction loss varies directly with the length of hose if all other variables are held constant.

    True or False?

17. If the flow of water triples and all other variables are held constant, the friction loss in the hose will increase by

    a. 3 times.
    b. 6 times.
    c. 9 times.
    d. 12 times.

18. When the flow remains constant, friction loss varies directly with hose diameter.

    True or False (inversely)?

19. For any given velocity, the friction loss in hose will be about the same regardless of water pressure.

    True or False?

20. The tendency of nozzles to move in the opposite direction of water flow is called

    a. nozzle recoil.
    b. nozzle pressure.
    c. nozzle velocity.
    d. nozzle reaction.

Short Answer

On a separate sheet of paper, answer/explain the following questions.

1. What is the difference between hydrodynamics and hydrostatics?

2. List several physical characteristics of water.

3. What does one pound of water weigh?

4. List the typical operating pressures for modern nozzles.

5. Explain the effect of pressure and temperature on vapor pressure and boiling point.

6. Explain the difference between weight and density.

7. List the formula for calculating the weight of a substance.

8. Explain how to calculate the weight of 1 gallon of water.

9. List the formula for calculating the number of gallons in a cylinder.

10. Explain the difference between weight and pressure.

11. List the formula for calculating pressure.

12. List and briefly define each of the five pressure principles.

13. What is the difference between psia and psig?

14. List the standard nozzle pressures for the majority of nozzles used in the fire service.

15. List and briefly define each of the four friction loss principles.

ACTIVITIES

1. Calculate the weight of water for a vessel containing 100 cubic feet of water.

2. A container measuring 10 feet by 10 feet by 3 feet is full of water. How much does the water weigh?

3. Calculate the weight of water contained in a 750-gallon onboard water tank.

4. How many tons of water are being delivered into a structure when three 2½-inch lines are flowing 250 gpm for a duration of 10 minutes?

5. Calculate the weight of a section of 3-inch hose filled with water and measuring 200 feet long.

6. Determine the nozzle reaction for the following smooth-bore lines:
   a. 3/4-inch tip handline   NR = _____
   b. 1-inch tip handline   NR = _____
   c. 1-inch tip monitor nozzle   NR = _____

7. Determine the nozzle reaction for the following combination nozzle lines:
   a. 100 gpm flowing   NR = _____
   b. 150 gpm flowing   NR = _____
   c. 200 gpm flowing   NR = _____
   d. 150 gpm flowing (low-   NR = _____
      pressure combination
      nozzle)

PRACTICE PROBLEMS

1. Calculate the weight of water for the onboard water supply on at least two apparatus.

2. Which of the vessels in Figure 10–14 has the greatest pressure at its base?

3. A gauge at the base of an open vessel indicates a pressure of 50 psi. How high is the water within the vessel?

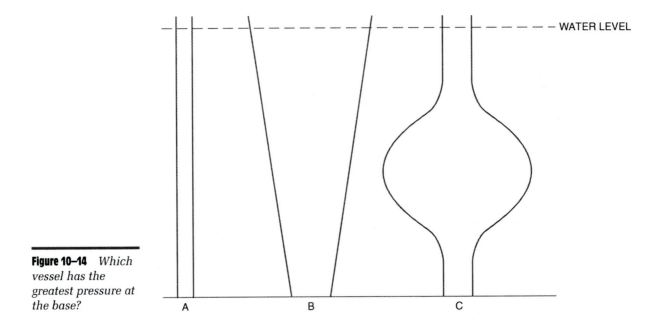

Figure 10–14 *Which vessel has the greatest pressure at the base?*

4. Determine the additional flow available from a hydrant given the following:

   static reading   = 80 psi

   residual reading = 60 psi, flowing 250 gpm through 2½-inch hose

5. Calculate nozzle reaction for a ¾-inch smooth-bore tip nozzle operating on a handline.

6. Calculate nozzle reaction for a combination nozzle operating on a handline flowing 300 gpm.

7. Calculate the number of Btus one gallon of water will absorb going from 60 degrees Fahrenheit in a liquid state to 212 degrees Fahrenheit in a vapor state.

BIBLIOGRAPHY

Hickey, Harry E. *Hydraulics for Fire Protection.* Boston, MA: National Fire Protection Association Publications, 1980.

Brock, Pat D. *Fire Protection Hydraulics and Water Supply Analysis.* Fire Protection Publications. Stillwater, OK: Oklahoma State University, 1990.

Fire Protection Handbook, 17th edition. Quincy, MA: National Fire Protection Association, 1991.

Chapter 11

FIREGROUND FLOW AND FRICTION LOSS CONSIDERATIONS

Learning Objectives

Upon completion of this chapter, you should be able to:

- List the three basic steps to efficient and effective pump operations.
- List and discuss the two basic fireground formulas used to determine needed flow.
- Calculate needed flow using both the Iowa State formula and the National Fire Academy formula.
- Define and discuss the concept of available water.
- Explain the two basic methods for developing discharge flows.
- Explain how flow meters simplify discharge flow development.
- Calculate friction loss for a 100-foot section of hose given the gpm and hose diameter.
- Calculate elevation gain and loss for a given hose lay.

NFPA 1002
Standard for Fire Apparatus Driver/Operator Professional Qualifications
(2003 Edition)

This chapter addresses parts of the following knowledge element within sections 5.2.1 and 5.2.2:

Hydraulic calculations for friction loss and flow using both written formulas and estimation methods

The following element within section 5.2.4 is also addressed:

Calculation of pump discharge pressure

INTRODUCTION

■ **Note**

Success in extinguishing a fire or protecting an exposure depends, in part, on the proper quantity of water delivered to the fire or exposure.

Moving water from a supply source through a pump to a discharge point defines fire pump operations. The main purpose of this activity is to extinguish a fire or protect an exposure. Simply moving water at any quantity and pressure will not guarantee success. Rather, success depends, in part, on the *proper quantity* of water delivered to the fire or exposure. Further, the *correct pressure* must occur at the nozzle to provide proper stream reach and shape. Efficient and effective fire pump operations ensure that proper quantity and pressure are provided to the discharge point.

The first step in efficient and effective pump operations is to determine the amount of water needed to extinguish a fire. Often, little consideration is given to determining needed flow. Rather, suppression efforts are usually initiated with available resources, even though the consideration of available resources (water supply, pump, hose, and nozzle capabilities) is actually the second step to efficient and effective pump operations. Determining available flow is simply identifying the limits of each component within fire pump operations, that is, the water supply, pump, hose, and nozzle. The final step is to develop discharge flows ensuring that proper pressures and volumes are provided to each discharge line.

This chapter focuses on concepts and calculations related to water flow on the fireground. First, needed and available flow are discussed, and then concepts related to developing discharge flows are presented. Finally, the discussion turns to loss of pressure within fire hose. The next chapter, Chapter 12, presents calculations for different hose line configurations.

NEEDED FLOW

needed flow

the estimated flow required to extinguish a fire

The estimated flow required to extinguish a fire is called the **needed flow**, also referred to as required flow. Determining the needed flow for a fire is a crucial process for several reasons, perhaps most obvious being that suppression efforts may be hindered to the point that avoidable damage may occur. For example, providing an inadequate water flow to the fire may allow the fire to continue to burn longer than necessary. In fact, suppression efforts may be so limited that they may not be able to bring the fire under control. Second, inadequate water flow and pressure may place firefighters in greater jeopardy from, for example, suppression efforts taking longer than necessary resulting in increased exposure to fireground hazards. Finally, figuring the needed fire flow allows for the determination of the resources necessary to provide the flow. In most cases, the initial commitment of resources determines the success of suppression efforts. The time it takes to disconnect, relocate, and reconnect hose lines may be longer than the time it takes for the structure to burn out of control.

Recall from Chapter 10 that water has a tremendous capacity to absorb heat. When water is converted to steam, it not only absorbs heat, it also helps smother a fire through dilution of air (oxygen). One means for determining the needed flow is to determine how much water is needed that, when converted from a liquid to a vapor, will fill an area with steam. The concept is that the fire will be extinguished through both heat absorption and smothering. With the information provided in Chapter 10, we can calculate the amount of water required to fill a room with steam. First, we determine the volume, in cubic feet, of the area. Next, we divide the total volume by the expansion ratio of water (1,700) to establish the cubic feet of water required to extinguish the fire. Finally, we convert feet to gallons. For example, what is the needed flow for a room measuring 50 feet wide, by 60 feet long, by 9 feet tall? First, the volume is 27,000 ft^3 (60 ft × 50 ft × 9 ft). Next, the cubic feet of water is 15.8 ft^3 (27,000 ft^3/1,700). Finally the needed flow in gallons is 118.8 (15.8 ft^2 × 7.48 gal/ft^3). The number 7.48 is the number of gallons in one cubic foot. Keep in mind that this calculation only determines the number of gallons of water that, when converted to steam, will fill an area with steam. This assumes a conversion of all water to steam and that the area is not vented. In addition, it does not address the fire load within the area. However, the concept of this calculation is the basis for several of the current needed flow formulas used in the fire service.

There are several formulas available for preplanning needed fire flows. One such formula was developed by the Insurance Service Office (ISO). Typically used for insurance grading purposes, the formula includes factors such as type of construction, occupancy class, exposures, and area. The formula is considered to yield good estimates of needed fire flow. The basic formula reads:

$$NFF = C \times O \times (X + P) \tag{11-1}$$

where NFF = needed fire flow
C = construction factor that is determined using a coefficient related to class of construction and the area of the structure
O = occupancy factor
$X + P$ = exposure factor

Another formula for determining needed water flow is contained in NFPA 1142, *Water Supplies for Suburban and Rural Fire Fighting.* The focus of this standard is on areas where water must be transported to the scene. Although these formulas are not complicated, they require information that takes time to gather. In addition, several charts and tables must be referenced in order to process the information gathered.

Inevitably, situations will arise when the needed flow has not been determined. In these cases, a simple method of calculating required flow on the fireground is needed. Factors such as building construction, occupancy, and exposures cannot be included. The two widely used formulas for calculating needed flow on the fireground are the Iowa State formula and the National Fire Academy formula.

Iowa State Formula

The Iowa State formula, developed by the Fire Service Extension at Iowa State University, estimates the needed flow, in gpm, for either an entire structure or section of a structure. This formula is also used for preplanning. When used for preplanning, typically the entire structure is calculated for needed flow. On the fireground, the needed flow for the specific area or areas involved is calculated. In either case, the estimated flow, if applied properly, should be able to efficiently and effectively control a typical enclosed-structure fire.

The Iowa formula was one of the first rational approaches to determining needed fire flow. The formula is based on research of enclosed fires consisting primarily of class A combustible material within structures common during the 1950s and 1960s. Consequently, higher heat release materials within today's structures require higher flow rates than yielded by the formula. In addition, the formula is based on fire growth within an enclosed structure after 10 minutes from ignition. Calculated flow rates should be increased when response delays occur or when increased oxygen is available as with ventilated structures or when openings exist such as hallways, connecting rooms, or windows. The calculated flow rate does not take into consideration fire spread into void spaces or exposure protection. In essence, the formula estimates the needed flow by dividing the volume of an affected area by 100. The formula can be expressed as:

$$NF = \frac{V}{100} \qquad (11\text{-}2)$$

where NF = needed flow in gpm
V = volume of the area in cubic feet
100 = is a constant in ft³/gpm

Consider a fire-involved floor area measuring 60 feet long by 50 feet wide with a 9-foot ceiling. The volume of the room is 27,000 ft³ (60 ft × 50 ft × 9 ft = 27,000 ft³). The needed flow is calculated as follows:

$$NF = \frac{V}{100}$$

$$= \frac{27{,}000}{100 \text{ ft}^3/\text{gpm}}$$

$$= 270 \text{ gpm}$$

The formula can also be expressed as:

$$NF = 0.01 \text{ gpm/ft}^3 \times V \qquad (11\text{-}3)$$

When the volume is determined, simply move the decimal point two places to the left. For example, the needed flow for room volume of 35,000 ft³ can be calculated as follows:

$$NF = 0.01 \text{ gpm/ft}^3 \times V$$
$$= 0.01 \text{ gpm/ft}^3 \times 35{,}000 \text{ ft}^3$$
$$= 350 \text{ gpm}$$

National Fire Academy Formula

The National Fire Academy (NFA) formula is a quick and easy calculation used to determine needed flow for a structure on the fireground. It utilizes the square footage of a structure and is expressed as:

$$NF = \frac{A}{3} \qquad (11\text{-}4)$$

where NF = needed flow in gpm
A = area of a structure in square feet (length × width)
3 = constant in ft²/gpm

Consider the structure in the previous example. The area of this structure is 3,000 ft² (60 × 50 = 3,000). The needed flow is calculated as follows:

$$NF = \frac{A}{3}$$

$$= \frac{3,000 \text{ ft}^2}{3 \text{ ft}^2/\text{gpm}}$$

$$= 1,000 \text{ gpm}$$

When more than one floor is involved, the square footage of each floor can be added together. The needed flow can also be adjusted to compensate for the percentage of fire. For example, if 50% of the structure is involved, the needed flow would be 500 gpm (1,000 gpm × .5 = 500 gpm).

This formula can also be expressed as:

$$NF = 0.333 \text{ gpm/ft}^2 \times A \qquad\qquad (11-5)$$

Another way of looking at this calculation is that the needed flow is equal to a third of the floor area.

Practice

Using the Iowa State formula and the NFA formula, calculate the needed flow for a structure that is 120 feet long by 60 feet wide by 10 feet tall.

Iowa State:

$$NF = \frac{V}{100 \text{ ft}^3/\text{gpm}}$$

$$= \frac{(120 \text{ ft} \times 60 \text{ ft} \times 10 \text{ ft})}{100 \text{ ft}^3/\text{gpm}}$$

$$= \frac{72,000 \text{ ft}^3}{100 \text{ ft}^3/\text{gpm}}$$

$$= 720 \text{ gpm}$$

NFA:

$$NF = \frac{A}{3 \text{ ft}^2/\text{gpm}}$$

$$= \frac{(120 \text{ ft} \times 60 \text{ ft})}{3 \text{ ft}^2/\text{gpm}}$$

$$= \frac{72,000 \text{ ft}^2}{3 \text{ ft}^2/\text{gpm}}$$

$$= 2,400 \text{ gpm}$$

Needed Flow Consideration

It must be kept in mind that formulas used to determine needed flows are only estimates. Consider the difference in needed flow derived from the previous two formulas. For the same building, the needed flow derived from the Iowa formula (270 gpm) is significantly less than that derived from the NFA formula (1,000 gpm). Each formula is based on a set of assumptions and conditions. When the assumptions and conditions are different, the formula may be less accurate. For example, the formulas provide needed flow for *assumed* fire suppression tactics. Differences in assumed tactics affect the accuracy of the formula.

Flows required in excess of calculated needed flow can easily occur for several reasons. First, conditions can change from the time the needed flow was calculated during preplanning to the time of the incident. For example, the occupancy can change, additions can be constructed, water supplies may change, and increased combustible storage can occur. In addition, the fireground calculation method does not include factors such as building construction and exposures. Even so, if needed flow is calculated, the chances of efficient and effective suppression efforts are increased considerably. Further, the differences between calculated flows and actual flows can be used to provide better calculations in the future.

AVAILABLE FLOW

available flow
the amount of water that can be moved from the supply to the fire scene

After the needed flow is determined, the next step is to determine the available flow. The amount of water that can be moved from the supply to the fire scene is called **available flow**. The available flow will be limited by the capabilities of each component within the pumping operation. The goal is to evaluate each of the following components to maximize the available flow.

Water Supply Determining available flow starts with an evaluation of the water supply. Obviously, water supplies should be selected that provide the needed flow. When the needed flow has not been calculated or when the needed flow is greater than the available supply, a water supply should be selected that allows the pump to flow capacity. Considerations for evaluating available water for each type of supply are discussed in Chapter 7.

Pump The size of the pump is also an important consideration for evaluating available water flow. Assuming other components are of adequate capability, available water will be limited by the size of the pump. Recall that pumps are expected to provide their rated capacity at 150 psi, 70% of rated capacity at 200 psi, and 50% of rated capacity at 250 psi (Figure 11–1). The pressure at which the pump is expected to operate will also limit available flow.

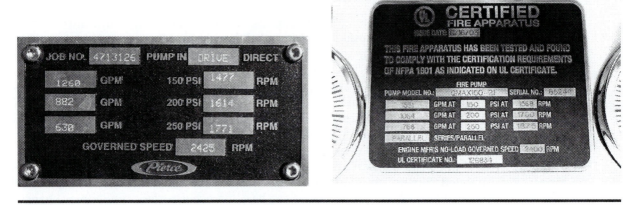

Figure 11–1   *Higher pressures during a pumping operation can reduce the available flow from a pump.*

Hose The size and length of hose used for both supply and discharge will affect available water. The size affects available flow in that different size hose have different flow capacities. In essence, large flow requires either large size hose or multiple lines. The length of hose also affects available water in that friction loss increases with longer hose lines. In this case, pressure is affected rather than the quantity of water. The longer the line, the greater the friction loss, and, in turn, the greater the pressure required to provide the nozzle with its proper operating pressure. The limiting factors are the pressure rating of the hose and the maximum pressure rating of the pump. Longer lays may be impossible because the pressure needed to adequately supply the nozzle is greater than the pressure rating of the hose or the pump.

Nozzles The type or types of nozzles used will also affect available flow. This component is perhaps the easiest to change yet the component least evaluated. In essence, nozzles must be provided with their designed operating pressure and flow (Figure 11–2). The total output from all operating nozzles will equal the available flow. For example, assuming all other components are capable of providing 500 gpm, if only four 95-gpm nozzles are available and are operating, the available flow will be 380 gpm. When automatic nozzles are used, the gpm flow, within the nozzles' range of flows, must be selected. When smooth-bore nozzles are used, the gpm must either be looked up on a table or calculated using the following formula:

$$Q = 29.7 \times d^2 \times \sqrt{NP} \tag{11–6}$$

where Q = gallons per minute
  29.7 = 30 is the constant most often used for fireground calculations
  d = nozzle diameter in inches
  NP = pressure in psi

Figure 11–2 *Nozzles are designed to operate at a specific pressure and flow (or range of flows).*

Note: For fireground calculations, the $\sqrt{NP}$ for hand lines, 7.07, is rounded to 7 and the $\sqrt{NP}$ for master streams, 8.94, is rounded to 9.

For example, what is the gallon per minute flow through a ¾-inch tip smooth-bore nozzle with 50 psi at the nozzle?

$$Q = 29.7 \times d^2 \times \sqrt{NP}$$
$$= 29.7 \times 0.75^2 \times \sqrt{50}$$
$$= 29.7 \times 0.56 \times 7.07$$
$$= 117.58 \text{ or rounded up to } 118$$

Note: Using the fireground calculation constants of 30 and 7 yields similar results of 117.6 or 118.

Practice

What is the flow through a 1-inch smooth-bore nozzle on a master stream device?

$$Q = 29.7 \times d^2 \times \sqrt{NP}$$
$$= 29.7 \times 01^2 \times \sqrt{80}$$
$$= 29.7 \times 1 \times 8.9$$
$$= 264.33 \text{ or } 264 \text{ gpm}$$

Note: Using the fireground calculation constants of 30 and 9 yields a slightly higher value of 270.

The available water flow is simply the amount of water that is capable of being provided to the scene. In some cases, the available water flow will be less than the needed flow. When this occurs, plans must be made to increase the available flow by securing additional water supplies or increasing the existing flow. In other cases, the available flow will be greater than the needed flow. In either case, after the supply has been secured, the number and size of discharge configurations can be manipulated to provide the available flow. In other words, the total flow a pumping operation is capable of providing is available at the discharge side of the pump. The actual discharge flow, however, may be equal to or less than the available flow.

DISCHARGE FLOW

After the needed water flow is determined and the available water flow secured, the next step is to develop discharge flows. The amount of water flowing from the discharge side of a pump through hose, appliances, and nozzles to the scene is called **discharge flows**. The most important factor in developing discharge flows is the determination of the quantity of water a specific line is expected to flow. In other words, the gallons per minute the line will be flowing. This determination is important for two reasons. First, identifying the gallons per minute a line is flowing is important for keeping track of the total flow provided to the scene as well as of the remaining flow available in the pumping operation. Second, knowing the flow of a line is essential for providing the proper operating pressure and flow required by the nozzle.

There are two methods used to ensure that nozzles are provided with the proper flow and pressure. One method uses flow meters while the other uses pressure gauges.

discharge flow
the amount of water flowing from the discharge side of a pump through the hose, appliances, and nozzles to the scene

■ **Note**
The most important factor in developing discharge flows is the determination of the quantity of water a specific line is expected to flow.

Flow Meters

Without question, the easiest method for ensuring that nozzles are provided with the proper flow and pressure is with flow meters (Figure 11–3). When flow meters are used, the major consideration is knowing the designed flow of the nozzle. Flow meters measure the quantity and rate (gpm) of water flowing through a line. Because matter cannot be created or destroyed, the same flow leaving the discharge

Figure 11–3 *Flow meters eliminate the need for friction loss calculations on the fireground. Courtesy Class 1.*

side of the pump will also travel through the hose and discharge from the nozzle. For example, when the flow meter on the pump panel reads 100 gpm, the nozzle will be discharging 100 gpm. Recall that pressure is a function of restriction. If the nozzle is designed to operate at 100 gpm, the restriction designed into the nozzle will create the proper operating pressure. The key, though, is knowing the required gpm for the nozzle.

The use and reliance of flow meters on pumping apparatus is relatively new. Although flow meters have been around for a number of years, increased accuracy and reliability of newer flow meters has increased their popularity and use on fire apparatus. In addition, they are gaining popularity because they dramatically simplify flow development and increase flow accuracy over estimated hydraulic calculations. Some flow meters provide both gpm and pressure. When flow meters provide only gpm, the operating pressure of the hose can easily be exceeded. When only one line is in operation, the master discharge gauge can be used to determine pressure. Should the flow meter fail, the master discharge gauge can again be used. When more than one line is in operation, this option is not available.

Pressure Gauges

The pressure gauge method for developing discharge flows is far more complicated (Figure 11–4) than the flow meter method. When pressure gauges are used, the pump operator must calculate any changes in pressure within the discharge line. Hypothetically, if the pressure did not change, the pump operator would simply

Figure 11–4 *The traditional use of pressure gauges on a pump panel require pump operators to estimate/calculate friction loss in hose. Courtesy Class 1.*

increase pump speed until the discharge pressure gauge equaled the operating pressure of the nozzle. In reality this cannot occur because friction loss in the hose and appliances as well as changes in elevation always affect the pressure. Therefore, the discharge pressure required to provide the nozzle with its designed operating pressure and flow must be calculated by determining friction loss in hose and appliances and pressure gains or losses due to elevation.

FRICTION LOSS CALCULATIONS

The loss of pressure within hose resulting from friction is inevitable. Scientific formulas have been developed and refined over the years that can accurately calculate friction loss. Unfortunately, most of the formulas are not practical for use on the fireground, because, perhaps obviously, they are too cumbersome and complex. In addition, the variables in the formulas cannot be adequately controlled or measured on the fireground.

Fireground formulas and rules of thumb have also been developed over the years to try to estimate friction loss, each with varying levels of accuracy. When the standard hose was 1½-inch and 2½-inch cotton-jacketed hose, the formulas and rules of thumb were accurate enough for fireground use. However, the gradual increase in different hose diameters as well as the number of manufacturers and construction methods have significantly confounded the issue of friction loss calculations. It is common for a fire department to use anywhere from three to five different sizes of various ages of hose manufactured by several different companies using different construction methods. When you add the dilemma of kinks and bends on the fireground, the problem of calculating friction loss becomes readily obvious.

In some cases, friction loss calculations are not required on the fireground. When several discharge hose layouts are consistently used, friction loss can be computed in advance. For example, most modern pumpers have preconnected attack lines. When the same diameter and length of hose and nozzle are used, the pump operator simply increases pump speed to the predesignated discharge pressure. In addition, tables and charts can be developed so that pump operators simply look up either the friction loss or the discharge pressure (Figure 11–5; see also in Appendix F). In addition, a variety of slide rules (Figure 11–6) and even hand-held friction loss calculators can be used (Figure 11–7). Finally, when flow meters are used, calculations are not necessary.

Even so, pump operators must still be able to manually calculate friction loss with a reasonable degree of accuracy. For one thing, such calculations are needed to develop tables and charts as for specific equipment used by a fire department. Likewise, calculations are required to predetermine pump pressures on commonly used discharge lays. In addition, calculations should be made to evaluate the variety of fireground friction loss formulas and rules of thumb. Finally, and perhaps most important, the fireground has a way of ruining even the best laid plans, the situation in which manual calculations will eventually occur.

As previously stated, a variety of fireground methods have been developed to calculate friction loss. The following are several of the more popular methods used. It should be noted that the last method discussed, cq^2L, is considered the most accurate.

■ Note

The fireground has a way of ruining even the best-laid plans.

201 Titusville Road — Union City, PA 16438
Ph: 814/438-7616 — Fax: 814/438-8163

G.P.M.	1-1/2"	1-3/4"	2"	2-1/2"	G.P.M.	1-3/4"	2"	2-1/2"	3"
60	8.6	4.9	2.2	.72	200	54.0	24.0	8.0	3.2
70	11.8	6.6	2.9	.98	220	65.3	29.0	9.7	3.9
80	15.4	8.6	3.8	1.3	240	77.8	34.6	11.5	4.6
90	19.4	10.9	4.9	1.6	250	84.4	37.5	12.5	5.0
100	24.0	13.5	6.0	2.0	280	105.8	47.0	15.7	6.3
110	29.0	16.3	7.3	2.4	300		54.0	18.0	7.2
120	34.6	19.4	8.6	2.9	320		61.4	20.5	8.2
130	40.6	22.8	10.2	3.4	340		69.4	23.1	9.2
140	47.0	26.5	11.8	3.9	350		77.8	24.5	9.8
150	54.0	30.4	13.5	4.5	380		86.6	28.9	11.6
160	61.4	34.6	15.4	5.1	400		96.0	32.0	12.8
180	77.8	43.7	19.4	6.5	420			35.3	14.1

HOSE FRICTION LOSS (PSI PER 100 FT)

DATA SHOWN IS THEORETICAL, ACTUAL LOSS MAY VARY SLIGHTLY BASED ON CONDITIONS
TRAINING MANUAL AVAILABLE

Figure 11–5 *Friction loss table can be used to estimate friction loss on the fireground.*

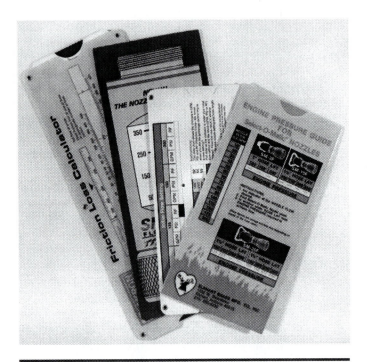

Figure 11–6   *Friction loss slide rules can be used to estimate friction loss on the fireground.*

Figure 11–7   *Friction loss calculators are also available to simplify fireground calculations. Courtesy Akron Brass Company.*

Hand Method

The hand method is a simple way to calculate friction loss in 100-foot sections of 2½-inch hose. Beginning with the thumb, each finger is assigned a number representing gpm flow (Figure 11–8). For example, the index finger has an assigned value of two, representing 200 gpm, while the ring finger has an assigned value of 4, representing 400 gpm. Next, the base of each finger is assigned an even number as indicated in Figure 11–8. To find the friction loss in 100-foot sections of 2½-inch hose, simply select the finger representing the gpm flow and multiply the two numbers assigned to the finger. For example, the friction loss for 100 feet of 2½-inch hose flowing 300 gpm is 18 psi (3 × 6 = 18). The friction loss for 100 feet of 2½-inch hose flowing 500 gpm is 50 psi (5 × 10 = 50). The space between the fingers can be assigned 1.5, 2.5, 3.5, and so on for the gpm values and 3, 5, 7 for the gaps between figures. The friction loss for 100 feet of 2½-inch hose flowing 150 gpm is 4.5 gpm (1.5 × 3 = 4.5). Several variations to this basic theme have evolved over the years and the system has been modified for use with various hose sizes.

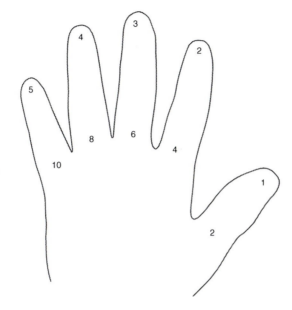

Figure 11–8   *The "Hand Method" can be used to help simplify fireground calculations.*

Practice

Calculate the friction loss for the following hose lines:

100 feet of 2½-inch flowing 400 gpm

$$FL = 32 \text{ psi } (4 \times 8)$$

100 feet of 2½-inch flowing 200 gpm

$$FL = 8 \text{ psi } (2 \times 4)$$

100 feet of 2½-inch flowing 250 gpm

$$FL = 12.5 \text{ psi } (2.5 \times 5)$$

200 feet of 2½-inch flowing 200 gpm

$$FL = 2 \times 4 \times 2 \text{ (number of 100 foot sections of hose)}$$
$$= 16 \text{ psi}$$

Drop Ten Method

The drop ten method is a simple rule of thumb that provides friction loss in 100-foot sections of hose, and, although less accurate than others, it can be used for more than just 2½-inch hose. This method simply subtracts 10 from the first two numbers of gpm flow. For example, the friction loss in 100 feet of hose when 500 gpm is flowing is 40 psi. (50 − 10 = 40). Likewise, the friction loss in 100 feet of hose when 250 gpm is flowing is 15 psi (25 − 10 = 15).

$2q^2 + q$

Developed by the National Board of Fire Underwriters and in use for many years, the full formula for this method is written as:

$$FL = 2q^2 + q \qquad (11\text{--}7)$$

where FL = friction loss in psi per 100-foot sections of hose
   2 = constant for 2½-inch cotton-jacketed hose
   q = flow in hundreds of gallons per minute

The amount of water flowing, q, is given in hundreds of gpm and is calculated using the following formula:

$$q = \frac{\text{gpm}}{100} \qquad (11\text{--}8)$$

where q = hundreds of gpm
   gpm = actual flow
   100 = constant

Consider 250 gpm flowing through a 2½-inch line 500 feet long. The flow in hundreds of gpm (q) can be calculated as follows:

$$q = \frac{250 \text{ gpm}}{100}$$

$$= 2.5 \text{ hundreds of gpm}$$

A simple method of calculating q is to move the decimal point two places to the left.
    The friction loss in 100 feet of 2½-inch hose flowing 500 gpm is calculated as follows:

$$FL = 2q^2 + q$$

$$= 2\left(\frac{500}{100}\right)^2 + 5$$

$$= 2\,(5)^2 + 5$$

$$= 2 \times 25 + 5$$

$$= 55$$

This formula provides results for 2½-inch cotton-jacketed hose for flows greater than 100 gpm. For flows less than 100 gpm, the formula is modified as follows:

$$FL = 2q^2 + \tfrac{1}{2}q \qquad (11\text{--}9)$$

To simplify calculations further, some departments use the following:

$$FL = 2q^2 \qquad (11\text{--}10)$$

The use of hose other than 2½ inches in diameter requires a conversion factor. Typically this is accomplished by multiplying the gpm by a conversion factor to get the equivalent flow through 2½-inch hose. The equivalent flow (gpm) is then plugged into the formula to determine the friction loss per 100 feet of hose.

It should be noted that this formula provides friction loss for cotton-jacketed hose. New hose construction methods reduce friction loss compared to older cotton-jacketed hose. The effects of newer hose construction on friction loss necessitate the use of a more accurate formula such as cq^2L.

Practice

Calculate the friction loss for 100 feet of 2½-inch hose flowing 200, 300, and 400 gpm.

200 gpm

$$
\begin{aligned}
FL &= 2q^2 + q \\
&= 2 \times (200/100)^2 + (200/100) \\
&= 2 \times 2^2 + 2 \\
&= 2 \times 4 + 2 \\
&= 10 \text{ psi}
\end{aligned}
$$

300 gpm

$$
\begin{aligned}
FL &= 2q^2 + q \\
&= 2 \times (300/100)^2 + (300/100) \\
&= 2 \times 3^2 + 3 \\
&= 2 \times 9 + 3 \\
&= 21 \text{ psi}
\end{aligned}
$$

400 gpm

$$
\begin{aligned}
FL &= 2q^2 + q \\
&= 2 \times (400/100)^2 + (400/100) \\
&= 2 \times 4^2 + 4 \\
&= 2 \times 16 + 4 \\
&= 36 \text{ psi}
\end{aligned}
$$

Condensed *q* Formula

This formula is an adaptation of the Fire Underwriters' formula ($FL = 2q^2 + q$) and is used specifically for 3-inch hose. The condensed q formula is written as:

$$FL = q^2 \tag{11–11}$$

For example, the friction loss in 100 feet of 3-inch hose flowing 500 gpm is calculated as follows:

$$
\begin{aligned}
FL &= q^2 \\
&= \left(\frac{500}{100}\right)^2 \\
&= 5^2 \\
&= 25 \text{ psi}
\end{aligned}
$$

cq^2L

According to the 19th edition of the *NFPA Fire Protection Handbook*, this formula is derived from a combination of Bernoulli's equation, the Darcy-Weibach equation, $FL = (fV^2L)/(2gD)$, and the Continuity equation, $Q = V \times A$. The Darcy-Weibach equation is used to calculate friction loss in long, straight pipes with uniform diameter and roughness. The Continuity equation expresses a conservation of mass law that states mass cannot be created or destroyed. In this case, it suggests that the amount of water entering and leaving a hose or pipe must be the same—the term most often used is conserved—as it exits the hose or pipe. If 500 gpm enters a hose line, then 500 gpm must exit. Consider a garden hose flowing water: If the end of the hose is squeezed, the velocity of the water increases. You might say the water is hurrying up to get out so that the same quantity exits the hose as is entering. The resultant formula, $FL = cq^2L$, is both simple and reasonably accurate for varying hose line diameters, so the formula can be used for both preplanning and fireground calculations. The formula is expressed as:

$$FL = c \times q^2 \times L \qquad\qquad (11\text{--}12)$$

where　　FL = friction loss in psi
　　　　　　c = constant for a specific hose diameter
　　　　　　q = gpm ÷ 100 (flow in hundreds of gpm)
　　　　　　L = length of hose in hundreds of feet

The constant, c, takes into account all the variables associated with friction loss except flow rate (q) and length (L). The average constants for several common hose diameters are provided in Table 11–1. These values are average values and can be replaced with more accurate values for specific manufacturers and hose sizes.

Table 11–1　*Average constants (c) for several common firefighting hose diameters when using the friction loss equation FL = cq²L.*

Hose Diameter (in.)	Friction Loss Constant (c)
¾ (booster)	1,100.00
1 (booster)	150
1½	24
1¾ (with 1½ in. couplings)	15.5
2	8
2½	2
3 (with 2½ in. couplings)	.8
3 (with 3 in. couplings)	.677
4	.2
5	.08
6	.05

Source: *Fire Protection Handbook*, 19th ed. vol. 1, 2003. National Fire Protection Association.

The length of hose, L, is given in hundreds of feet and is calculated using the following formula:

$$L = \frac{\text{hose length}}{100}$$ (11–13)

where

$$
\begin{aligned}
L &= \text{hundreds of feet} \\
\text{hose length} &= \text{actual length of hose} \\
100 &= \text{constant}
\end{aligned}
$$

For example, the value of L for a 500-foot hose lay is calculated as follows:

$$L = \frac{500 \text{ ft}}{100}$$

$$= 5 \text{ hundreds of feet}$$

Again, a simple method of calculating this formula is to move the decimal point two places to the left.

The following steps, then, can be used to calculate friction loss using this formula:

Step 1: Identify the constant for the diameter hose being used (refer to Table 11–1).

Step 2: Determine the hundreds of gallons of flow using the formula $q = $ gpm/100.

Step 3: Determine the hundreds of feet of hose using the formula $L = $ hose length/100.

Step 4: Use the answers to the previous steps in the friction loss formula, $FL = c \times q^2 \times L$.

For example, what is the friction loss in 500 feet of 2½-inch hose flowing 250 gpm?

Step 1: c = 2 (from Table 11 – 1)

Step 2: q $= \dfrac{250}{100}$

$\qquad = 2.5$

Step 3: L $= \dfrac{500}{100}$

$\qquad = 5$

Step 4: $FL = 2 \times (2.5)^2 \times 5$

$\qquad = 2 \times 6.25 \times 5$

$\qquad = 62.5$

To compare friction loss in different size hose, use the same flow and length of hose as in the previous example and calculate the friction loss in 3- and 4-inch hose.

3-inch hose

$$FL = c \times q^2 \times L$$
$$= 0.8 \times \left(\frac{250}{100}\right)^2 \times \left(\frac{500}{100}\right)$$
$$= 0.8 \times (2.5)^2 \times 5$$
$$= 0.8 \times 6.25 \times 5$$
$$= 25$$

4-inch hose

$$FL = c \times q^2 \times L$$
$$= 0.2 \times \left(\frac{250}{100}\right)^2 \times \left(\frac{500}{100}\right)$$
$$= 0.2 \times (2.5)^2 \times 5$$
$$= 0.2 \times 6.25 \times 5$$
$$= 6.25, \text{ or rounded off } 6$$

Practice

Calculate the friction loss for the following hose dimensions with each line flowing 150 gpm.

	100 feet	**250 feet**	**500 feet**
1½ inch	_____	_____	_____
1¾ inch	_____	_____	_____
2½ inch	_____	_____	_____
3 inch	_____	_____	_____

1½-inch lines

100 feet $FL = 24 \times (150/100)^2 \times (100/100)$
$$= 24 \times 1.5^2 \times 1$$
$$= 24 \times 2.25 \times 1$$
$$= 54 \text{ psi}$$

250 feet $FL = 24 \times 2.25 \times 2.5$
$$= 135 \text{ psi}$$

500 feet $FL = 24 \times 2.25 \times 5$
$$= 270 \text{ psi}$$

$1\frac{3}{4}$-inch lines

100 feet $FL = 15.5 \times (150/100)^2 \times (100/100)$
$= 15.5 \times 1.5^2 \times 1$
$= 15.5 \times 2.25 \times 1$
$= 34.8$ or 35 psi

250 feet $FL = 15.5 \times 2.25 \times 2.5$
$= 87.1$ or 87 psi

500 feet $FL = 15.5 \times 2.25 \times 5$
$= 174.3$ or 174 psi

$2\frac{1}{2}$-inch lines

100 feet $FL = 2 \times (150/100)^2 \times (100/100)$
$= 2 \times 1.5^2 \times 1$
$= 2 \times 2.25 \times 1$
$= 4.5$ or 5 psi

250 feet $FL = 2 \times 2.25 \times 2.5$
$= 11.25$ or 11 psi

500 feet $FL = 2 \times 2.25 \times 5$
$= 22.5$ or 23 psi

3-inch lines

100 feet $FL = .8 \times (150/100)^2 \times (100/100)$
$= .8 \times 1.5^2 \times 1$
$= .8 \times 2.25 \times 1$
$= 1.8$ or 2 psi

250 feet $FL = .8 \times 2.25 \times 2.5$
$= 4.5$ or 5 psi

500 feet $FL = .8 \times 2.25 \times 5$
$= 9$ psi

	100 feet	250 feet	500 feet
$1\frac{1}{2}$ inch	54	135	270
$1\frac{3}{4}$ inch	35	87	174
$2\frac{1}{2}$ inch	5	11	23
3 inch	2	5	9

appliance friction loss
the reduction in pressure resulting from increased turbulence caused by the appliance

APPLIANCE FRICTION LOSS

Recall that friction loss is a result of water rubbing against the interior of the hose. **Appliance friction loss** is the reduction in pressure resulting from increased turbulence caused by the appliance. When the interior is smooth and continuous, less

Table 11–2   *Estimated friction loss in common appliances.**

Type of Appliance	Friction Loss (psi)
2½-in. to 2½-in. wye	5
2½-in. to 1½-in. wye	10
1½-in. to 1½-in. wye	15
2½-in. to 1½-in. siamese	10
1½-in. to 1½-in. siamese	15
Reducer	5
Increaser	5
Monitor	15
Four-way valve	15
Stand pipe	25
Sprinkler	150
Combination	175
Foam eductor	200

*A variety of friction loss pressures are used for these appliances. Each appliance will have a specific loss in pressure based in part on age and manufacture.

friction loss will occur than when the interior is rough and with varying changes in the interior lining. Hose appliances such as wyes, siamese, increasers, reducers, manifolds, and master stream devices change the interior lining by causing slight protrusions and debits. These changing surfaces within the lining cause increased friction loss that must be accounted for when determining the total loss in pressure from friction. Table 11–2 should be used only as a guide. Friction loss can vary with the same type of appliance based on flow and manufacture. More accurate friction information can be obtained from the manufacturer or from tests conducted by a department.

ELEVATION GAIN AND LOSS

Recall that when hose lines are operated at elevation above the pump, greater pressures are required to provide the required pressure at the nozzle. In turn, when hose lines are operated at elevations below the pump, less pressure is required to provide the required nozzle pressure. Two methods are used on the fireground to calculate this gain or loss of pressure: calculations using feet above or below the pump level and using number of floor levels above or below the pump.

Elevation Formula in Feet

The following formula can be used to calculate pressure gain and loss when the elevation in feet above or below the pump is known.

$$EL = 0.5 \times H \qquad\qquad (11-14)$$

where EL = the gain or loss of elevation in psi
         0.5 = pressure exerted at base of 1-in^3 column of water 1 foot
                 high, expressed as psi/ft (actual pressure is .433 in^2/ft)
           H = height in feet

Elevation Gain   Consider a hose line operating 50 feet above the pump. The estimated pressure gain can be calculated as follows:

$$EL = 0.5 \times H$$
$$= 0.5 \text{ psi/ft} \times 50 \text{ ft}$$
$$= 25 \text{ psi}$$

Elevation Loss   Consider a hose line operating 30 feet below the pump. The pressure loss can be calculated as follows:

$$EL = 0.5 \times H$$
$$= 0.5 \text{ psi/ft} \times -30 \text{ ft}$$
$$= -15 \text{ psi}$$

Elevation Formula by Floor Level

When hose lines are operated within structures, elevation gain or loss can be calculated using the number of floor levels above or below the pump, as follows:

$$EL = 5 \times H \qquad\qquad (11-15)$$

where EL = the gain or loss of elevation in psi
           5 = gain or loss in pressure for each floor level
          H = height in number of floor levels above or below the pump

Elevation Gain   Consider a hose line operating on the fifth story of a structure when the pump is on the same elevation as the first floor. The pressure gain can be calculated as follows:

$$EL = 5 \times H$$
$$= 5 \times 4$$
$$= 20 \text{ psi}$$

Elevation Loss Consider a hose line operating in the basement of a structure three floor levels below ground. The pressure loss can be calculated as follows:

$$EL = 5 \times H$$
$$= 5 \times -3$$
$$= -15 \text{ psi}$$

Practice

Calculate the elevation gain or loss for the following:

Hose line 75 feet above the pump.

$$EL = 0.5 \text{ psi/ft} \times 75 \text{ ft}$$
$$= 37.5 \text{ or } 38 \text{ psi}$$

Hose line 35 feet below the pump.

$$EL = 0.5 \text{ psi/ft} \times -35 \text{ ft}$$
$$= 17.5 \text{ or } 18 \text{ psi}$$

Hose line taken to the third floor of a structure; the first floor is at ground level.

$$EL = 5 \text{ psi/fl} \times 2 \text{ fl}$$
$$= 10 \text{ psi}$$

Hose line operating on the second level below ground; the first level is at ground level.

$$EL = 5 \text{ psi/fl} \times -1 \text{ fl}$$
$$= -5 \text{ psi}$$

SUMMARY

Efficient and effective pump operations require that pump operators determine the amount of needed water flow. Ideally, this figuring is done during preplanning. When needed flow is not known, the pump operator must estimate it. The two most common formulas for estimating needed flow are the Iowa State formula and the NFA formula. Next, pump operators determine available water by considering the limits of each component within fire pump operations: the water supply, pump, hose, and nozzle. Finally, pump operators must develop appropriate discharge flows.

Proper discharge flows require that nozzles be provided with their designed operating pressure and gpm flow. The easiest method is to use flow meters. When pressure gauges are used, friction loss in hose must be estimated. Friction loss can be estimated using tables, slide rules, friction loss calculators, rules of thumb, and friction loss formulas. In addition, the pump operator must account for other losses in pressure such as losses in pressure from appliance friction loss and changes in elevation.

Chapter Formulas

Needed Fire Flow Formulas

11–1 Insurance Service Office (ISO)

$NFF = C \times O \times (X + P)$.

11–2 Iowa State formula

$$NF = \frac{V}{100}$$

11–3 Iowa State formula in multiplication format

$NF = 0.01 \ \text{gpm/ft}^3 \times V$

11–4 National Fire Academy Formula

$$NF = \frac{A}{3}$$

11–5 National Fire Academy Formula in multiplication format

$NF = 0.333 \ \text{gpm/ft}^2 \times A$

11–6 Determine GPM through smooth-bore nozzles

$Q = 30 \times d^2 \times \sqrt{NP}$

Friction Loss Formulas

11–7 National Board of Fire Underwriters

$FL = 2q^2 + q$

11–8 To determine the number of hundreds of gpm flowing

$$q = \frac{\text{gpm}}{100}$$

11–9 Flows less than 100 gpm

$FL = 2q^2 + \frac{1}{2}q$

11–10 Simplified formula

$FL = 2q^2$

11–11 Condensed q formula for 3-inch hose

$FL = q^2$

11–12 Popular friction loss formula

$FL = c \times q^2 \times L$

11–13 Determine number of 100 lengths of hose

$$L = \frac{\text{hose length}}{100}$$

11–14 Determine elevation gain/loss using feet

$EL = 0.5 \times H$ (in units of feet)

11–15 Determine elevation gain/loss using floors

$EL = 0.5 \times H$ (number of floor levels above or below the pump)

REVIEW QUESTIONS

Key Terms and Concepts

On a separate sheet of paper, identify and/or define each of the following.

1. Needed flow
2. Available flow
3. Discharge flow
4. Appliance friction loss

Multiple Choice and True/False

Select the most appropriate answer.

1. Which of the following is the correct NFA formula for determining needed flow?

   a. $NF = \dfrac{A}{3}$

   b. $NF = 0.333 \times A$

   c. Both a and b are correct.

   d. None of the above are correct.

2. The amount of water that can be moved from the supply to the scene is called

   a. needed flow.

   b. available flow.

   c. discharge flow.

   d. None of the above are correct.

3. A flow meter on the pump panel reads 200 gpm when flowing water through a discharge line consisting of 500 feet of 1¾-inch hose operating on the third floor of a structure. How much water is the nozzle discharging?

   a. 200 gpm

   b. 210 gpm

   c. 215 gpm

   d. None of the above are correct.

4. In the majority of friction loss formulas, q is typically considered to represent

   a. total gpm flow.

   b. quantity of water flowing.

   c. flow in hundreds of gpm.

   d. rate flow of the nozzle.

5. Because of the manner in which water acts on the inside of most appliances, the standard friction loss for appliances is 5 psi.

   True or False?

Short Answer

On a separate sheet of paper, answer/explain the following questions.

1. Explain the role of needed flow, available flow, and discharge flow in efficient and effective pump operations.

2. Using both the Iowa State formula and the NFA formula, calculate the needed flow for a two-story single family dwelling measuring 60 feet long by 45 feet wide.

3. Explain the difference between discharge flow development using flow meters and pressure gauges.

4. Calculate the friction in the following hose lines:

   a. 400 ft of 2½-in. flowing 250 gpm

   b. 400 ft of 3-in. flowing 250 gpm

   c. 150 ft of 1½-in. flowing 125 gpm

   d. 150 ft of 1¾-in. flowing 125 gpm

   e. 1,000 ft of 4-in. flowing 500 gpm

5. Calculate the gain or loss in pressure for the following elevations, referring to Figure 11–9:

   a. hose line operating 25 ft above the pump

   b. hose line operating 15 ft below the pump

   c. hose line operating on the 7th floor

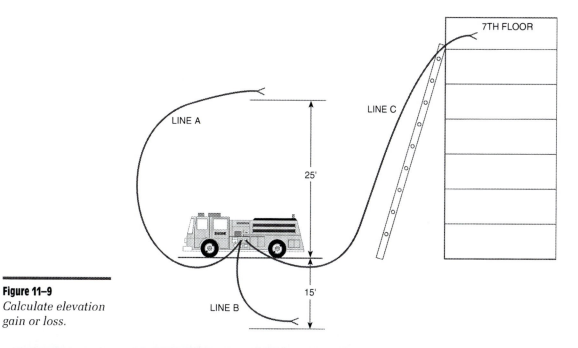

Figure 11–9
Calculate elevation gain or loss.

ACTIVITY

1. Contact the manufacturer(s) of several appliances to determine the actual friction loss of the component.

PRACTICE PROBLEM

1. You have been asked to develop a friction loss table for 100-foot sections of 1½-inch and 1¾-inch hose for the following flows: 95, 100, 125, 150, 175, 200, and 250. Provide the table.

BIBLIOGRAPHY

Hickey, Harry E. *Hydraulics for Fire Protection*. Boston, MA: National Fire Protection Association Publications, 1980.

Burns, Edward, and Phelps, Burton W. "Redefining Needed Fire Flow for Structure Firefighting." *Fire Engineering*, November 1994.

Fire Protection Handbook, 19th edition, vol. 1, 2003. Quincy, MA: National Fire Protection Association.

ISO Guide for Determination of Required Fire Flow. New York: Insurance Services Office, 1974.

Wiseman, John D. Jr. How Valid Are NFF Formulas? *Firefighter's News*, June/July (1996) pp. 78–84.

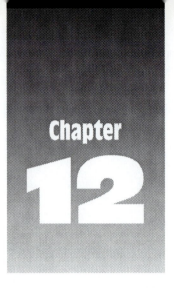

PUMP DISCHARGE PRESSURE CALCULATIONS

Learning Objective

Upon completion of this chapter, you should be able to:

■ Calculate pump discharge pressures for a variety of hose line configurations using the formula $PDP = NP + FL + AFL + EL$.

NFPA 1002
Standard the Fire Apparatus Driver/Operator Professional Qualifications (2003 Edition)

This chapter addresses parts of the following knowledge element within sections 5.2.1 and 5.2.2:

Hydraulic calculations for friction loss and flow using both written formulas and estimation methods.

The following knowledge element within section 5.2.4 is also addressed:

Calculation of pump discharge pressure.

INTRODUCTION

In many respects, the sum total of this textbook boils down to this chapter: calculating pump discharge pressures. Almost every point discussed in the text can be logically tracked to the point of flowing water from a nozzle. From driving the apparatus to preventive maintenance to operating the pump, the ultimate achievement of fire pump operations is providing the proper flow and pressure to a nozzle.

As stated in Chapter 11, the use of flow meters simplifies the process of fire pump operations by eliminating the need to calculate friction loss. Pump operators simply increase the pump speed until the correct gallons per minute for a nozzle are indicated on the flow meter. In turn, the same flow is delivered to the nozzle. When supplied with the correct gpm, the nozzle, by design, automatically develops its proper operating pressure. When pressure gauges are used, the pump operator provides the nozzle with the correct operating pressure. In turn, the design of the nozzle automatically ensures the correct gpm are flowing. The major difference between the use of pressure gauges and flow meters is the requirement to calculate pressure changes within a hose configuration.

The focus of this chapter is on calculating pump discharge pressure. The sample calculations begin with simple single lines and progress to more complicated multiple-line configurations. Friction loss constants and appliance friction loss can be obtained from either Table 11–1 or Table 11–2.

PUMP DISCHARGE PRESSURE CALCULATIONS

■ **Note**

The calculated changes in pressure coupled with the nozzle pressure are the basis for determining pump discharge pressure.

The preceding chapters in this section of the text provide the framework for calculating pump discharge pressure (PDP). Recall that nozzles are designed to operate at a specific pressure. Further recall that friction loss in hose and appliances as well as elevation will affect pressure in hose. These changes in pressure must be compensated for to ensure that the nozzle is provided with the proper pressure. The calculated changes in pressure coupled with the nozzle pressure are the basis for determining pump discharge pressure. Pump discharge pressure is the pressure at the pump panel for a specific hose configuration and can be calculated using the following formula:

$$PDP = NP + FL + AFL + EL$$

where PDP = pump discharge pressure
NP = nozzle pressure
FL = friction loss in hose (Any of the friction loss calculations discussed in Chapter 11 can be used. However, for the purpose of accuracy and current use, the formula $c \times q^2 \times L$ is used.)

AFL = appliance friction loss
EL = elevation gain or loss = $0.5 \times H$ or ($EL = 5 \times$ number of floor levels above ground level)

Calculations Considerations

Although pump discharge pressure can be calculated in a variety of ways, it is best to develop a consistent method to ensure that variables are not left out. For the purpose of illustration, calculations in this chapter are presented in the basic steps discussed in the following. Toward the end of the chapter, calculations are condensed in the interest of space and tedious minor calculations.

Step 1:   Determine the operating pressure and flow for the nozzle. One of the first steps in any pump discharge pressure calculation is to determine the gpm to flow. When fixed- or variable-flow combination (fog) nozzles are used, the gpm will often be identified on the nozzle. Automatic nozzles can provide a range of flows while maintaining proper operating pressures. In this case, the pump operator must simply choose the flow for the nozzle. When smooth-bore (straight-stream) nozzles are used, the flow can either be looked up on a chart (see Appendix F) or calculated. Recall from Chapter 11 that the formula for determining gpm from a smooth-bore nozzle is as follows:

$$\text{gpm} = 30 \times d^2 \times \sqrt{NP}$$

In addition to knowing the flow for a specific nozzle, the operating pressure of the nozzle must also be known. Most nozzles have a designed operating pressure (Table 12–1). When calculating gpm for smooth bore nozzles on the fireground, the square root of the nozzle pressure calculations can be replaced with 7 for handlines and 9 for master streams.

Step 2:   After the gpm is determined, the friction loss in the hose can be calculated. Recall that friction loss is affected by the diameter of a hose. Therefore, friction loss must be calculated separately for each diameter hose within the lay. Several examples of this concept are presented later in this chapter.

Step 3:   Determine the appliance friction loss within the hose lay. Keep in mind that friction loss varies for specific appliances.

Step 4:   Calculate the loss or gain in pressure from changes in elevation.

Step 5:   The last step is to use the numbers obtained from the previous steps in the pump discharge pressure formula. The result is the pressure required at the pump panel to provide the nozzle with its proper operating pressure and flow.

Table 12–1 *Operating pressures for typical nozzle types.*

Type of Nozzle	Operating Pressure (psi)
Smooth-bore, hand line	50
Combination (fog), low-pressure	75
Smooth-bore, master stream	80
Combination (fog) and automatic	100

SINGLE LINES

The easiest discharge configuration is that of a single line. Keep in mind, though, that different sizes of hose, types of nozzles, and elevation must be factored into the calculations. The following are several examples of single-line pump discharge pressure calculations.

Nozzle Comparisons

Combination Nozzle Consider a 150-foot section of 1¾-inch hose flowing 125 gpm with a combination (fog) nozzle (Figure 12–1). What is the pump discharge pressure?

Step 1: Nozzle operating pressure and flow, NP, = 100 and gpm = 125

Step 2: Hose friction loss:

$$FL = c \times q^2 \times L$$
$$= 15.5 \times \left(\frac{125}{100}\right)^2 \times \left(\frac{150}{100}\right)$$

$$= 15.5 \times (1.25)^2 \times 1.5$$
$$= 15.5 \times 1.56 \times 1.5$$
$$= 36.27$$

Step 3: Appliance friction loss, AFL, = no appliance

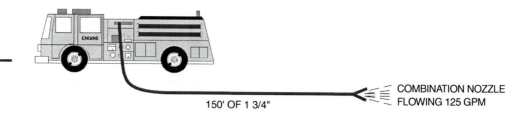

Figure 12–1 *Single line: combination nozzle.*

150' OF 1 3/4"

COMBINATION NOZZLE
FLOWING 125 GPM

Step 4:　Elevation pressure change, *EL*, = no change in elevation

Step 5:　Pump discharge pressure:

$$PDP = NP + FL + AFL + EL$$
$$= 100 + 36 + 0 + 0$$
$$= 136$$

Smooth-Bore Nozzle Calculate the pump discharge pressure for a 150-foot section of 1¾-inch handline equipped with a ¾-inch tip smooth-bore nozzle (Figure 12–2).

Step 1:　Nozzle operating pressure and flow, *NP*, = 50 psi and

$$gpm = 30 \times d^2 \times \sqrt{NP}$$
$$= 30 \times (¾'')^2 \times \sqrt{50}$$
$$= 30 \times (0.75)^2 \times \sqrt{50}$$
$$= 30 \times 0.56 \times 7$$
$$= 117.6 \text{ or } 118$$

Step 2:　Hose friction loss:

$$FL = c \times q^2 \times L$$
$$= 15.5 \times \left(\frac{118}{100}\right)^2 \times \left(\frac{150}{100}\right)$$
$$= 15.5 \times (1.18)^2 \times 1.5$$
$$= 15.5 \times 1.39 \times 1.5$$
$$= 32.37 \text{ or } 32$$

Step 3:　Appliance friction loss, *AFL*, = no appliance

Step 4:　Elevation pressure change, EL, = no change in elevation

Step 5:　Pump discharge pressure:

$$PDP = NP + FL + AFL + EL$$
$$= 50 + 32 + 0 + 0$$
$$= 82$$

Figure 12–2　*Single line: smooth bore nozzle (hand line).*

150' OF 1 3/4"

SMOOTH BORE NOZZLE WITH A 3/4 " TIP

Practice

Calculate *PDP* for both an automatic nozzle and a 1-inch tip smooth-bore using the friction loss formulas (1) $2q^2+q$, (2) hand method, and (3) cq^2L for the hand line: 300 feet of 2½-inch hose flowing 200 gpm for the automatic nozzle.

As mentioned earlier, a systematic approach to calculating PDP may help ensure critical steps or calculations are not overlooked. The basic steps proposed in this text are: *NP* (step 1) + *FL* (step 2) + *AFL* (step 3) + *EL* (step 4) = *PDP* (step 5).

		Automatic Nozzle	**Smooth-Bore**
PDP	**1.** $2q^2+q$	_____	_____
	2. hand method	_____	_____
	3. cq^2L	_____	_____

Calculations

Automatic Nozzle

1. $PDP = NP + FL\ [(2q^2+q) \times L] + AFL + EL$

Step 1: NP = 100

gpm = 200

Step 2: FL = $(2q^2+q) \times$ L
   = $[2 \times (200/100)^2 + (200/100)] \times (300/100)$
   = $[2 \times (2)^2 + 2] \times 3$
   = $(2 \times 4 + 2) \times 3$
   = 10×3
   = 30 psi

Step 3:   No appliance friction loss

Step 4:   No elevation gain or loss

Step 5: $PDP = NP + FL\ [(2q^2+q) \times L] + \cancel{AFL} + EL$
   = 100 + 30
   = 130 psi

2. $PDP = NP + FL$ (hand method, see Figure 11–8) + $AFL + EL$

Step 1: NP = 100

gpm = 200

Step 2: FL = hand method $\times$ L
   = 2 (top of index finger) $\times$ 4 (base of index finger) $\times$ (300/100)
   = 8×3
   = 24 psi

Step 3:   No appliance friction loss

Step 4:   No elevation gain or loss

Step 5: $PDP = NP + FL$ (hand method, see Figure 11–8) $+ \cancel{AFL} + \cancel{EL}$
$= 100 + 24$
$= 124$ psi

3. $PDP = NP + FL\ (cq^2L) + AFL + EL$

Step 1: $NP = 100$ and gpm $= 200$

Step 2: $FL = c \times q^2 \times L$
$= 2 \times (200/100)^2 \times (300/100)$
$= 2 \times (2)^2 \times 3$
$= 2 \times 4 \times 3$
$= 24$ psi

Step 3:   No appliance friction loss

Step 4:   No elevation gain or loss

Step 5: $PDP = NP + FL\ (cq^2L) + \cancel{AFL} + \cancel{EL}$
$= 100 + 24$
$= 124$ psi

Smooth-Bore Nozzle

1. $PDP = NP + FL\ [(2q^2+q) \times L] + AFL + EL$

Step 1: $NP = 50$
gpm $= 29.7 \times d^2 \times \sqrt{NP}$
$= 29.7 \times 1^2 \times \sqrt{50}$
$= 29.7 \times 1 \times 7.07$
$= 209.97$ or 210 gpm

Note: Using the common foreground figures in the formula ($30 \times d^2 \times 7$) provides virtually the same answer: 210 gpm.

Step 2: $FL = (2q^2+q) \times L$
$= [2 \times (210/100)^2 + (210/100)] \times (300/100)$
$= [2 \times (2.1)^2 + 2.1] \times 3$
$= (2 \times 4.41 + 2.1) \times 3$
$= 10.92 \times 3$
$= 32.76$ or 33 psi

Step 3:   No appliance friction loss

Step 4:   No elevation gain or loss

Step 5: $PDP = 50 + 33$
$= 83$ psi

2. $PDP = NP + FL$ (hand method, see Figure 11–8) $+ AFL + EL$

Step 1: NP = 50

gpm = 210

Step 2: FL = hand method $\times L$

= 2 (top of index finger top) $\times$ 4 (base of index finger) $\times$ (300/100)

= 8 $\times$ 3

= 24 psi

Step 3:   No appliance friction loss

Step 4:   No elevation gain or loss

Step 5: $PDP = 50 + 24$

= 74 psi

3. $PDP = NP + FL\,(cq^2L) + AFL + EL$

Step 1: NP = 50

gpm = 210

Step 2: FL = $c \times q^2 \times L$

= $2 \times (210/100)^2 \times (300/100)$

= $2 \times (2.1)^2 \times 3$

= $2 \times 4.41 \times 3$

= 26.46 or 27 psi

Step 3:   No appliance friction loss

Step 4:   No elevation gain or loss

Step 5: $PDP = 50 + 27$

= 77 psi

		Automatic Nozzle	Smooth-Bore
PDP	1. $2q^2+q$	130	88
	2. hand method	124	74
	3. cq^2L	124	77

Hose Diameter Comparisons

The following calculations for 2½-inch and 3-inch lines, with all other variables held constant, provide a good illustration of the difference hose size makes on the pump discharge pressure.

2½-inch Line Pumper A is flowing 350 gpm through 500 feet of 2½-inch hose equipped with an automatic nozzle (Figure 12–3). Calculate the discharge pressure.

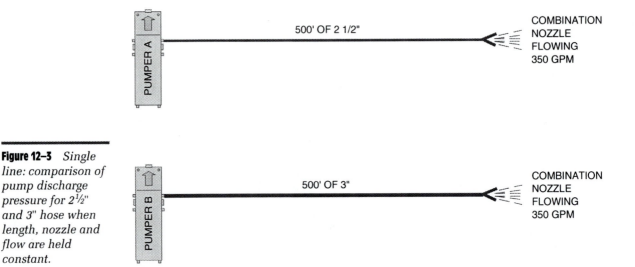

Figure 12–3 *Single line: comparison of pump discharge pressure for 2½" and 3" hose when length, nozzle and flow are held constant.*

Step 1: Nozzle operating pressure and flow, NP, = 100 and gpm = 350

Step 2: Hose friction loss:

$$FL = c \times q^2 \times L$$
$$= 2 \times \left(\frac{350}{100}\right)^2 \times \left(\frac{500}{100}\right)$$
$$= 2 \times (3.5)^2 \times 5$$
$$= 2 \times 12.25 \times 5$$
$$= 122.5 \text{ or } 123$$

Step 3: Appliance friction loss, AFL, = no appliance

Step 4: Elevation pressure change, EL, = no change in elevation

Step 5: Pump discharge pressure:

$$PDP = NP + FL + AFL + EL$$
$$= 100 + 123 + 0 + 0$$
$$= 223$$

3-inch Line Pumper B has the same hose lay configuration except that 3-inch hose is being used instead of 2½-inch hose (refer again to Figure 12–3). What is the pump discharge pressure?

Step 1: Nozzle operating pressure and flow, NP, = 100 and gpm = 350

Step 2: Hose friction loss:

$$FL = c \times q^2 \times L$$

$$= .8 \times \left(\frac{350}{100}\right)^2 \times \left(\frac{500}{100}\right)$$

$$= .8 \times (3.5)^2 \times 5$$

$$= .8 \times 12.25 \times 5$$

$$= 49$$

Step 3: Appliance friction loss, AFL, = no appliance

Step 4: Elevation pressure change, EL, = no change in elevation

Step 5: Pump discharge pressure:

$$PDP = NP + FL + AFL + EL$$

$$= 100 + 49 + 0 + 0$$

$$= 149$$

Note the significant difference in pump discharge pressure between pumper A and B. Because of the high discharge pressure, pumper A can deliver less than 70% of its rated capacity, but pumper B is capable of delivering 100% of its rated capacity.

Practice

Calculate PDP for a $2\frac{1}{2}$-inch hose using the FL formulas in the following table for the master stream line: 200 feet of hose with a $1\frac{1}{2}$-inch tip smooth-bore master stream nozzle. Also calculate the same 3-inch line except with hose using the condensed Q and cq^2L formula (no monitor nozzle).

		$2\frac{1}{2}$-inch	3-inch
PDP	**1.** $2q^2+q$	_____	
	2. Hand method	_____	
	3. Drop 10	_____	
	4. cq^2L	_____	_____
	5. Condensed Q		_____

$2\frac{1}{2}$-inch hose line

1. $PDP = NP + FL\ [(2q^2+q) \times L] + AFL + EL$

Step 1: $NP = 80$

$$gpm = 29.7 \times d^2 \times \sqrt{NP}$$

$$= 29.7 \times (1.5)^2 \times \sqrt{80}$$

$$= 29.7 \times 2.25 \times 8.94$$

$$= 597.4 \text{ or } 597 \text{ psi}$$

Note: Using the rounded figures in the formula ($30 \times d^2 \times 9$) provides a slightly higher value of 607.5 gpm.

Step 2: FL $= (2q^2+q) \times L$
$= (2 \times (597/100)^2 + (597/100)) \times (200/100)$
$= (2 \times (5.97)^2 + 5.97) \times 2$
$= (2 \times 35.6 + 5.97) \times 2$
$= 77.17 \times 2$
$= 154.34$ or 154 psi

Step 3:   No appliance friction loss

Step 4:   No elevation gain or loss

Step 5: $PDP = NP + FL\ [(2q^2+q) \times L] + \cancel{AFL} + \cancel{EL}$
$= 80 + 154$
$= 234$ psi

2. $PDP = NP + FL$ (hand method, see Figure 11–8) $+ AFL + EL$

Step 1: NP $= 80$

gpm $= 597$

Step 2: FL $=$ hand method $\times L$
$= 6$ (top of first finger on second hand) $\times 12$ (base of first finger on second hand) $\times (200/100)$
$= 6 \times 12 \times 2$
$= 144$ psi

Step 3:   No appliance friction loss

Step 4:   No elevation gain or loss

Step 5: $PDP = 80 + 144$
$= 224$ psi

3. $PDP = NP + FL$ (Drop 10) $+ AFL + EL$

Step 1: NP $= 80$

gpm $= 597$

Step 2: FL $= 59 - 10 \times L\ (200/100)$
$= 49 \times 2$
$= 98$ psi

Step 3:   No appliance friction loss

Step 4:   No elevation gain or loss

Step 5: $PDP = 80 + 98$
$= 178$ psi

4. $PDP = NP + FL\ (cq^2L) + AFL + EL$

Step 1: NP $= 80$

gpm $= 597$

Step 2: $FL = c \times q^2 \times L$
$$= 2 \times (597/100)^2 \times (200/100)$$
$$= 2 \times (5.97)^2 \times 2$$
$$= 2 \times 35.6 \times 2$$
$$= 142.4 \text{ or } 142 \text{ psi}$$

Step 3:   No appliance friction loss

Step 4:   No elevation gain or loss

Step 5: $PDP = 80 + 142$
$$= 222 \text{ psi}$$

3-inch line

1. $PDP = NP + FL\ (cq^2L) + AFL + EL$

Step 1: $NP = 80$

gpm $= 597$

Step 2: $FL = c \times q^2 \times L$
$$= .8 \times (597/100)^2 \times (200/100)$$
$$= .8 \times (5.97)^2 \times 2$$
$$= .8 \times 35.6 \times 2$$
$$= 56.9 \text{ or } 57 \text{ psi}$$

Step 3:   No appliance friction loss

Step 4:   No elevation gain or loss

Step 5: $PDP = 80 + 57$
$$= 137 \text{ psi}$$

2. $PDP = NP + FL\ (q^2L) + AFL + EL$

Step 1: $NP = 80$

gpm $= 597$

Step 2: $FL = q^2 \times L$
$$= (597/100)^2 \times (200/100)$$
$$= (5.97)^2 \times 2$$
$$= 35.6 \times 2$$
$$= 71.2 \text{ or } 71 \text{ psi}$$

Step 3:   No appliance friction loss

Step 4:   No elevation gain or loss

Step 5: $PDP = 80 + 71$
$$= 151 \text{ psi}$$

		2½-inch	3-inch
PDP	**1.** $2q^2+q$	234	
	2. Hand method	224	
	3. Drop 10	178	
	4. cq^2L	222	137
	5. Condensed Q		151

Elevation Comparisons

Recall from Chapter 11 that changes in elevation affect the pressure within hose. The following calculations illustrate elevation calculations using both the change in grade method and the floor level method.

Elevation Gain, Change in Grade Consider 300 feet of 1¾-inch line with a combination nozzle flowing 100 gpm when the line is taken up a hill to an elevation of 50 feet above the pump (Figure 12–4). What is the pump discharge pressure?

Step 1:   Nozzle operating pressure and flow, *NP*, = 100 psi and gpm = 100

Step 2:   Hose friction loss:

$$FL = c \times q^2 \times L$$
$$= 15.5 \times \left(\frac{100}{100}\right)^2 \times \left(\frac{300}{100}\right)$$

$$= 15.5 \times 1^2 \times 3$$
$$= 15.5 \times 1 \times 3$$
$$= 46.5 \text{ or } 47$$

Step 3:   Appliance friction loss, *AFL*, = no appliance

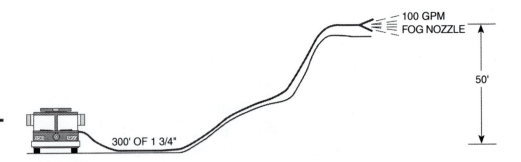

Figure 12–4 *Single line: elevation gain.*

100 GPM
FOG NOZZLE

50'

300' OF 1 3/4"

Step 4:   Elevation pressure change:

$$EL = 0.5 \times h$$
$$= 0.5 \times 50 \text{ ft}$$
$$= 25$$

Step 5:   Pump discharge pressure:

$$PDP = NP + FL + AFL + EL$$
$$= 100 + 47 + 0 + 25$$
$$= 172$$

Elevation Loss, Change in Grade Consider 300 feet of 1¾-inch line with a combination nozzle flowing 100 gpm when the line is taken down a hill to an elevation of 50 feet below the pump (Figure 12–5). What is the pump discharge pressure?

Step 1:   Nozzle operating pressure and flow, NP, = 100 psi and gpm = 100

Step 2:   Hose friction loss:

$$FL = c \times q^2 \times L$$
$$= 15.5 \times \left(\frac{100}{100}\right)^2 \times \left(\frac{300}{100}\right)$$
$$= 15.5 \times 1^2 \times 3$$
$$= 15.5 \times 1 \times 3$$
$$= 46.5, \text{ or } 47$$

Step 3:   Appliance friction loss, AFL, = no appliance

Step 4:   Elevation pressure change:

$$EL = 0.5 \times h$$
$$= 0.5 \times -50 \text{ ft}$$
$$= -25$$

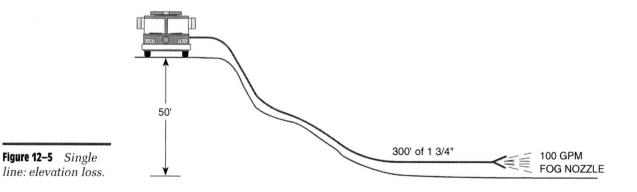

Figure 12–5 *Single line: elevation loss.*

50'

300' of 1 3/4"

100 GPM
FOG NOZZLE

Step 5:   Pump discharge pressure:

$$PDP = NP + FL + AFL + EL$$
$$= 100 + 47 + 0 + -25$$
$$= 122$$

Elevation Gain, Floor Level Consider 200 feet of 1½-inch line with a ¾-inch smooth-bore nozzle taken to the third floor of a structure (Figure 12–6). What is the pump discharge pressure?

Step 1:   Nozzle operating pressure and flow, NP, = 50 psi and

$$gpm = 30 \times d^2 \times \sqrt{NP} \text{ (using fireground formula)}$$
$$= 30 \times (0.75)^2 \times \sqrt{50}$$
$$= 30 \times 0.56 \times 7$$
$$= 117.6, \text{ or } 118$$

Step 2:   Hose friction loss:

$$FL = c \times q^2 \times L$$
$$= 24 \times \left(\frac{118}{100}\right)^2 \times \left(\frac{200}{100}\right)$$

$$= 24 \times (1.18)^2 \times 2$$
$$= 24 \times 1.39 \times 2$$
$$= 66.72 \text{ or } 67$$

Step 3:   Appliance friction loss, AFL, = no appliance

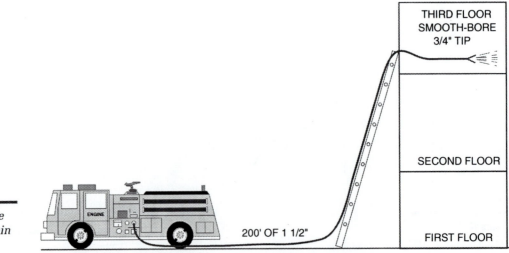

Figure 12–6 *Single line: elevation gain in a structure.*

THIRD FLOOR
SMOOTH-BORE
3/4" TIP

SECOND FLOOR

FIRST FLOOR

ENGINE

200' OF 1 1/2"

Step 4:   Elevation pressure change:

$$EL = 5 \times \text{number of levels}$$
$$= 5 \times 2$$
$$= 10$$

Step 5:   Pump discharge pressure:

$$PDP = NP + FL + AFL + EL$$
$$= 50 + 67 + 0 + 10$$
$$= 127$$

Practice

Calculate PDP using the FL formulas and elevations in the following table for 300 feet of 3-inch hose flowing 250 gpm through an automatic nozzle.

		+15 ft	5th Floor (first level = 1)	– 30 feet
PDP	**1.** cq^2L	_____	_____	_____
	2. Condensed Q	_____	_____	_____

+15 ft (elevation gain)

1. $PDP = NP + FL\,(cq^2L) + AFL + EL$

Step 1: $NP = 100$

   $gpm = 250$

Step 2: $FL = c \times q^2 \times L$
$$= .8 \times (250/100)^2 \times (300/100)$$
$$= .8 \times (2.5)^2 \times 3$$
$$= .8 \times 6.25 \times 3$$
$$= 15 \text{ psi}$$

Step 3:   No appliance friction loss

Step 4: $EL = 0.5 \times h$
$$= 0.5 \times 15$$
$$= 7.5 \text{ or } 8 \text{ psi}$$

Step 5: $PDP = NP + FL\,(cq^2L) + \cancel{AFL} + EL$
$$= 100 + 15 + 8$$
$$= 123 \text{ psi}$$

2. $PDP = NP + FL\,(q^2L) + AFL + EL$

Step 1: $NP = 100$

   $gpm = 250$

Step 2: $FL = q^2 \times L$
$= (250/100)^2 \times (300/100)$
$= (2.5)^2 \times 3$
$= 6.25 \times 3$
$= 18.75$ or 19 psi

Step 3: No appliance friction loss

Step 4: $EL = 0.5 \times h$
$= 0.5 \times 15$
$= 7.5$ or 8 psi

Step 5: $PDP = NP + FL\ (q^2L) + \cancel{AFL} + EL$
$= 100 + 19 + 8$
$= 127$ psi

5th Floor (first floor = 1)

1. $PDP = NP + FL\ (cq^2L) + AFL + EL$

Step 1: $NP = 100$
gpm $= 250$

Step 2: $FL = c \times q^2 \times L$
$= .8 \times (250/100)^2 \times (300/100)$
$= .8 \times (2.5)^2 \times 3$
$= .8 \times 6.25 \times 3$
$= 15$ psi

Step 3: No appliance friction loss

Step 4: $EL = 5 \times$ number of levels
$= 5 \times 4$
$= 20$ psi

Step 5: $PDP = NP + FL\ (cq^2L) + \cancel{AFL} + EL$
$= 100 + 15 + 20$
$= 135$

2. $PDP = NP + FL\ (q^2L) + AFL + EL$

Step 1: $NP = 100$
gpm $= 250$

Step 2: $FL = q^2 \times L$
$= (250/100)^2 \times (300/100)$
$= (2.5)^2 \times 3$
$= 6.25 \times 3$
$= 18.75$ or 19 psi

Step 3:   No appliance friction loss

Step 4: EL $= 5 \times$ number of levels
$= 5 \times 4$
$= 20$ psi

Step 5: $PDP = NP + FL\,(q^2L) + \cancel{AFL} + EL$
$= 100 + 19 + 20$
$= 139$ psi

–30 feet (elevation loss)

1. $PDP = NP + FL\,(cq^2L) + AFL + EL$

   Step 1: NP $= 100$

   gpm $= 250$

   Step 2: FL $= c \times q^2 \times L$
   $= .8 \times (250/100)^2 \times (300/100)$
   $= .8 \times (2.5)^2 \times 3$
   $= .8 \times 6.25 \times 3$
   $= 15$ psi

   Step 3:   No appliance friction loss

   Step 4: EL $= 0.5 \times h$
   $= 0.5 \times -30$
   $= -15$ psi

   Step 5: $PDP = NP + FL\,(cq^2L) + \cancel{AFL} + EL$
   $= 100 + 15 + -15$
   $= 100$ psi

2. $PDP = NP + FL\,(q^2L) + AFL + EL$

   Step 1: NP $= 100$

   gpm $= 250$

   Step 2: FL $= q^2 \times L$
   $= (250/100)^2 \times (300/100)$
   $= (2.5)^2 \times 3$
   $= 6.25 \times 3$
   $= 18.75$ or 19 psi

   Step 3:   No appliance friction loss

   Step 4: EL $= 0.5 \times h$
   $= 0.5 \times -30$
   $= -15$ psi

Step 5: $PDP = NP + FL\,(q^2L) + \cancel{AFL} + EL$
$= 100 + 19 + -15$
$= 104 \text{ psi}$

		+15 ft	5th Floor (first level = 1)	– 30 feet
PDP	1. cq^2L	123 psi	135 psi	100 psi
	2. Condensed Q	127 psi	139 psi	104 psi

Different Hose Sizes

When different size hose diameters are used within the same lay, friction loss must be calculated separately for each diameter of hose. This type of lay typically uses medium-diameter hose to overcome distance and then is reduced to a smaller attack line. Following is an example of how to calculate single line lays utilizing two different sizes of hose: 3-inch to 1½-inch hose. Consider a 700-foot lay consisting of 550 feet of 3-inch line and 150 feet of 1½-inch line. The line is equipped with an automatic nozzle flowing 150 gpm (Figure 12–7). What is the pump discharge pressure?

Step 1:   Nozzle operating pressure and flow, NP, = 100 and gpm = 150.

Step 2:   Hose friction loss, in this case, must be calculated separately for the 3-inch line and the 1½-inch line. FL_s represents the 3-inch supply line, while FL_a represents the 1½-inch attack line.

$FL_s = c \times q^2 \times L$ $FL_a = c \times q^2 \times L$
$= 0.8 \times (1.5)^2 \times 5.5$ $= 24 \times (1.5)^2 \times 1.5$
$= 0.8 \times 2.25 \times 5.5$ $= 24 \times 2.25 \times 1.5$
$= 9.9, \text{ or } 10 \text{ psi}$ $= 81 \text{ psi}$

Step 3:   Appliance friction loss, AFL, = 5 psi (reducer)

Step 4:   Elevation pressure change, EL, = no change in elevation

Step 5:   Pump discharge pressure:

$PDP = NP + FL_s + FL_a + AFL + EL$
$= 100 + 10 + 81 + 5 + 0$
$= 196$

550' OF 3"   150' OF 1 1/2"   150 GPM AUTOMATIC NOZZLE

REDUCER

Figure 12–7   *Single line: change of hose diameter.*

Practice

Calculate PDP using the FL formula cq^2L for 500 feet of 2½-inch hose reduced to 100 feet of 1½-inch hose with a ¾-inch tip smooth-bore nozzle.

$$PDP = NP + FL + AFL + EL$$

Step 1: NP = 50

$$gpm = 29.7 \times d^2 \times \sqrt{NP}$$

$$= 29.7 \times .75^2 \times \sqrt{50}$$
$$= 29.7 \times .56 \times 7.07$$
$$= 117.58 \text{ or } 118 \text{ gpm}$$

Note: Using the rounded figures in the formula ($30 \times d^2 \times 7$) provides virtually the same answer: 117.6 or 118.

Step 2: FL FL_a: 500 feet of 2½-inch hose
$$= 2 \times (118/100)^2 \times (500/100)$$
$$= 2 \times (1.18)^2 \times 5$$
$$= 2 \times 1.39 \times 5$$
$$= 13.9 \text{ or } 14 \text{ psi}$$

FL_b: 100 feet of 1½-inch hose
$$= 24 \times (118/100)^2 \times (100/100)$$
$$= 24 \times (1.18)^2 \times 1$$
$$= 24 \times 1.39 \times 1$$
$$= 33.36 \text{ or } 33 \text{ psi}$$

FL_{a+b} = 14 + 33
$$= 47 \text{ psi}$$

Step 3:   Reducer = 5 psi

Step 4:   No elevation gain or loss

Step 5: $PDP = NP + FL + AFL + \cancel{EL}$
$$= 50 + 47 + 5$$
$$= 102 \text{ psi}$$

MULTIPLE LINES

Multiple-line calculations range from simple to complex. When the pump is supplying more than one line, lower pressure lines must be feathered. The following are examples of multiple-line pump discharge pressure calculations.

Multiple Like Lines

When the pump is supplying two or more lines that are the same in size and flow, only one line is calculated. Because the other line is the same, the results of the calculations would be the same. When initiating flow for multiple like lines, the pump operator need only increase the pump discharge pressure to the calculated pressure of the one line.

For the remaining calculations in this chapter, formulas are gradually omitted and calculations gradually condensed by omitting mention of obvious steps.

Two Like Lines Consider two sections of 3-inch hose 1,000 feet long each with a 1¼-inch tip handheld smooth-bore nozzle (Figure 12–8). What is the pump discharge pressure for each line?

Step 1:   Nozzle operating pressure and flow, NP, = 50 and

$$gpm = 30 \times d^2 \times \sqrt{NP}$$
$$= 30 \times (1.25)^2 \times \sqrt{50}$$
$$= 30 \times 1.56 \times 7$$
$$= 327.6, \text{ or } 328$$

Step 2:   Hose friction loss:

$$FL = c \times q^2 \times L$$
$$= .8 \times (3.28)^2 \times 10$$
$$= .8 \times 10.76 \times 10$$
$$= 86.08 \text{ or } 86$$

Step 3:   Appliance friction loss, AFL, = no appliances

Step 4:   Elevation pressure change, EL, = no change in elevation

Step 5:   Pump discharge pressure:

$$PDP = NP + FL + AFL + EL$$
$$= 50 + 86$$
$$= 136$$

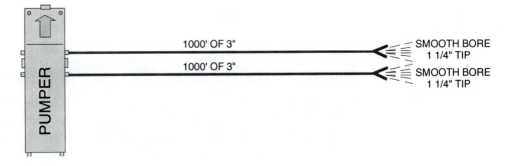

Figure 12–8 *Multiple lines: two like lines.*

Both lines should be pumped at 136 psi to obtain a nozzle pressure of 50 psi. Regardless of how many lines are flowing, as long as they are all the same size and flow, the discharge pressure will be the same. Keep in mind that the elevation must also be the same.

Multiple Lines of Different Sizes and Flows

When the pump is supplying multiple lines of different sizes and flows, each line must be calculated separately.

Two Lines of Different Size Consider a pump supplying two lines. Line A is 500 feet of 3-inch hose flowing 250 gpm through a fog nozzle. Line B is 200 feet of 2½-inch hose flowing 300 gpm through an automatic nozzle (Figure 12–9). What is the pump discharge pressure for both lines?

Step 1: Nozzle operating pressure and flow, NP, = 100 psi for both Line A and Line B, and gpm = 250 for line A and 300 for Line B

Step 2: Hose friction loss:

Line A
$$FL = .8 \times (2.5)^2 \times 5$$
$$= .8 \times 6.25 \times 5$$
$$= 25$$

Line B
$$FL = 2 \times 3^2 \times 2$$
$$= 2 \times 9 \times 2$$
$$= 36$$

Step 3: Appliance friction loss, AFL, = no appliances in either line

Step 4: Elevation pressure change, EL, = no change in elevation in either line

Step 5: Pump discharge pressure:

Line A
$$PDP = 100 + 25$$
$$= 125$$

Line B
$$PDP = 100 + 36$$
$$= 136$$

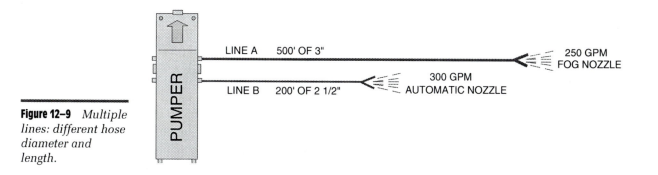

Figure 12–9 *Multiple lines: different hose diameter and length.*

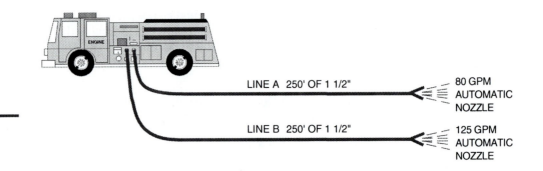

Figure 12–10
*Multiple lines:
different flow.*

LINE A   250' OF 1 1/2"

80 GPM
AUTOMATIC
NOZZLE

LINE B   250' OF 1 1/2"

125 GPM
AUTOMATIC
NOZZLE

Two Lines of Different Flow Consider a pump supplying two lines of equal length (250 ft) and diameter (1½ in.). However, line A is flowing 80 gpm and line B is flowing 125 gpm both through automatic nozzles (Figure 12–10). What is the pump discharge pressure for each line?

Step 2:   Hose friction loss:

Line A

$$FL = 24 \times .8^2 \times 2.5$$
$$= 24 \times .64 \times 2.5$$
$$= 38.4, \text{ or } 38$$

Line B

$$FL = 24 \times (1.25)^2 \times 2.5$$
$$= 24 \times 1.56 \times 2.5$$
$$= 93.6, \text{ or } 94$$

Step 5:   Pump discharge pressure:

Line A

$$PDP = 100 + 38$$
$$= 138$$

Line B

$$PDP = 100 + 94$$
$$= 194$$

Three Lines of Various Configurations Determine the pump discharge pressure for each of the lines shown in Figure 12–11.

Step 2:   Hose friction loss:

Line A

$$FL = .8 \times 2.5^2 \times 4$$
$$= 8 \times 6.25 \times 4$$
$$= 20$$

Line B

$$FL_s = .8 \times (.95)^2 \times 2.5$$
$$= .8 \times .9 \times 2.5$$
$$= 1.8, \text{ or } 2$$
$$FL_a = 24 \times (.95)^2 \times 1$$
$$= 24 \times .9 \times 1$$
$$= 21.6, \text{ or } 22$$

Line C

$$FL = 15 \times (.95)^2 \times 4$$
$$= 15 \times .9 \times 4$$
$$= 54$$

Step 3:   Appliance friction loss: line A $AFL = 0$; line B $AFL = 5$; and line C $AFL = 0$

Step 4:   Elevation pressure change:

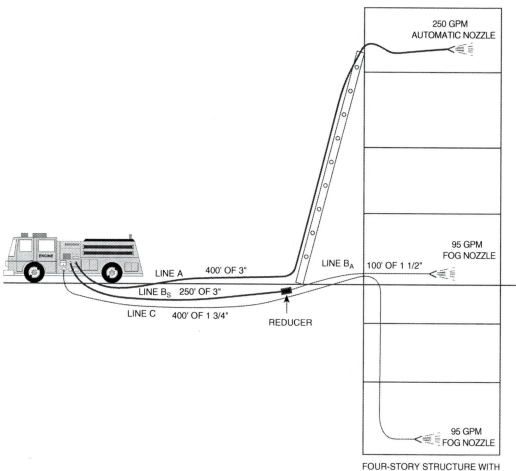

Figure 12–11   *Multiple lines (3): various configurations.*

Line A	Line B	Line C
$EL = 5 \times 3$	$EL = 0$	$EL = 5 \times (-3)$
$= 15$		$= -15$

Step 5:   Pump discharge pressure:

Line A

$PDP = NP + FL + EL$
$= 100 + 20 + 15$
$= 135$

Line B

$PDP = NP + FL_s + FL_a + AFL$
$= 100 + 2 + 22 + 5$
$= 129$

Line C

$PDP = NP + FL + EL$
$= 100 + 54 + {-15}$
$= 139$

Practice

Calculate PDP for the following pumping operation using cq^2L. Two pumpers are operating on the fire scene. Pumper 1 is attacking the fire with two 200-feet $1\frac{3}{4}$-inch lines with automatic nozzles flowing 125 gpm and operating on the third level of a structure. Pumper 2 is covering exposures with three lines. The first master stream line$_a$ is 400 feet in length, 200 feet of 3-inch reduced to 200 feet of $2\frac{1}{2}$-inch with a combination nozzle flowing 300 gpm and operating on the second level. The second master stream line$_b$ is 600 feet of 3-inch with a $1\frac{1}{2}$-inch tip on the smooth-bore master stream which is 10 feet below the pumper. The third line$_c$ is the same as line$_b$ with the exception that it is 25 feet below the pumper. All master streams lines are monitors.

Pumper 1: PDP = NP + FL + AFL + EL

    Step 1:   Equal lines, calculate for one line

$$NP = 100$$
$$gpm = 125$$

    Step 2: $FL = 15.5 \times (125/100)^2 \times (200/100)$
$$= 15.5 \times (1.25)^2 \times 2$$
$$= 15.5 \times 1.56 \times 2$$
$$= 48.36 \text{ or } 48 \text{ psi}$$

    Step 3:   AFL, Monitor nozzle = 15 psi

    Step 4: $EL = 5 \times 2$
$$= 10 \text{ psi}$$

    Step 5: $PDP = NP + FL + AFL + EL$
$$= 100 + 48 + 15 + 10$$
$$= 173 \text{ psi}$$

Pumper 2: PDP = NP + FL + AFL + EL

    Step 1:   (NP and gpm)

        Line$_a$ $NP = 100$
                      gpm = 300

        Line$_b$ $NP = 80$
                      gpm $= 30 \times 1.5^2 \times 9$
                      gpm = 607.5 or 608

        Line$_c$ $NP = 80$
                      gpm = 608

    Step 2: FL

        Line$_a$ 3-inch line $FL = .8 \times (300/100)^2 \times (200/100)$
                                            $= .8 \times (3)^2 \times 2$

$$= .8 \times 9 \times 2$$
$$= 14.4 \text{ or } 14 \text{ psi}$$

Line$_a$	2½-inch line	$FL = 2 \times (300/100)^2 \times (200/100)$
		$= 2 \times (3)^2 \times 2$
		$= 2 \times 9 \times 2$
		$= 36 \text{ psi}$
Line$_a$		$= 14 + 36$
		$= 50 \text{ psi}$
Line$_b$		$= .8 \times (608/100)^2 \times (600/100)$
		$= .8 \times (6.08)^2 \times 6$
		$= .8 \times 36.96 \times 6$
		$= 177.4 \text{ or } 177 \text{ psi}$
Line$_c$		$= 177 \text{ psi}$

Step 3:   Appliance *FL*

Line$_a$ = 5 psi (reducer) + 15 psi (monitor nozzle)
              = 20 psi

Lines$_{b \text{ and } c}$ = 15 psi (monitor nozzles)

Step 4:   Elevation

Line$_a$ = 5 × 2
              = 10 psi

Line$_b$ = .5 × 10
              = −5 psi

Line$_c$ = .5 × 25
              = −12.5 or −13 psi

Step 5:   *PDP = NP + FL + AFL + EL*

Line$_a$ = 100 + 50 + 20 + 10
          = 180 psi

Line$_b$ = 80 + 177 + 15 + −5
          = 267 psi

Line$_c$ = 80 + 177 + 15 + −13
          = 259 psi

Pumper 1 *PDP* would be 173 psi (like lines pumped at same pressure because they are like lines).

Pumper 2 *PDP* would be 267 psi (line$_b$) and the other two lines would be gated down to 180 psi (line$_a$) and 259 psi (line$_c$).

WYED LINES

Wyed lines are hose configurations in which one hose line supplies two or more separate lines. Typically, one larger line supplies two or more smaller lines, for example, a 2½- or 3-inch supply line wyed to two or more 1½- or 1¾-inch attack lines. When calculating wyed lines, the supply line is calculated separately from the attack lines. Because the supply line feeds each of the wyed attack lines, the flow through the supply line will be the sum total of the flow through each of the wyed attack lines. If the attack lines are of equal size and flow, simply calculate friction loss for one of the lines.

Simple Wyed Configuration

Calculate the pump discharge pressure for the wyed line shown in Figure 12–12.

Step 1:   Nozzle operating pressure and flow, NP, = 100 psi.

The flow through each of the 1¾-inch attack lines is 125 gpm, while the flow through the 3-inch supply line is 250 gpm. The attack line flow is designated as gpm_a, and the flow through the supply line will be gpm_s. This distinction is necessary for separately calculating the friction loss in the supply and attack lines. Because both attack lines are the same (like lines), only one line is calculated.

$$gpm_s = 250 \text{ and } gpm_a = 125$$

Step 2:   Hose friction loss, in this case, must be calculated separately for the 3-inch line, FL_s, and the 1¾-inch line, FL_a. L_s represents the length

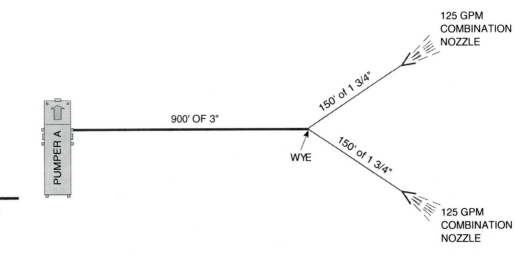

Figure 12–12   *Wye line: simple configuration.*

of the 3-inch supply line, while L_a represents the length of the $1\tfrac{3}{4}$-inch attack line.

$$FL_s = c \times \left(\frac{gpm_s}{100}\right)^2 \times L_s \qquad\qquad FL_a = c \times \left(\frac{gpm_a}{100}\right)^2 \times L_a$$

$$\begin{aligned} &= 0.8 \times (2.5)^2 \times 9 \\ &= 0.8 \times 6.25 \times 9 \\ &= 45 \end{aligned} \qquad\qquad \begin{aligned} &= 15.5 \times (1.25)^2 \times 1.5 \\ &= 15.5 \times 1.56 \times 1.5 \\ &= 36.27, \text{ or } 36 \end{aligned}$$

Step 3:   Appliance friction loss, AFL, = 10 psi ($2\tfrac{1}{2}$ inches to $1\tfrac{1}{2}$ inches). Note that 3-inch hose typically utilizes $2\tfrac{1}{2}$-inch couplings, and $1\tfrac{3}{4}$-inch hose typically uses $1\tfrac{1}{2}$-inch couplings.

Step 5:   Pump discharge pressure:

$$\begin{aligned} PDP &= NP + FL_s + FL_a + AFL \\ &= 100 + 45 + 36 + 10 \\ &= 191 \end{aligned}$$

Practice

Calculate PDP for the following pumping operation using cq^2L. A pumper is supporting two wyed lines as follows:

Line 1 =   400 feet of 3-inch line wyed to two 150 feet of $1\tfrac{3}{4}$-inch line with an automatic nozzle flowing 125 gpm.

Line 2 =   400 feet of $2\tfrac{1}{2}$-inch line wyed to two 150 feet of $1\tfrac{1}{2}$-inch line with an automatic nozzle flowing 125 gpm.

$$PDP = NP + FL + AFL + EL$$

Line 1

Step 1: NP = 100

gpm_a = 125

gpm_s = 250 (combined flow for the two attack lines)

Step 2: FL_a 150 feet of $1\tfrac{3}{4}$-inch line

$$\begin{aligned} &= 15.5 \times (125/100)^2 \times (150/100) \\ &= 15.5 \times (1.25)^2 \times 1.5 \\ &= 15.5 \times 1.56 \times 1.5 \\ &= 36.27 \text{ or } 36 \text{ psi} \end{aligned}$$

FL_s 400 feet of 3-inch line

$$\begin{aligned} &= .8 \times (250/100)^2 \times (400/100) \\ &= .8 \times (2.25)^2 \times 4 \\ &= .8 \times .06 \times 4 \\ &= 16.2 \text{ or } 16 \text{ psi} \end{aligned}$$

$$FL_{a+s} = 36 + 16$$
$$= 52 \text{ psi}$$

Step 3:   10 psi ($2\frac{1}{2}$ to $1\frac{1}{2}$ wye)

Step 4:   No elevation gain or loss

Step 5: $PDP = NP + FL + AFL + \cancel{EL}$
$$= 100 + 52 + 10$$
$$= 162 \text{ psi}$$

Line 2

Step 1: NP $= 100$

$\text{gpm}_a = 125$

$\text{gpm}_s = 250$

Step 2: FL_a 150 feet of $1\frac{1}{2}$-inch line
$$= 24 \times (125/100)^2 \times (150/100)$$
$$= 24 \times (1.25)^2 \times 1.5$$
$$= 24 \times 1.56 \times 1.5$$
$$= 56.16 \text{ or } 56 \text{ psi}$$

FL_b 400 feet of $2\frac{1}{2}$-inch line
$$= 2 \times (250/100)^2 \times (400/100)$$
$$= 2 \times (2.25)^2 \times 4$$
$$= 2 \times 5.06 \times 4$$
$$= 4.48 \text{ or } 40 \text{ psi}$$

FL $= FL_a + FL_s$
$$= 56 + 40$$
$$= 96 \text{ psi}$$

Step 3: $2\frac{1}{2}$ to $1\frac{1}{2}$ wye = 10 psi

Step 4:   No elevation gain or loss

Step 5: $PDP = NP + FL + AFL + \cancel{EL}$
$$= 100 + 96 + 10$$
$$= 206 \text{ psi}$$

PDP

Line 1   162 psi

Line 2   206 psi

Complicated Wyed Configuration

Complicated wye configurations include variables such as elevation and some-times unequal lines on the downstream side of the wye. When calculating PDP for complicated configurations, be sure to clearly indicate each line and recheck fig-ures to help reduce the chance of error.

Practice

Calculate the pump discharge pressure for the wyed line shown in Figure 12–13.

Step 1:   Nozzle operating pressure and flow, NP, = 100 psi (each of the nozzles are combination (fog) nozzles) and

Line A gpm_s = 200 (supply line), gpm_a = 100 (wyed attack like line)

Line B gpm_s = 400 (supply line), gpm_a = 250 (wyed attack line B1), gpm_a = 150 (wyed attack line B2)

Note that the two lines downstream from the wye are unequal (different size and flow), therefore, friction loss calculations must be made for both lines.

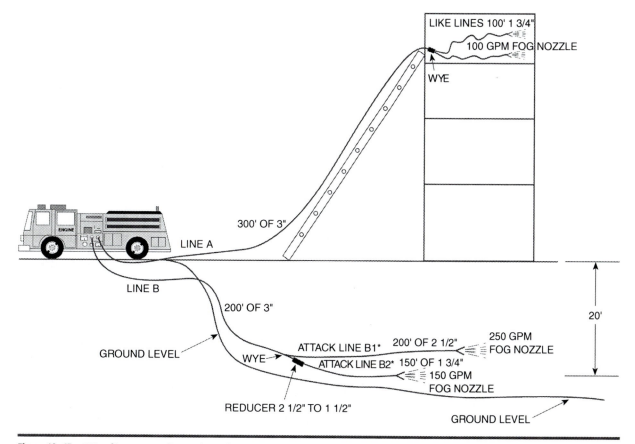

Figure 12–13   *Wye line: complicated configuration.*

Step 2:   Hose friction loss:

Line A

Supply Line   **Attack Line**

$$FL_s = c \times \left(\frac{gpm_s}{100}\right)^2 \times L_s \qquad FL_a = c \times \left(\frac{gpm_a}{100}\right)^2 \times L_a$$

$$= 0.8 \times 2^2 \times 3 \qquad\qquad = 15.5 \times 1^2 \times 1$$
$$= 0.8 \times 4 \times 3 \qquad\qquad = 15.5 \times 1 \times 1$$
$$= 9.6, \text{ or } 10 \qquad\qquad\quad = 15.5, \text{ or } 16$$

Line B

Supply   **Attack Line B1***   **Attack Line B2***

$FL_s = .8 \times 4^2 \times 2$ $FL_a = 2 \times (2.5)^2 \times 2$ $FL_a = 15.5 \times (1.5)^2 \times 1.5$
$\quad = 0.8 \times 16 \times 2$ $\quad = 2 \times 6.25 \times 2$ $\quad = 15.5 \times 2.25 \times 1.5$
$\quad = 25.6, \text{ or } 26$ $\quad = 25$ $\quad = 52.3, \text{ or } 53$

Step 3:   Appliance friction loss:

Line A: $AFL = 10$ psi ($2\frac{1}{2}$ inches to $1\frac{1}{2}$ inches wye)

Line B: $AFL = 10$ ($2\frac{1}{2}$ inches to $2\frac{1}{2}$ inches wye and a $2\frac{1}{2}$ inches to $1\frac{1}{2}$ inches reducer for the second attack line)

Step 4:   Elevation pressure change:

Line A   **Line B**

EL $= 5 \times$ number of stories   EL $= h \times 5$
$\quad\ = 5 \times 3$ $\quad\ = -20 \times .5$
$\quad\ = 15$ $\quad\ = -10$

Step 5:   Pump discharge pressure:

Line A   **Line B**

$PDP = NP + FL_s + FL_a + AFL + EL$ $PDP = NP + FL_s + {}^*FL_a + AFL = EL$
$\quad\ = 100 + 10 + 16 + 10 + 15$ $\quad\ = 100 + 26 + 53 + 10 - 10$
$\quad\ = 151$ $\quad\ = 179$

Note that the two attack lines downstream of the wye in Figure 12–13 (designated by the asterisks) require two different pressures. The highest pressure line is included in the pump discharge pressure calculation. The second attack line is gated down (feathered) to the lower required pressure. This is similar to having two lines of different pressure at the pump panel. Obviously, this would require a pressure gauge on, or just aft of, the wye. Since gauges on wyes are typically not common, uneven lines downstream of the wye should be avoided.

SIAMESE LINES

Siamese lines are hose configurations in which two or more separate lines supply one line, monitor nozzle, fixed system, and a pump in a relay or similar situation. Typically, two smaller or equal lines supply one larger or equal line, for example, several 2½-inch supply lines siamesed to a 3-inch line. When calculating siamese lines, the supply line is calculated separately from the attack line. The flow through the line downstream of the siamese will be divided among the supply lines. If the supply lines are of equal size and flow, simply calculate friction loss for one of the lines.

Calculate the pump discharge pressure for the siamese line shown in Figure 12–14.

Step 1:   Nozzle operating pressure and flow, NP, = 80 psi and

$$\text{gpm}_a = 30 \times (1.375)^2 \times 9$$
$$= 30 \times 1.89 \times 9$$
$$= 510.3, \text{ or } 510 \text{ (total flow)}$$

$$\text{gpm}_s = 255 \text{ (flow through each of the supply lines)}$$

Note that flow through the attack line will be divided between the supply lines.

Step 2:   Hose friction loss:

$$FL_s = c \times \left(\frac{\text{gpm}_s}{100}\right)^2 \times L_s \qquad\qquad FL_a = c \times \left(\frac{\text{gpm}_a}{100}\right)^2 \times L_a$$

$$\begin{array}{ll} = 2 \times 2.55^2 \times 6 & \qquad = .8 \times 5.1^2 \times 2 \\ = 2 \times 6.5 \times 6 & \qquad = .8 \times 26 \times 2 \\ = 78 & \qquad = 41.6 \text{ or } 42 \end{array}$$

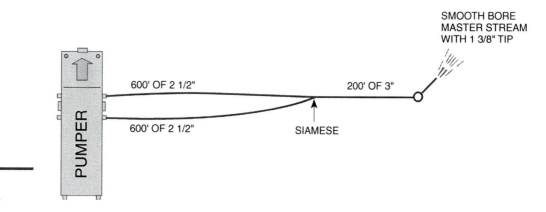

Figure 12–14
Siamese lines.

Step 3:   Appliance friction loss, AFL_a, = 15 psi (siamese) and AFL_b = 15 (monitor nozzle)

Step 4:   Pump discharge pressure:

$$PDP = NP + FL_s + FL_a + AFL_a + AFL_b$$
$$= 80 + 78 + 42 + 15 + 15$$
$$= 230$$

Practice

Calculate PDP for a siamese hose configuration with three 500-foot sections of 2½-inch line leading to a 200-foot section of 3-inch line with a 1½-inch tip on a smooth-bore monitor master stream nozzle.

$$PDP = NP + FL + AFL + EL$$

Step 1: NP $= 80$

$gpm_a = 30 \times d^2 \times 9$
$= 30 \times 1.5^2 \times 9$
$= 30 \times 2.25 \times 9$
$= 607.5$ or 608

$gpm_s = 203$ (combined flow divided by the number lines supplying the siamese)

Step 2: FL FL_a: 200 feet of 3-inch line
$= .8 \times (608/100)^2 \times (200/100)$
$= .8 \times (6.08)^2 \times 2$
$= .8 \times 36.96 \times 2$
$= 59$ psi

FL_s: 500 feet of 2½-inch line
$= 2 \times (203/100)^2 \times (500/100)$
$= 2 \times (2.03)^2 \times 5$
$= 2 \times 4.12 \times 5$
$= 41.2$ or 41 psi

$FL_{a+s} = 59 + 41$
$= 100$ psi

Step 3: $AFL = 10$ (siamese) + 15 (monitor)
$= 25$ psi

Step 4:   No elevation gain or loss

Step 5: $PDP = NP + FL + AFL + \cancel{EL}$
$= 80 + 100 + 25$
$= 205$ psi

A handy fireground rule of thumb used for siamese lines is:

Two-line siamese: use 25% of the friction loss for one line.

Three-line siamese: use 10% of the friction loss for one line.

STANDPIPE SUPPORT

Calculate the pump discharge pressure required to support the standpipe configuration illustrated in Figure 12–15.

Step 1:   Nozzle operating pressure and flow, NP, = 100 psi and gpm = 100

Step 2:   Hose friction loss:

$$FL_A = c \times q^2 \times L$$
$$= 24 \, (100/100)^2 \times (150/100)$$
$$= 24 \times 1 \times 1.5$$
$$= 36 \text{ psi}$$
$$FL_s = .8 \times (100/100)^2 \times (300/100)$$
$$= .8 \times 1 \times 3$$
$$= 2.4 \text{ or } 2 \text{ psi}$$

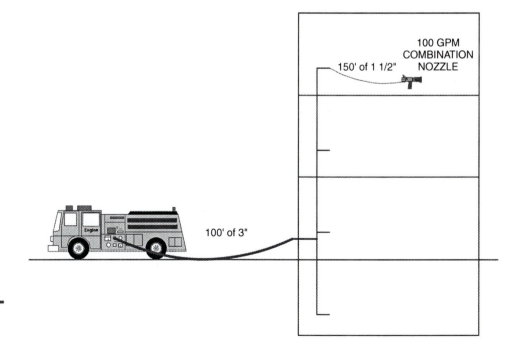

Figure 12–15
Standpipe support.

Step 3:   Appliance friction loss, AFL, = 25

Step 4:   Elevation pressure changes:

$$EL = 5 \times \text{number of levels}$$
$$= 5 \times 2$$
$$= 10 \text{ psi}$$

Step 5:   Pump discharge pressure:

$$PDP = NP + (FL_A + FL_S) + AFL + EL$$
$$= 100 + 36 + 2 + 25 + 10$$
$$= 173 \text{ psi}$$

Recall from Chapter 9 that, unless otherwise indicated, the supply line for sprinkler systems should be pumped at 150 psi.

SUMMARY

Hose lay configurations can be as simple as a single line or as complicated as three lines with wyes, siamese, and elevation gain or loss. Regardless of the method used, pump operators must strive to provide nozzles with the proper flow and pressure by calculating the appropriate pump discharge pressure.

REVIEW QUESTIONS

Short Answer

For each of the following, review the figure and provide the requested information on a separate sheet of paper.

1. Single line with smooth-bore nozzle, no elevation, see Figure 12–16.

   a. Quantity of water flowing:

   b. Nozzle pressure:

   c. Friction loss per 100 feet:

   d. Pump discharge pressure:

2. Two lines of same flow with different lengths, no elevation, see Figure 12–17.

   a. Total friction loss for Line A:

   b. Total friction loss for Line B:

   c. *PDP* for Line A:

   d. *PDP* for Line B:

300' OF 1 3/4"   1/2" SMOOTH BORE (HANDLINE)

Figure 12–16   *Single line with smooth-bore nozzle.*

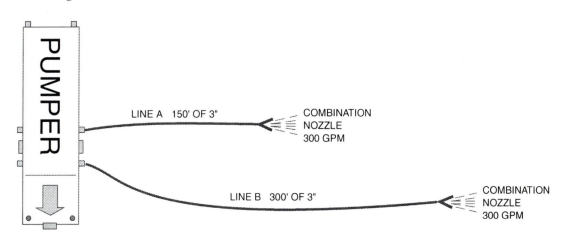

PUMPER

LINE A   150' OF 3"   COMBINATION NOZZLE 300 GPM

LINE B   300' OF 3"   COMBINATION NOZZLE 300 GPM

Figure 12–17   *Two lines of same flow with different lengths.*

3. Two lines of same length with different flows, no elevation, see Figure 12–18.

   a. *FL* Line A:

   b. *FL* Line B:

   c. *PDP* Line A:

   d. *PDP* Line B:

4. Multiple lines of same length and flow with different size hose, no elevation, see Figure 12–19.

   a. Friction Loss (*FL*):

     Line A =

     Line B =

     Line C =

     Line D =

   b. Pump Discharge Pressure (*PDP*):

     Line A =

     Line B =

     Line C =

     Line D =

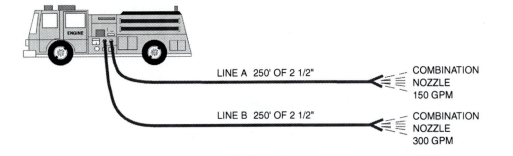

Figure 12–18   *Two lines of same length with different flows.*

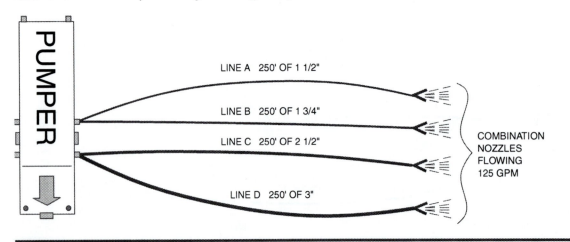

Figure 12–19   *Multiple lines of same length and flow with different size hose.*

5. Single line wyed to two like lines, no elevation, see Figure 12–20.

   a. *FL* in the 3-inch hose:

   b. *FL* in the 1¾-inch hose:

   c. *PDP:*

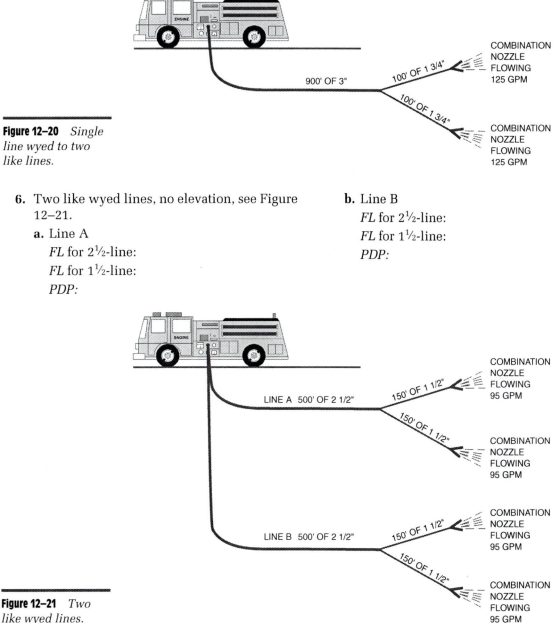

Figure 12–20 *Single line wyed to two like lines.*

6. Two like wyed lines, no elevation, see Figure 12–21.

   a. Line A

   FL for 2½-line:

   FL for 1½-line:

   PDP:

   b. Line B

   FL for 2½-line:

   FL for 1½-line:

   PDP:

Figure 12–21 *Two like wyed lines.*

7. Siamese line, no elevation, see Figure 12–22.

   a. Friction loss (FL_s) for the lines supplying the siamese:

   b. Friction loss (FL_a) for the line supplied by the siamese:

   c. Pump discharge pressure for the hose configuration:

8. Two lines, one a siamese and the other a single line, each of the same total length and flow, no elevation, see Figure 12–23.

   a. *PDP* Line A:

   b. *PDP* Line B:

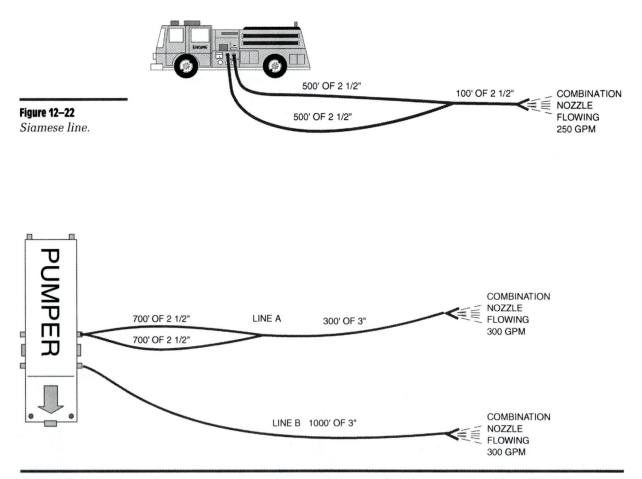

Figure 12–22
Siamese line.

500' OF 2 1/2"

500' OF 2 1/2"

100' OF 2 1/2"

COMBINATION
NOZZLE
FLOWING
250 GPM

PUMPER

700' OF 2 1/2"   LINE A   300' OF 3"

700' OF 2 1/2"

COMBINATION
NOZZLE
FLOWING
300 GPM

LINE B   1000' OF 3"

COMBINATION
NOZZLE
FLOWING
300 GPM

Figure 12–23   *Two lines, one a siamese and the other a single line, each with the same total length and flow.*

9. Two siamese lines, no elevation, see Figure 12–24.

    a. Which siamese line has the lower *PDP?*

10. Monitor nozzle supplied by three lines, no elevation, see Figure 12–25.

    a. *PDP:*

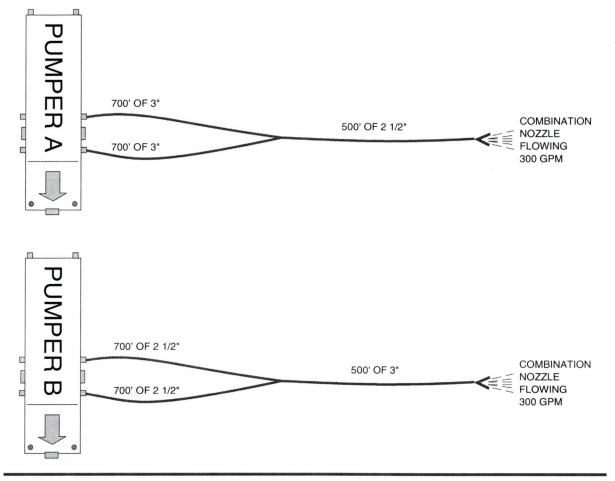

Figure 12–24   *Two siamese lines.*

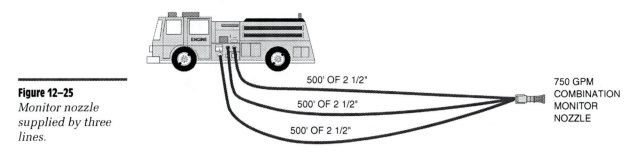

Figure 12–25
Monitor nozzle supplied by three lines.

11. One three-inch handline operating on the third floor, see Figure 12–26.

    a. Pressure change caused by elevation:

    b. Flow through nozzle (gpm):

    c. *PDP:*

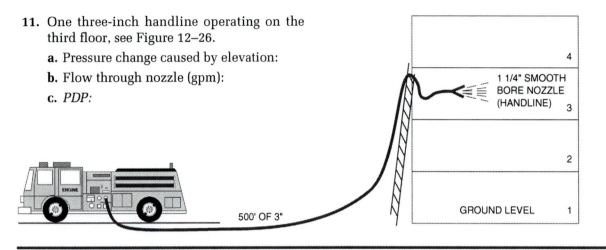

Figure 12–26   *One 3-inch handline operating on the 3rd floor.*

12. Multiple lines with elevation, see Figure 12–27.

    a. *PDP* Line A:

    b. *PDP* Line B:

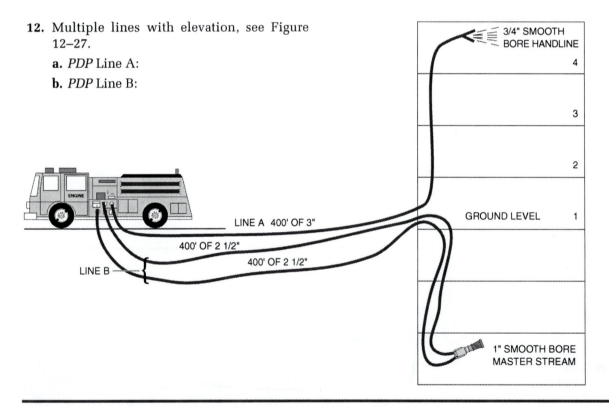

Figure 12–27   *Multiple lines with elevation.*

13. Complicated hose lay, see Figure 12–28.

  a. *PDP* Line A:

  b. *PDP* Line B:

On a separate sheet of paper, draw the described hose lay and determine the pump discharge pressures for each of the following, being sure to properly identify each component.

14. 500 feet of 2½-inch line flowing 325 gpm through an automatic nozzle.

15. Two lines, one consisting of 300 feet of 3-inch hose flowing 250 gpm through a fog nozzle, and the second consisting of 300 feet of 1½-inch hose flowing 125 gpm through an automatic nozzle.

16. One 900-foot line of 3-inch hose wyed to two 150-foot sections of 1¾-inch hose, each flowing 200 gpm through an automatic nozzle.

17. A siamese lay consisting of two 750-foot sections of 3-inch hose to 150 feet of 3-inch hose flowing 500 gpm through a master stream combination nozzle.

18. Three 400-foot lines of 3-inch hose attached to a master stream flowing 1,000 gpm through a combination nozzle.

19. Two lines consisting of a wyed line (line A) and a siamese line (line B). Line A is 400 feet of 3-inch hose wyed to two 200-foot lines of 1¾-inch hose with ¾-inch tips taken to the sixth floor of a structure. Line B is two 500-foot lines of 3-inch hose siamesed to 350 feet of 3-inch hose with a 1-inch tip taken to the second basement level in a structure.

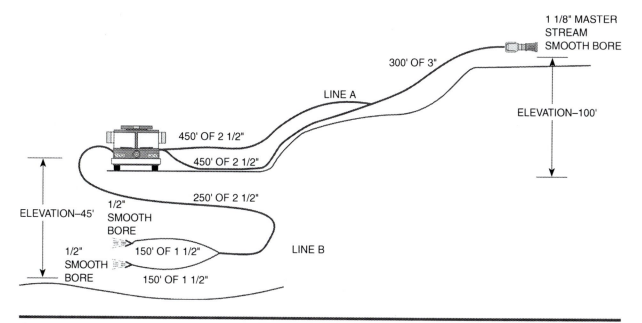

Figure 12–28 *Complicated hose lay.*

ACTIVITY

1. Review a response district and identity four structures. For each of the structures, determine four likely hose lay configurations, two of which are simple and two of which are complex. Finally, determine pump discharge pressures for each hose lay configuration.

PRACTICE PROBLEM

1. For a given structure, develop a pumping operation in which the hose line configurations include all the variables presented in this chapter.

Appendixes

Appendix

A

NFPA STANDARDS RELATED
TO PUMP OPERATIONS

NFPA 11   *Standard for Low-, Medium-, and High-Expansion Foam Systems*
NFPA 11A   *Standard for Medium- and High-Expansion Foam Systems*
NFPA 13   *Standard for the Installation of Sprinkler Systems*
NFPA 13D   *Standard for the Installation of Sprinkler Systems in One- and Two-Family Dwellings and Manufactured Homes*
NFPA 13E   *Recommended Practice for Fire Department Operations in Properties Protected by Sprinkler and Standpipe Systems*
NFPA 13R   *Standard for the Installation of Sprinkler Systems in Residential Occupancies up to and Including Four Stories in Height*
NFPA 14   *Standard for the Installation of Standpipe and Hose Systems*
NFPA 18   *Standard on Wetting Agents*
NFPA 291   *Recommended Practice for Fire Flow Testing and Marking of Hydrants*
NFPA 414   *Standard for Aircraft Rescue and Fire-Fighting Vehicles*
NFPA 1001   *Standard for Fire Fighter Professional Qualifications*
NFPA 1002   *Standard for Fire Apparatus Driver/Operator Professional Qualifications, 2003 edition*
NFPA 1142   *Standard on Water Supplies for Suburban and Rural Fire Fighting*

NFPA 1145	*Guide for the Use of Class A Foams in Manual Structural Fire Fighting*
NFPA 1150	*Standard on Fire-Fighting Foam Chemicals for Class A Fuels in Rural, Suburban, and Vegetated Areas*
NFPA 1201	*Standard for Developing Fire Protection Services for the Public*
NFPA 1403	*Standard on Live Fire Training Evolutions*
NFPA 1410	*Standard on Training for Initial Emergency Scene Operations*
NFPA 1451	*Standard for a Fire Service Vehicle Operations Training Program*
NFPA 1500	*Standard on Fire Department Occupational Safety and Health Program*
NFPA 1561	*Standard on Emergency Services Incident Management System*
NFPA 1582	*Standard on Comprehensive Occupational Medical Program for Fire Departments*
NFPA 1901	*Standard for Automotive Fire Apparatus*
NFPA 1906	*Standard for Wildland Fire Apparatus*
NFPA 1911	*Standard for Service Tests of Fire Pump Systems on Fire Apparatus*
NFPA 1914	*Standard for Testing Fire Department Aerial Devices*
NFPA 1915	*Standard for Fire Apparatus Preventive Maintenance Program*
NFPA 1961	*Standard on Fire Hose*
NFPA 1962	*Standard for the Inspection, Care, and Use of Fire Hose, Couplings, and Nozzles and the Service Testing of Fire Hose*
NFPA 1963	*Standard for Fire Hose Connections*
NFPA 1964	*Standard for Spray Nozzles (Shutoff and Tip)*
NFPA 1965	*Standard for Fire Hose Appliances*

Appendix

B

SAMPLE GENERIC STANDARD OPERATING PROCEDURES

FIRE DEPARTMENT NAME

Generic SOP

Safe Operation & Movement of Emergency Vehicles

PURPOSE: The purpose of this standard operating procedure is to establish responsibility and procedures for the safe operation and movement of vehicles assigned to the department. This standard is intended to meet the requirements of the National Fire Protection Association (NFPA) 1500, *Standard on Fire Department Occupational Safety and Health Program*. This policy does not imply a comprehensive list of items that could occur.

RESPONSIBILITIES: The chief, officers and all drivers assigned to the department are responsible for ensuring the safe operation and movement of emergency vehicles.

PROCEDURES: All drivers must maintain a healthy vigilance to safety and awareness of the hazards associated with driving emergency vehicles. **The following procedures will be followed to ensure the safe operation and movement of the department vehicles:**

1. During emergency response, installed audible and visual warning devices shall be operated until arrival on scene. Speed will be governed by weather and traffic conditions. Drivers are responsible for maintaining control of emergency vehicles at all times by operating emergency vehicles no faster than conditions permit.
2. During emergency response, drivers will bring the vehicle to a complete stop: when directed by a law enforcement officer, at red traffic light or stop sign (may proceed when safe), when encountering stopped school bus with flashing warning light (may proceed when school bus driver indicates it is safe), at unguarded railroad grade crossing (may proceed after verifying tracks are clear). Never, ever assume you have the right of way because you are using the emergency warning lights and sirens.
3. Operation of any department vehicle under other than emergency situations will be in accordance with existing state, county, and city traffic regulations. All personnel shall be seated with safety devices properly engaged prior to vehicle movement.
5. When backing any department vehicle, at least one spotter is required, two when personnel are available or conditions require. Spotters shall wear hearing protection devices when vehicles utilize loud backup warning devices. Vehicles will be backed at slow speeds. If eye contact is lost between the driver and spotter, the vehicle will be stopped immediately.
6. When parking a vehicle with the engine running, the operator will ensure the emergency brake is set and chocks are properly placed at the wheel. When parking the vehicle with the engine off, place the vehicle in gear for standard transmissions or in park for automatic transmission and set the emergency brake.

FIRE DEPARTMENT NAME

Generic SOP

Securing Tools and Equipment

PURPOSE: The purpose of this standard operating procedure is to establish responsibility and procedures for securing tools and equipment on department vehicles. This standard is intended to meet the requirements of the National Fire Protection Association (NFPA) 1500, *Standard on Fire Department Occupational Safety and Health Program.* This policy does not imply a comprehensive list of items that could occur.

RESPONSIBILITIES: All firefighters are responsible for compliance with the procedures as outlined in this operating instruction.

PROCEDURES: The following procedures shall be followed to ensure that tools and equipment are properly secured.

1. All hand tools and equipment will be secured in a compartment or appropriate location on each vehicle. All compartment doors will be completely closed and/or latched prior to any vehicle being moved.
2. A walk-around of the vehicle will be completed after each vehicle checkout and before returning from a response to ensure that all compartments are properly latched.
3. All exterior mounted equipment will be secured to the vehicles by safety straps or latches. Breathing apparatus and spare air bottles will be secured by safety straps.
4. All personnel will ensure that ground ladders are securely mounted in the brackets provided on the vehicles.
5. All vehicle operators will make a check of all safety straps during vehicle checkout and when returning from a response. If straps are not serviceable, immediately notify the (officer).

FIRE DEPARTMENT NAME
Generic SOP
Emergency Vehicles Inspections

PURPOSE: The purpose of this standard operating procedure is to establish responsibility and procedures for the inspection of vehicles assigned to the department for the safety and health of members and other persons involved in those activities. This standard is intended to meet the requirements of the National Fire Protection Association (NFPA) 1500, *Standard on Fire Department Occupational Safety and Health Program.* This policy does not imply a comprehensive list of items that could occur.

RESPONSIBILITIES: The (officer) and all firefighters assigned to the department are responsible for ensuring that the inspection of all fire equipment is performed on at least a weekly basis, within 24 hours after use, and when any major repair and/or modification is made.

PROCEDURES: The (officer) and the fire chief are required to ensure that vehicles and equipment of the department are in full operational readiness. This requirement necessitates constant vigilance to detail and prompt attention to the effectiveness of the inspection program.

1. Preventive maintenance inspections will be performed in accordance with the vehicle inspection form (checklist) for each department vehicle. Discrepancies found during the inspection shall be documented and reported immediately to the (officer). These forms shall be maintained as documentation for open discrepancies (items not corrected) as well as to document closed discrepancies (items corrected).
2. The (officer) will ensure that all vehicle inspection forms are properly maintained and that on the first of each month open entries from previous months are transcribed to the new monthly form. The (officer) will routinely review vehicle inspection forms to determine trends and preventive maintenance intervention, if necessary.
3. Only qualified individuals shall perform vehicle inspections and document inspection results. Safety will be maintained as a priority during the inspection process.

Appendix C

SAMPLES OF STATE LAWS FOR EMERGENCY VEHICLE DRIVERS

Vehicle Code

Division 1   **Words and Phrases Defined** (Section 100-680)

100.   Unless the provision or context otherwise requires, these definitions shall govern the construction of this code.

165.   An authorized emergency vehicle is:

(a) Any publicly owned and operated ambulance, lifeguard, or lifesaving equipment or any privately owned or operated ambulance licensed by the Commissioner of the California Highway Patrol to operate in response to emergency calls.

(b) Any publicly owned vehicle operated by the following persons, agencies, or organizations:

(1) Any federal, state, or local agency, department, or district employing peace officers as that term is defined in Chapter 4.5 (commencing with Section 830) of Part 2 of Title 3 of the Penal Code, for use by those officers in the performance of their duties.

(2) Any forestry or fire department of any public agency or fire department organized as provided in the Health and Safety Code.

(c) Any vehicle owned by the state, or any bridge and highway district, and equipped and used either for fighting fires, or towing or servicing other vehicles, caring for injured persons, or repairing damaged lighting or electrical equipment.

(d) Any state-owned vehicle used in responding to emergency fire, rescue or communications calls and operated either by the Office of Emergency Services or by any public agency or industrial fire department to which the Office of Emergency Services has assigned the vehicle.

(e) Any vehicle owned or operated by any department or agency of the United States government when the vehicle is used in responding to emergency fire, ambulance, or lifesaving calls or is actively engaged in law enforcement work.

(f) Any vehicle for which an authorized emergency vehicle permit has been issued by the Commissioner of the California Highway Patrol.

165.5.   No act or omission of any rescue team operating in conjunction with an authorized emergency vehicle as defined in Section 165, while attempting to resuscitate any person who is in immediate danger of loss of life, shall impose any liability upon the rescue team or the owners or operators of any authorized emergency vehicle, if good faith is exercised. For the purposes of this section, "rescue team" means a special group of physicians and surgeons, nurses, volunteers, or employees of the owners or operators of the authorized emergency vehicle who have been trained in cardiopulmonary resuscitation and have been designated by the owners or operators of the emergency vehicle to attempt to resuscitate persons who are in immediate danger of loss of life in cases of emergency.

This section shall not relieve the owners or operators of any other duty imposed upon them by law for the designation and training of members of a rescue team or for any provisions regarding maintenance of equipment to be used by the rescue team.

Members of a rescue team shall receive such training in a program approved by, or conforming to, standards prescribed by an emergency medical care committee established pursuant to Article 3 (commencing with Section 1797.270) of Chapter 4 of Division 2.5 of the Health and Safety Code, or a voluntary area health planning agency established pursuant to Section 437.7 of the Health and Safety Code.

676.5.   A "water tender vehicle" is a vehicle designed to carry not less than 1,500 gallons of water and used primarily for transporting and delivering water to be applied by other vehicles or pumping equipment at fire emergency scenes.

Division 11   **Rules of the Road** (Section 21050-21062)

Chapter 1   Obedience to and Effect of Traffic Laws

Article 2   Effects of Traffic Laws

21055.   The driver of an authorized emergency vehicle is exempt from:

Chapter 2 Traffic Signs, Signals, and Markings (commencing with Section 21350),
Chapter 3 Driving, Overtaking, and Passing (commencing with Section 21650),
Chapter 4 Right-Of-Way (commencing with Section 21800),
Chapter 5 Pedestrians' Rights and Duties (commencing with Section 21950),
Chapter 6 Turning and Stopping and Turning Signals (commencing with 22100),
Chapter 7 Speed Laws (commencing with Section 22348),
Chapter 8 Special Stops Required (commencing with Section 22450),
Chapter 9 Stopping, Standing, and Parking (commencing with Section 22500), and
Chapter 10 Removal of Parked and Abandoned Vehicles (commencing with Section 22650) of this division, and
Division 16.5 Off-Highway Vehicles, Chapter 5 Off-Highway Vehicle Operating Rules, Article 3 Speed Laws (commencing with Section 38305) and Article 4 Turning and Starting (commencing with Section 38312),

under all of the following conditions:

(a) If the vehicle is being driven in response to an emergency call or while engaged in rescue operations or is being used in the immediate pursuit of an actual or suspected violator of the law or is responding to, but not returning from, a fire alarm, except that fire department vehicles are exempt whether directly responding to an emergency call or operated from one place to another as rendered desirable or necessary by reason of an emergency call and operated to the scene of the emergency or operated from one fire station to another or to some other location by reason of the emergency call.

(b) If the driver of the vehicle sounds a siren as may be reasonably necessary and the vehicle displays a lighted red lamp visible from the front as a warning to other drivers and pedestrians.
A siren shall not be sounded by an authorized emergency vehicle except when required under this section.

21056.   Section 21055 does not relieve the driver of a vehicle from the duty to drive with due regard for the safety of all persons using the highway, nor protect him from the consequences of an arbitrary exercise of the privileges granted in that section.

MAINE REVISED STATUES ANNOTATED

Title 29-A.   Motor Vehicles

Chapter 19.   Operation, Subchapter I. Rules of the Road

§ 2054. Emergency and auxiliary lights; sirens: privileges

1. **Definitions.**   As used in this section, unless the context otherwise indicates, the following terms have the following meanings.
   A. "Ambulance" means any vehicle designed, constructed and routinely used or intended to be used for the transportation of ill or injured persons and licensed by Maine Emergency Medical Services pursuant to Title 32, chapter 2-B.
   B. "Authorized emergency vehicle" means any one of the following vehicles:
      (1)   An ambulance;
      (5)   A Department of Conservation vehicle used for forest fire control;
      (9)   An emergency medical service vehicle;
      (10)   A fire department vehicle;
      (11)   A hazardous material response vehicle;
      (16)   A vehicle operated by a municipal fire inspector, a municipal fire chief, an assistant or deputy chief or a town forest fire warden;
   D. "Emergency light" means a light, other than standard equipment lighting such as headlights, taillights, directional signals, brake lights, clearance lights, parking lights and license plate lights, that is displayed on a vehicle and used to increase the operator's visibility of the road or the visibility of the vehicle to other operators and pedestrians.
   F. "Fire vehicle" means any vehicle listed under paragraph B, subparagraph (5) or (16).
   G. "Hazardous material response vehicle" means a vehicle equipped for and used in response to reports of emergencies resulting from actual or potential releases, spills or leaks of, or other exposure to, hazardous substances that is authorized by a mutual aid agreement pursuant to Title 37-B, section 795, subsection 3 and approved by the local emergency planning committee or committees whose jurisdiction includes the area in which the vehicle operates.

2. **Authorized lights.**   Authorized lights are governed as follows.
   A. Only an ambulance, an emergency medical service vehicle, a fire department vehicle . . . may be equipped with a device that provides for alternate flashing of the vehicle's headlights.
   F. Only vehicles listed in this paragraph . . . may be equipped with, display or use a red auxiliary or emergency light.
      (1)   Emergency lights used on an ambulance, and emergency medical service vehicle, a fire department vehicle, a fire vehicle or a hazardous

material response vehicle must emit a red light or a combination of red and white lights.

(2)   The municipal officers or a municipal official designated by the municipal officers, with the approval of the fire chief, may authorize an active member of a municipal or volunteer fire department to use a flashing red signal light not more than 5 inches in diameter on a vehicle. The light may be displayed but may be used only while the member is en route to or at the scene of a fire or other emergency. The light must be mounted as near as practicable above the registration plate on the front of the vehicle or on the dashboard. A light mounted on the dashboard must be shielded so that the emitted light does not interfere with the operator's vision.

(3)   Members of an emergency medical service licensed by Maine Emergency Medical Services may display and use on a vehicle a flashing red signal light of the same proportion, in the same location and under the same conditions as those permitted municipal and volunteer firefighters, when authorized by the chief official of the emergency medical service.

3.   **Sirens.**   A bell or siren may not be installed or used on any vehicle, except an authorized emergency vehicle.

4.   **Right-of-way.**   An authorized emergency vehicle operated in response to, but not returning from, a call or fire alarm . . . has the right-of-way when emitting a visual signal using an emergency light and an audible signal using a bell or siren.

5.   **Exercise of privileges.**   The operator of an authorized emergency vehicle when responding to, but not upon returning from, an emergency call or fire alarm . . . may exercise the privileges set forth in this subsection. The operator of an authorized emergency vehicle may:

A.   Park or stand, notwithstanding the provisions of this chapter;

B.   Proceed past a red signal, stop signal or stop sign, but only after slowing down as necessary for safe operation;

C.   Exceed the maximum speed limits as long as life or property is not endangered, except that employees of the Department of Corrections may not exercise this privilege;

D.   Disregard regulations governing direction of movement or turning in specified directions; and

E.   Proceed with caution past a stopped school bus that has red lights flashing only:

(1)   After coming to a complete stop; and

(2)   When signaled by the school bus operator to proceed.

6.   **Emergency lights and audible signals.**   The operator of an authorized emergency vehicle who is exercising the privileges granted under subsection 5 shall use an emergency light authorized by subsection 2. The operator of an

authorized emergency vehicle who is exercising the privileges granted under subsection 5, paragraphs B, C, D, and E shall sound a bell or siren when reasonably necessary to warn pedestrians and other operators of the emergency vehicle's approach.

7. **Duty to drive with due regard for safety.**   Subsections 4, 5, and 6 do not relieve the operator of an authorized emergency vehicle from the duty to drive with due regard for the safety of all persons, nor do those subsections protect the operator from the consequences of the operator's reckless disregard for the safety of others.

MINNESOTA STATUTES

Transportation 160-174A

169   Traffic Regulations

169.01 **Definitions**

Subdivision 1.   Terms.

For the purposes of this chapter, the terms defined in this section shall have the meanings ascribed to them.

Subdivision 5. Authorized emergency vehicle.

"Authorized emergency vehicle" means any of the following vehicles when equipped and identified according to law:

(1) a vehicle of a fire department;

(2) a publicly owned police vehicle or a privately owned vehicle used by a police officer for police work under agreement, express or implied, with the local authority to which the officer is responsible;

(3) a vehicle of a licensed land emergency ambulance service, whether publicly or privately owned;

(4) an emergency vehicle of a municipal department or a public service corporation, approved by the commissioner of public safety or the chief of police of a municipality;

(5) any volunteer rescue squad operating pursuant to Laws 1959, chapter 53;

(6) a vehicle designated as an authorized emergency vehicle upon a finding by the commissioner of public safety that designation of that vehicle is necessary to the preservation of life or property or to the execution of emergency governmental functions.

169.03 **Emergency vehicles; exemptions; application.**

Subdivision 1.   Scope.

The provisions of this chapter applicable to the drivers of vehicles upon the highways shall apply to the drivers of all vehicles owned or operated by the United States, this state, or any county, city, town, district, or any other political subdivision of the state, subject to such specific exemptions as are set forth in this chapter with reference to authorized emergency vehicles.

Subdivision 2.   Stops.

The driver of any authorized emergency vehicle, when responding to an emergency call, upon approaching a red or stop signal or any stop sign shall slow down as necessary for safety, but may proceed cautiously past such red or stop sign or signal after sounding siren and displaying red lights.

Subdivision 3. One-way roadways.

The driver of any authorized emergency vehicle, when responding to any emergency call, may enter against the run of traffic on any one-way street, or highway where there is authorized division of traffic, to facilitate traveling to the area in which an emergency has been reported; and the provisions of this section shall not affect any cause of action arising prior to its passage.

Subdivision 4. Parking at emergency scene.

An authorized emergency vehicle, when at the scene of a reported emergency, may park or stand, notwithstanding any law or ordinance to the contrary.

Subdivision 5. Course of duty.

No driver of any authorized emergency vehicle shall assume any special privilege under this chapter except when such vehicle is operated in response to any emergency call or in the immediate pursuit of an actual or suspected violator of the law.

169.17 Emergency vehicles.

The speed limitations set forth in sections 169.14 to 169.17 do not apply to authorized emergency vehicles when responding to emergency calls, but the drivers thereof shall sound audible signal by siren and display at least one lighted red light to the front. This provision does not relieve the driver of an authorized emergency vehicle from the duty to drive with due regard for the safety of persons using the street, nor does it protect the driver of an authorized emergency vehicle from the consequence of a reckless disregard of the safety of others.

Appendix

D

SAMPLE PUMP MANUFACTURERS' TROUBLESHOOTING GUIDES

Waterous Company

TROUBLESHOOTING

CONDITION	POSSIBLE CAUSE	SUGGESTED REMEDY
Pump fails to prime or loses prime	Air leaks	Clean and tighten all intake connections. Make sure intake hoses and gaskets are in good condition. Use the following procedure to locate air leaks: 1. Connect intake hose to pump and attach intake cap to end of hose. 2. Close all pump openings. 3. Open priming valve and operate primer until vacuum gauge indicates 22 in. Hg. (If primer fails to draw specified vacuum, it may be defective, or leaks are too large for primer to handle.) 4. Close priming valve and shut off primer. If vacuum drops more than 10 in. Hg in 5 minutes, serious air leaks are indicated. With engine stopped, air leaks are frequently audible. If leaks cannot be heard, apply engine oil to suspected points and watch for break in film or oil being drawn into pump. Completely fill water tank (if so equipped). Connect intake hose to hydrant or auxiliary pump. Open one discharge valve and run in water until pump is completely filled and all air is expelled. Close discharge valve, apply pressure to system and watch for leaks or overflowing water tank. A pressure of 100 psi is sufficient. DO NOT EXCEED RECOMMENDED PRESSURE. If pump has not been operated for several weeks, packing may be dried out. Close discharge and drain valves and cap intake openings. Operate primer to build up a strong vacuum in pump. Run pump slowly and apply oil to impeller shaft near packing gland. Make sure packing is adjusted properly.
	Dirt on intake strainer	Remove all leaves, dirt and other foreign material from intake strainer. When drafting from shallow water source with mud, sand, or gravel bottom, protect intake strainer in one of the following ways: 1. Suspend intake strainer from a log or other floating object to keep it off the bottom. Anchor float to prevent it from drifting into shallow water. 2. Remove top from a clean barrel. Sink barrel so open end is below water surface. Place intake strainer inside barrel. 3. Make an intake box, using fine mesh screen. Suspend intake strainer inside box.
	No oil in priming tank	With rotary primer, oil is required to maintain a tight rotor seal. Check priming tank oil supply and replenish, if necessary.

Waterous Company

F–1031, Section 1000

Page 7 of 10

TROUBLESHOOTING

CONDITION	POSSIBLE CAUSE	SUGGESTED REMEDY
Pump fails to prime or loses prime (cont'd)	Defective priming valve	A worn or damaged priming valve may leak and cause pump to lose prime. Consult primer instructions for priming valve repair.
	Improper clearance in rotary gear or vane primer	After prolonged service, wear may increase primer clearance and reduce efficiency. Refer to primer instructions for adjusting primer clearance.
	Engine speed too low	Refer to instructions supplied with primer for correct priming speeds. Speeds much higher than those recommended do not accelerate priming, and may actually damage priming pump.
	Bypass line open	If a bypass line is installed between the pump discharge and water tank to prevent pump from overheating with all discharge valves closed, look for a check valve in the line. If valve is stuck open, clean it, replace it or temporarily block off line until a new valve can be obtained.
	Lift too high	Do not attempt lifts exceeding 22 feet except at low altitudes and with equipment in new condition.
	End of intake hose not submerged deep enough	Although intake hose might be immersed enough for priming, pumping large volumes of water may produce whirlpools, which will allow air to be drawn into intake hose. Whenever possible, place end of intake hose at least two feet below water source.
	High point in intake line	If possible, avoid placing any part of intake hose higher than pump inlet. If high point cannot be prevented, close discharge valve as soon as pressure drops, and prime again. This procedure will usually eliminate air pockets in intake line, but it may have to be repeated several times.
	Primer not operated long enough	Refer to instructions supplied with primer for required priming time. The maximum time for priming should not exceed 45 seconds for lifts up to 10 feet.
Insufficient capacity	Insufficient engine power	Engine requires maintenance. Check engine in accordance with manufacturer's instructions supplied with truck.
A. Engine and pump speed too low at full throttle		Engine operated at high altitudes and/or high air temperatures. Engine power decreases with an increase in altitude or air temperature, except for turbo charged engines. Adjusting carburetor or changing carburetor jets (or injector nozzles) may improve engine performance. Consult with engine manufacturer.
	Discharge relief valve set improperly	If relief valve is set to relieve below desired operating pressure, water will bypass and reduce capacity. Adjust relief valve in accordance with instructions supplied with valve.

Waterous Company

TROUBLESHOOTING

CONDITION	POSSIBLE CAUSE	SUGGESTED REMEDY
Insufficient capacity A. Engine and pump speed too low at full throttle (continued)	Transfer valve set improperly (Does not apply to single stage pumps.)	Place transfer valve in VOLUME (parallel) position when pumping more than two thirds rated capacity.
		When shifting transfer valve, make sure it travels all the way into new position. Failure of transfer valve to move completely into new position will seriously impair pump efficiency.
	Truck transmission in too high a gear	Consult vehicle instructions for correct pump gear. Pump usually works best with transmission in direct drive. If truck is equipped with an automatic transmission, be sure transmission is in pumping gear.
Insufficient capacity B. Engine and pump speed higher than specified for desired pressure and volume	Transfer valve set improperly (Does not apply to single stage pumps.)	Place transfer valve in VOLUME (parallel) position when pumping more than two-thirds rated capacity.
		When shifting transfer valve, make sure it travels all the way into new position. Failure of transfer valve to move completely into new position will seriously impair pump efficiency.
	Pump impeller(s) or wear rings badly worn	Install undersize wear rings if impeller to wear ring clearance is within limits indicated in MAINTENANCE INSTRUCTIONS. If not, install new impeller(s) and wear rings.
	Intake strainer, intake screens or impeller vanes fouled with debris	Remove intake strainer and hose, and clear away all debris. Pressure backwash (preferably in parallel or "volume" position) will usually clear impeller vanes when pump is stopped.
	Intake hose defective	On old intake hoses, the inner liner sometimes becomes so rough it causes enough friction loss to prevent pump from drawing capacity. Sometimes, the liner will separate from the outer wall and collapse when drafting. It is usually impossible to detect liner collapse, even with a light. Try drafting with a new intake hose; if pump then delivers capacity, it may be assumed that previous hose was defective.
	Intake hose too small	When pumping at higher than normal lifts, or at high altitudes, use a larger or additional intake hoses.

Waterous Company

TROUBLESHOOTING

CONDITION	POSSIBLE CAUSE	SUGGESTED REMEDY
Insufficient capacity C. Engine speed higher than specified for desired pressure and volume	Truck transmission in too low a gear	Consult vehicle instructions for correct pumping gear. Pump usually works best with transmission in direct drive. (Check both engine and pump speed, if possible, to be sure transmission is in "direct".)
Insufficient pressure	Pump speed too low	In general, the above causes and remedies for low pump capacity will also apply to low pump pressure. Check pump speed with a tachometer. If pump speed is too low, refer to engine manufacturer's instructions for method of adjusting engine speed governor.
Insufficient pressure (continued)	Pump capacity limits pump pressure	Do not attempt to pump greater volume of water at the desired pressure than the pump is designed to handle. Exceeding pump capacity may cause a reduction in pressure. Exceeding maximum recommended pump speed will produce cavitation, and will seriously impair pump efficiency.
	Flap valve stuck open	When pump is in PRESSURE (series), discharge will bypass to first stage intake. Operate pump at 75 psi pressure, and rapidly switch transfer valve back and forth between positions. If this fails, try to reach valve with a stick or wire and work it free.
Relief Valve Malfunction A. Pressure not relieved when discharge valves are closed	Sticky pilot valve	Disassemble and clean. Replace noticeably worn parts.
	Plugged tube lines	Disconnect lines and inspect.

Waterous Company

TROUBLESHOOTING

F–1031, Section 1000
Page 10 of 10

CONDITION	POSSIBLE CAUSE	SUGGESTED REMEDY
Relief Valve Malfunction	Sticky pilot valve	Disassemble and clean. Replace noticeably worn parts.
B. Pressure will not return to original setting after discharge valves are reopened	Sticky main valve	Disassemble and clean. Replace noticeably worn parts.
	Incorrect installation	Check all lines to be sure installation instructions have been followed.
Relief Valve Malfunction

C. Fluctuating pressure | Sticky pilot valve | Disassemble and clean. Replace noticeably worn parts. |
| | Water surges (relief valve) | Pressure fluctuation can result from a combination of intake and discharge conditions involving the pump, relief valve and engine. When the elasticity of the intake and discharge system and the response rate (reaction time) of the engine, pilot valve, and relief valve are such that the system never stabilizes, fluctuation results. With the proper combination of circumstances, fluctuation can occur regardless of the make or type of equipment involved. Changing one or more of these factors enough to disrupt this timing should eliminate fluctuation. |
| Relief Valve Malfunction

D. Slow response | Plugged filter or line | Clean lines and filter. |

4. CORRECTIVE MAINTENANCE

TROUBLE ANALYSIS

Table 4-1 lists the symptoms of some common problems and possible corrective measures. Before calling Hale for assistance, eliminate problem causes using Table 4-1. If you cannot correct a problem, before calling Hale Customer Service (215/825-6300) for assistance, please have the following information ready:

- Pump model and serial number

- Pump configuration information

- Observed symptoms and under what circumstances the symptoms occur.

Table 4-1. Hale Midship Pump Trouble Analysis

CONDITION	POSSIBLE CAUSE	SUGGESTED CORRECTION
PUMP WILL NOT ENGAGE.		
Standard Transmission with Manual Pump Shift	Clutch not fully disengaged or malfunction in shift linkage	Check clutch disengagement. Drive shaft must come to a complete stop before attempting pump shift.
Automatic Transmission with Manual Pump Shift	Automatic transmission not in neutral position	Repeat recommended shift procedures with transmission in neutral position
Standard Transmission with Power Shift System	Insufficient air or vacuum supply in shift system	Repeat recommended shift procedures. Check system for loss of vacuum or air supply. **Turn the engine off**, and employ shift override procedures as follows: • Hole is provided in shifting shaft to accomplish emergency shifting • Complete shift of control in cab to neutral, and proceed to complete shift of lower control manually.

DO NOT LEAVE THE CAB AFTER PUMP SHIFTING UNLESS THE SHIFT INDICATOR LIGHT IS ON, OR A SPEEDOMETER READING IS NOTED.

(continued)

Table 4-1. Hale Midship Pump Trouble Analysis *(continued)*

CONDITION	POSSIBLE CAUSE	SUGGESTED CORRECTION
PUMP WILL NOT ENGAGE (continued) Automatic Transmission with Power Shift System	Automatic transmission not in neutral position	Repeat recommended shift procedures with transmission in neutral position.
	Pump shift attempted before vehicle was completely stopped	Release braking system momentarily. Then reset and repeat recommended shifting procedures.
	Premature application of parking brake system (before truck comes to a complete stop)	Release braking system momentarily. Then reset and repeat recommended shifting procedures.
	Insufficient air or vacuum in shift system	Repeat recommended shift procedures. Check system for loss of air or vacuum. Check for leak in system. Employ manual override procedures if necessary. See Standard Transmission with Power Shift System.
	Air or vacuum leaks in shift system	Attempt to locate and repair leak(s). Leakage, if external, may be detected audibly. Leakage could be internal and not as easily detected.
PUMP LOSES PRIME OR IT WILL NOT PRIME NOTE: Weekly priming pump operation is recommended to promote good operation.	No lubricant in priming lubricant tank	Fill priming lubricant tank with Hale-approved lubricant
	Electric Priming System	No recommended engine speed is required to operate the electric primer; however, 1000 engine RPM will maintain truck electrical system while providing enough speed for initial pumping operation.
	Defective Priming System	Check priming system by performing "Dry Vacuum Test" per NFPA standards. If pump is tight, but primer pulls less than 22 inches of vacuum, it could indicate excessive wear in the primer.
	Defective Priming Valve (Electric)	Replace the sealing rings if defective. Lubricate the rings. Priming valve stuck open will allow loss of prime. Also, it will permit unnecessary running of electric priming motor. Ensure complete priming valve closure; dismantle and lubricate if necessary.
	Suction lifts too high	Do not attempt lifts exceeding 22 feet except at low elevations.
	Blocked suction strainer	Remove obstruction from suction hose strainer. Do not allow suction hose and strainer to rest on bottom of water supply.
	Suction connections	Clean and tighten all suction connections. Check suction hose and hose gaskets for possible defects.
	Primer not operated long enough	Proper priming procedures should be followed. Do not release the primer control before assuring a complete prime. Open the discharge valve slowly during completion of prime to ensure same. NOTICE: Do not run the primer over 45 seconds. If prime is not achieved in 45 seconds, stop and look for causes (for example, air leaks or blocked suction).

(continued)

Table 4-1. Hale Midship Pump Trouble Analysis *(continued)*

CONDITION	POSSIBLE CAUSE	SUGGESTED CORRECTION
PUMP LOSES PRIME OR IT WILL NOT PRIME (continued)	Air trap in suction line	Avoid placing any part of the suction hose higher than the suction intake. Suction hose should be laid with continuous decline to water supply. If trap in hose is unavoidable, repeated priming may be necessary to eliminate air pocket in suction hose.
	Pump pressure too low when nozzle is opened	Prime the pump again, and maintain higher pump pressure while opening discharge valve slowly.
	Air leaks	Attempt to locate and correct air leaks using the following procedure. 1. Perform dry vacuum test on pump per NFPA standards with 22 inches minimum vacuum required with loss not to exceed 10 inches of vacuum in 5 minutes. 2. If a minimum of 22 inches of vacuum cannot be achieved, the priming device or system may be defective, or the leak is too big for the primer to overcome (such as an open valve). The loss of vacuum indicates leakage and could prevent priming or cause loss of prime. 3. Attempt above dry prime and shut engine off. Audible detection of a leak is often possible. 4. Connect the suction hose from the hydrant or the discharge of another pumper to pressurize the pump with water, and look for visible leakage, and correct. A pressure of 100 PSI should be sufficient. Do not exceed pressure limitations of pump, pump accessories, or piping connections. 5. Check pump packing during attempt to locate leakage. If leakage is in excess of recommendations, adjust accordingly, following instructions in Section 3. 6. The suction side relief valves can leak. Plug the valve outlet connection, and re-test.
INSUFFICIENT PUMP CAPACITY.	Insufficient engine power	Engine power check or tune up may be required for peak engine and pump performance.
	Transfer Valve not in proper "Volume" position	Two-stage pumps only. Place transfer valve in "Volume" position (parallel) when pumping more than 2/3 rated capacity. For pressure above 200 PSI, pump should be placed in "Pressure" (series) position.
	Relief Valve improperly set	If relief valve control is set for too low a pressure, it will allow relief valve to open and bypass water. Reset Relief Valve control per the procedures in Section 3. Other bypass lines (such as foam system or inline valves) may reduce pump capacity or pressure.
	Engine Governor set incorrectly	Engine governor, if set for too low a pressure when on automatic, will decelerate engine speed before desired pressure is achieved. Reset the governor per the manufacturer's procedures.
	Truck transmission in wrong gear or clutch is slipping	Recheck the pumping procedure for the recommended transmission or gear range; see Section 3 for assistance. Use mechanical speed counter on the pump panel to check speed against possible clutch or transmission slipping or inaccurate tachometer. (Check the truck manual for the proper speed counter ratio.)
	Air leaks	See Air leaks under "PUMP LOSES PRIME OR IT WILL NOT PRIME."

(continued)

Table 4-1. Hale Midship Pump Trouble Analysis *(continued)*

CONDITION	POSSIBLE CAUSE	SUGGESTED CORRECTION
INSUFFICIENT PRESSURE	Check similar causes for insufficient capacity.	Recheck pumping procedures for recommended transmission gear or range. Use mechanical speed counter on pump panel to check actual speed against possible clutch or transmission slippage or inaccurate tachometer. (Check the truck manual for proper speed counter ratio.)
	Transfer Valve not in "Pressure" position	Two-stage pumps only. For desired pump pressure above 200 PSI, transfer valve should be in "Pressure" position.
ENGINE SPEEDS TOO HIGH FOR REQUIRED CAPACITY OR PRESSURE	Impeller blockage	Blockage in the impeller can prevent loss of both capacity and pressure. Back flushing of pump from discharge to suction may free blockage. Removal of one half of the pump body may be required; this is considered a major repair.
	Worn pump impeller(s) and clearance rings	Installation of new parts required.
	Blockage of suction hose entry	Clean suction hose strainer of obstruction, and follow recommended practices for laying suction hose – keep off the bottom of the water supply but at least 2 feet below the surface of the water.
	Defective suction hose	Inner line of suction hose may collapse when drafting and is usually undetectable. Try a different suction hose on same pump test mode for comparison against original hose and results.
	Lift too high, suction hose too small	Higher than normal lift (10 feet) will cause higher engine speeds, high vacuum, and rough operation. Larger suction hose will assist above condition.
	Truck transmission in wrong range or gear	Check recommended procedures for correct transmission selection; see Section 3 and truck manual.
RELIEF VALVE DOES NOT RELIEVE PRESSURE WHEN VALVES ARE CLOSED.	Incorrect setting of Control (Pilot) Valve	Check and repeat proper procedures for setting relief valve system; see Section 3.
	Relief valve inoperative	Possibly in need of lubrication. Remove relief valve from pump; dismantle; clean and lubricate. Weekly use of the Relief Valve is recommended.
RELIEF VALVE DOES NOT RECOVER AND RETURN TO ORIGINAL PRESSURE SETTING AFTER OPENING VALVES.	Dirt in system causing sticky or slow reaction	Relief Valve dirty or sticky. Follow instructions for disassembling, cleaning, and lubricating the Relief Valve.
		Blocked Relief Valve. Clean the valve with a small wire or straightened paper clip.
RELIEF VALVE OPENS WHEN CONTROL VALVE IS LOCKED OUT.	Drain hole in housing, piston, or sensing valve blocked.	Clean the hole with a small wire or straightened paper clip.
		Dismantle and clean the sensing valve.

(continued)

Table 4-1. Hale Midship Pump Trouble Analysis *(continued)*

CONDITION	POSSIBLE CAUSE	SUGGESTED CORRECTION
UNABLE TO ATTAIN PROPER SETTING ON RELIEF VALVE.	Wrong procedures	Check instructions for setting the Relief Valve; reset the valve.
	Blocked strainer	Check and clean the strainer in the supply line from the pump discharge to the control valve. Check the truck manual for the exact location. Check and clean tubing lines related to the Relief Valve and Control Valve.
	Foreign matter in the Control Valve	Remove the Control Valve, and clean it.
	Hunting condition	Insufficient water supply coming from the pump to the Control Valve. Check the strainer in the Relief Valve system.

Foreign matter in the Control Valve. Remove the Control Valve, and clean it. |
| LEAK AT PUMP PACKING | Packing out of adjustment or worn | Adjust the pump packing per the procedure in Section 3 of this manual (8 to 10 drops per minute leakage preferred).

Replace pump packing per Section 3 of this manual. Packing replacement is recommended every 2 or 3 years, depending on usage. |
| WATER IN PUMP GEARBOX | Leak coming from above pump | Check all piping connections and tank overflow for possible spillage falling directly on the pump gearbox.

Follow the procedures in Section 3 of this manual for adjustment or replacement of packing. Excess packing leakage permits the flushing of water over the gearbox casing to the input shaft area. Induction of this excessive water may occur through the oil seal or speedometer connection.

Inspect the oil seal; replace it if necessary. |
| DISCHARGE VALVES DIFFICULT TO OPERATE | Lack of lubrication | Recommended weekly lubrication of discharge and suction valve. Use a good grade of petroleum-base or silicon grease. |
| | Valve in need of more clearance | Add gasket to the valve cover (per the truck manual). Multigasket design allows additional gaskets for more clearance and free operation.

NOTE: Addition of too many gaskets to the valve will permit leakage. |
| REMOTE CONTROL DIFFICULT TO OPERATE | Lack of lubrication | Lubricate the remote control linkages and collar with oil. |

Appendix

E

SAMPLE VEHICLE INSPECTION FORMS

NASHVILLE FIRE DEPARTMENT
DAILY APPARATUS REPORT

Caution: Any Engine noise, overheating, malfunction of brakes or steering, check out of service and notify shop immediately. **ON 24 HOUR CALL.**

09 _____ Company _____ Make _____ Month _____

Beginning of Month Mileage: _____   Beginning Engine Hours: _____

Date Last Oil Change _____   Mileage Last Oil Change _____

DAILY CHECKLIST

TO BE CHECKED DAILY _____	Power Tools _____
Oil _____	Air Pressure _____
Coolant Level _____	Air - Horn - Siren _____
Fan and Alternator Belts or All Belts _____	Gauges - Dash Lights _____
Fuel _____	Pump Panel Gauges and Light _____
Transmission Fluid _____	Valves - Handles _____
Power Steering Fluid _____	Leaks - Oil - Fuel _____
Batteries - Clamps _____	Unusual Noises _____
Tires - Lug Nuts _____	Generator _____
Wipers - Blades _____	Hydraulic _____
Windshield - Windows _____	Ladder Rungs _____
Doors - Handles _____	Cables _____
Headlights - Turn Signals _____	Intercom _____
Warning Lights - Spots _____	Fly Nozzle _____
Seat Belts _____	Outriggers _____

Engineers/Relief Drivers responsible for checking the above items. Sign date Below.

CHECKED BY

Day	Driver	Day	Driver	Day	Driver	Day	Driver
1		9		17		25	
2		10		18		26	
3		11		19		27	
4		12		20		28	
5		13		21		29	
6		14		22		30	
7		15		23		31	
8		16		24			

MONTHLY REPORT

Actual Fire Runs	_____	**HOSE LAID OUT**		
1st Responder Emt Runs	_____	5" Hose Used	_____	Feet
Special Duty Runs	_____	3" Hose Used	_____	Feet
Miscellaneous Runs	_____	2 ½" Hose Used	_____	Feet
TOTAL ALL RUNS	_____	1 ¾" Hose Used	_____	Feet

TOTAL PUMP TIME _____   **TOTAL HOURS** _____
TOTAL MILES _____

COMPANY OFFICER SIGNATURE

_____   _____   _____
A   B   C

FUEL REPORT

Date	Operator	Fuel	Oil		Date	Operator	Fuel	Oil
TOTALS					TOTALS			

TOTAL FUEL   **TOTAL OIL**

APPARATUS SERVICE REPORT

Report Date:	Person Notified	Comments	Repairs Completed By:	Date:

Equipment Lost or In Need of Repair

Date	Item	Status Replace/Fixed	Reported By	Received By

• Fill out information at first of month at top front.   • Report stays with apparatus at all times
• End of month, complete and send to shop

Engineers/Relief Drivers - It is your responsibility to see that this report is
filled out properly and completely.

SOUTH PORTLAND FIRE DEPARTMENT
VEHICLE INSPECTION REPORT

VEHICLE ID #:_____   WEEK ENDING:____/___/____

CHECK MARK = NO PROBLEM   X = PROBLEM   O = REPAIRED

DAILY VEHICLE INSPECTIONS

	SUN.	MON.	TUES.	WEDS.	THURS.	FRI.	SAT.
OPERATOR NAME: 1		1	1	1	1	1	1
FUEL LOG -- # OF GALS: 2		2	2	2	2	2	2
OIL LEVEL: 3		3	3	3	3	3	3
TRANSMISSION FLUID: 4		4	4	4	4	4	4
FLUID LEAKS: 5		5	5	5	5	5	5
COOLANT / HOSES: 6		6	6	6	6	6	6
BATTERIES: 7		7	7	7	7	7	7
ELECTRICAL SYSTEM: 8		8	8	8	8	8	8
LIGHTS: 9		9	9	9	9	9	9
SIREN/HORNS: 10		10	10	10	10	10	10
RADIOS REC/TRANS: 11		11	11	11	11	11	11
AIR SYSTEM: 12		12	12	12	12	12	12
BRAKES: 13		13	13	13	13	13	13
EMERG/MAXI BRAKE: 14		14	14	14	14	14	14
WHEELS/TIRES: 15		15	15	15	15	15	15
GLASS/WIPERS: 16		16	16	16	16	16	16
WATER TANK: 17		17	17	17	17	17	17
PORTABLE EQUIPMENT: 18		18	18	18	18	18	18
SCBA: 19		19	19	19	19	19	19
MIRRORS: 20		20	20	20	20	20	20
BODY DAMAGE: 21		21	21	21	21	21	21

SATURDAY WEEKLY INSPECTION

ENDING MILAGE:_____   ENDING HOURS - ENGINE:_____   ENDING HOURS - AERIAL:_____

UNDER CARRIAGE INSPECTION

SPRINGS _____   AIRLINES _____   FUEL LINES _____

SHOCKS _____   EXHAUST SYSTEM_____   OIL / FLUID LEAKS _____

PUMP CHECK   **OPEN & CLOSE ALL GATES AND INTAKE VALVES**

PUMP SHIFT _____   PRIMING PUMP _____   CHANGE OVER VALVE _____   PRESSURE RELIEF _____   GUAGES _____

THROTTLE CONTROL _____   PRIMING PUMP TANK _____   TANK TO PUMP _____   TANK FILL _____

MAIN PUMP DRAIN _____   LEAKING VALVES _____   BOOSTER REEL _____   PUMP PACKING LEAKING   Y   N

TANK LEVEL INDICATOR - WATER _____ FOAM - _____   FOAM PUMP _____   FOAM SYSTEM_____

AERIAL DEVICE INSPECTION   **PERFORM ALL OPERATIONS**

HYDRAULIC LEVEL _____   CONTROLS _____   HYDRAULIC PRESSURE _____   PTO _____   LADDER GLIDES _____

HYDRAULIC LINES _____   CABLES _____   OUTRIGGERS / JACKS _____   HYDRAULIC LEAKS _____

PLEASE DATE AND INITIAL ALL COMMENTS TO ALLOW FOR FOLLOW-UP ------- PLEASE USE REVERSE SIDE FOR COMMENTS

KNOXVILLE FIRE DEPARTMENT

DAILY AND WEEKLY VEHICLE CHECKLIST

FIRE COMPANY _____   APPARATUS _____   DATE _____

ANY ENGINE NOISE, OVERHEATING, MALFUNCTION OF BRAKES OR STEERING, CHECK OUT
OF SERVICE IMMEDIATELY. NOTIFY THE SHOP.

Apparatus weekly check to be done each Monday.
Any minor repairs, call the shop and get on work list.
Any major repairs, call the shop immediately.

Marking code: - OK   Repairs needed - 0   Adjustment made - X

ENGINE.

	Sun.	Mon.	Tue.	Wed.	Thu.	Fri.	Sat.
1. Fuel level							
2. Oil level							
3. Radiator water level							
4. Unusual noises							
5. Engine (clean)							
6. Leaks (water, fuel, oil)							

ELECTRICAL SYSTEM:

	Sun.	Mon.	Tue.	Wed.	Thu.	Fri.	Sat.
1. Lights (all)							
2. Gauges (all)							
3. Windshield wipers							
4. Switches (all)							
5. Battery and cables (clean)							
6. Battery water level							

BATTERY CHECK: NO. 1

	Specific Gravity	Charge Time	Amps.
Sun.	____ ____ ____ ____ ____	_____	_____
Mon.	____ ____ ____ ____ ____	_____	_____
Tue.	____ ____ ____ ____ ____	_____	_____
Wed.	____ ____ ____ ____ ____	_____	_____
Thu.	____ ____ ____ ____ ____	_____	_____
Fri.	____ ____ ____ ____ ____	_____	_____
Sat.	____ ____ ____ ____ ____	_____	_____

BATTERY CHECK: NO. 2

	Specific Gravity	Charge Time	Amps.
Sun.	____ ____ ____ ____ ____	_____	_____
Mon.	____ ____ ____ ____ ____	_____	_____
Tue.	____ ____ ____ ____ ____	_____	_____
Wed.	____ ____ ____ ____ ____	_____	_____
Thu.	____ ____ ____ ____ ____	_____	_____
Fri.	____ ____ ____ ____ ____	_____	_____
Sat.	____ ____ ____ ____ ____	_____	_____

TIRES:

	Sun.	Mon.	Tue.	Wed.	Thu.	Fri.	Sat.
Visual conditions (cuts, tread irregularity, pressure, and lugs)							

BRAKES:

	Sun.	Mon.	Tue.	Wed.	Thu.	Fri.	Sat.
1. Air Brakes (pressure drop 15 lbs. per application, report to the shop)							
2. Air Brakes (effectiveness)							
3. Air Brakes (drain water from cylinder)							
4. Hydraulic Brakes (pedal travel)							
5. Hydraulic Brakes (fluid level)							
6. Parking Brake							

	Sun.	Mon.	Tue.	Wed.	Thu.	Fri.	Sat.

AERIAL EQUIPMENT:

1. Auxiliary generator (oil, fuel, operation) --------------------------

2. Cables and ladders and towers -------

3. Hydraulic hoses (leaks and condition) --------------------------

4. Bolts and nuts (loose or missing) ---

5. Physical defects --------------------

DRIVE TRAIN:
1. Clutch action and free travel 1" to 2½" --------------------------

2. Gear shift or noise in transmission-

3. Differential noise-------------------

MISCELLANEOUS:

1. Exterior mounted equipment and brackets---------------------------

2. Interior equipment and brackets-----

3. Check water level in booster tank---

4. Clean all tools and equipment-------

5. Check body, door latches, compartments----------------------

6. Clean apparatus--------------------

7. Operate nozzles--------------------

8. Check breathing equipment operation --------------------------

9. Check radio operation --------------

NOTE: When any item is marked O explain under remarks.

REMARKS: Condition - Date - Who notified - By Whom

Current speedometer reading _____ Tachometer _____

Speedometer reading when lubricated _____ Tachometer _____

OFFICER IN CHARGE:　　　　　　　　　　DRIVER IN CHARGE:

A Shift_____　　　_____

B Shift_____　　　_____

C Shift_____　　　_____

FD85-447(1000)

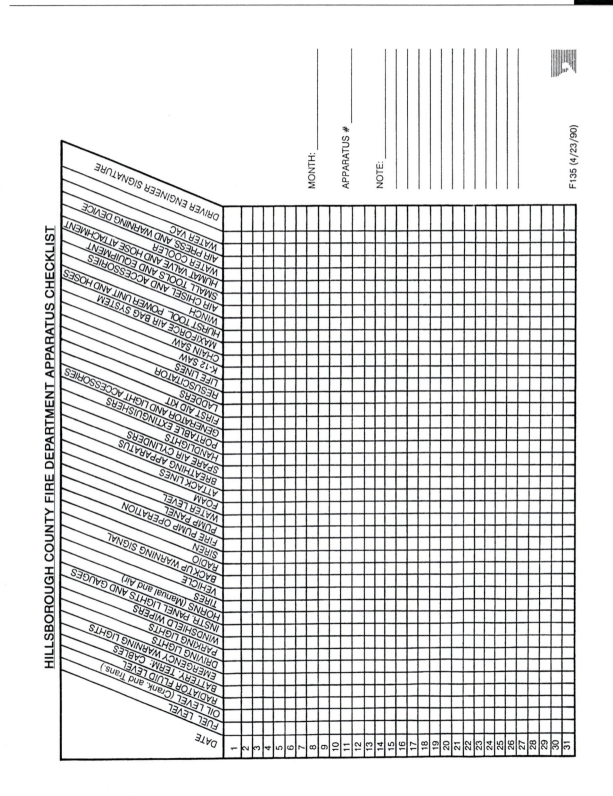

HILLSBOROUGH COUNTY FIRE DEPARTMENT APPARATUS CHECKLIST

MONTH:

APPARATUS #

NOTE:

F135 (4/23/90)

CHARTS FOR FRICTION LOSS, NOZZLE REACTION, AND NOZZLE DISCHARGE

Friction Loss in Rubber-Lined Fire Hose

Loss of Pressure in psi per 100' of Fire Hose

Flow in gpm	Size of Hose 1½"	1¾"	2½"	Flow in gpm	Size of Hose 2½"	*3"	4"	5"	6"
20	1	1		260	14	5	1		
30	2	1		270	15	6	1	1	
40	4	2		280	16	6	2	1	
50	6	4	1	290	17	7	2	1	
60	9	6	1	300	18	7	2	1	
70	12	8	1	310	19	8	2	1	
80	15	10	1	320	20	8	2	1	1
90	19	13	2	330	22	9	2	1	1
100	24	16	2	340	23	9	2	1	1
110	29	19	2	350	25	10	2	1	1
120	35	22	3	360	26	10	3	1	1
130	41	26	3	370	27	11	3	1	1
140	47	30	4	380	29	12	3	1	1
150	54	35	5	390	30	12	3	1	1
160	61	40	5	400	32	13	3	1	1
170	69	45	6	425	36	14	4	1	1
180	78	50	6	450	41	16	4	2	1
190	87	56	7	475	45	18	5	2	1
200	96	62	8	500	50	20	5	2	1
210		68	9	525	55	22	6	2	1
220		75	10	550	61	24	6	2	2
230		82	11	575	66	26	7	3	2
240		89	12	600	72	29	7	3	2
250		97	13	650	85	34	8	3	2

*3 inch with 2½-inch couplings

$FL = c \times g^2$
    where
      FL = friction loss per 100' of hose
      c = coefficient
      g = gallons per minute divided by 100

Note: Actual friction loss will vary depending on type of manufacturer and age/condition of hose.

Nozzle Reaction

Combination Nozzles				Smooth-Bore Nozzles		
gpm	75 psi	100 psi		Nozzle Diameter	50 psi	80 psi
20	9	10		1/8″	1	2
40	17	20		1/4″	5	8
60	26	30		3/8″	11	18
80	35	40		7/16″	15	24
95	42	48		1/2″	20	31
100	44	51		5/8″	31	49
120	52	61		3/4″	44	71
125	55	63		7/8″	60	96
140	61	71		15/16″	69	110
160	70	81		1″	79	126
180	79	91		1⅛″	99	159
200	87	101		1¼″	123	196
220	96	111		1⅜″	148	237
240	105	121		1½″	177	283
250	109	126		1⅝″	207	332
260	114	131		1¾″	240	385
280	122	141		1⅞″	276	442
300	131	152		2″	314	502

$NR = 0.0505 \times g \times \sqrt{NP}$
where
NR = nozzle reaction
0.0505 = constant
g = gallons per minute (gpm)
NP = nozzle pressure in psi

$NR = 1.57 \times d^2 \times NP$
where
NR = nozzle reaction
1.57 = constant
d = nozzle diameter in inches
NP = nozzle pressure in psi

Smooth-Bore Discharge Table

Discharge given in gallons per minute (gpm)

Nozzle Pressure in psi	1/8"	1/4"	3/8"	7/16"	1/2"	5/8"	3/4"	7/8"	15/16"	1"	1⅛"	1¼"	1⅜"	1½"	1⅝"	1¾"	1⅞"	2"
20	2	8	19	25	33	52	75	102	117	133	168	208	251	299	351	407	467	531
22	2	9	20	27	35	54	78	107	122	139	176	218	263	313	368	427	490	557
24	2	9	20	28	36	57	82	111	128	145	184	227	275	327	384	446	512	582
26	2	9	21	29	38	59	85	116	133	151	192	237	286	341	400	464	532	606
28	2	10	22	30	39	61	88	120	138	157	199	246	297	354	415	481	553	629
30	3	10	23	31	41	64	92	125	143	163	206	254	308	366	430	498	572	651
32	3	11	24	32	42	66	95	129	148	168	213	263	318	378	444	515	591	672
34	3	11	24	33	43	68	97	133	152	173	219	271	327	390	457	530	609	693
36	3	11	25	34	45	70	100	136	157	178	226	278	337	401	471	546	626	713
38	3	11	26	35	46	72	103	140	161	183	232	286	346	412	483	561	644	732
40	3	12	26	36	47	73	106	144	165	188	238	293	355	423	496	575	660	751
42	3	12	27	37	48	75	108	147	169	192	244	301	364	433	508	589	677	770
44	3	12	28	38	49	77	111	151	173	197	249	308	372	443	520	603	693	788
46	3	13	28	39	50	79	113	154	177	201	255	315	381	453	532	617	708	806
48	3	13	29	39	51	80	116	158	181	206	260	322	389	463	543	630	723	823
50	3	13	30	40	53	82	118	161	185	210	266	328	397	473	555	643	738	840
52	3	13	30	41	54	84	120	164	188	214	271	335	405	482	566	656	753	857
54	3	14	31	42	55	85	123	167	192	218	276	341	413	491	576	668	767	873
56	3	14	31	43	56	87	125	170	195	222	281	347	420	500	587	681	781	889
58	4	14	32	43	57	88	127	173	199	226	286	353	428	509	597	693	795	905

Smooth-Bore Discharge Table (Continued)

Nozzle Pressure in psi	Nozzle Diameter																	
	1/8"	1/4"	3/8"	7/16"	1/2"	5/8"	3/4"	7/8"	15/16"	1"	1⅛"	1¼"	1⅜"	1½"	1⅝"	1¾"	1⅞"	2"
60	4	14	32	44	58	90	129	176	202	230	291	359	435	518	607	705	809	920
62	4	15	33	45	58	91	132	179	206	234	296	365	442	526	618	716	822	935
64	4	15	33	45	59	93	134	182	209	238	301	371	449	535	627	728	835	950
68	4	15	34	47	61	96	138	188	215	245	310	383	463	551	647	750	861	980
70	4	16	35	48	62	97	140	190	218	248	314	388	470	559	656	761	874	994
72	4	16	35	48	63	98	142	193	221	252	319	394	476	567	665	772	886	1008
74	4	16	36	49	64	100	144	196	225	255	323	399	483	575	675	782	898	1022
76	4	16	36	50	65	101	146	198	228	259	328	405	490	583	684	793	910	1036
78	4	16	37	50	66	102	148	201	231	262	332	410	496	590	693	803	922	1049
80	4	17	37	51	66	104	149	203	233	266	336	415	502	598	701	814	934	1063
82	4	17	38	51	67	105	151	206	236	269	340	420	508	605	710	824	946	1076
84	4	17	38	52	68	106	153	208	239	272	345	425	515	612	719	834	957	1089
86	4	17	39	53	69	108	155	211	242	275	349	430	521	620	727	843	968	1102
88	4	17	39	53	70	109	157	213	245	279	353	435	527	627	736	853	979	1114
90	4	18	40	54	70	110	158	216	248	282	357	440	533	634	744	863	991	1127
92	4	18	40	55	71	111	160	218	250	285	361	445	539	641	752	872	1002	1139
94	4	18	40	55	72	112	162	220	253	288	364	450	544	648	760	882	1012	1152
96	5	18	41	56	73	114	164	223	256	291	368	455	550	655	768	891	1023	1164
98	5	18	41	56	74	115	165	225	258	294	372	459	556	662	776	900	1034	1176
100	5	19	42	57	74	116	167	227	261	297	376	464	562	668	784	910	1044	1188

$$gpm = 29.7 \times d^2 \times \sqrt{NP}$$

where

gpm = gallons per minute
29.7 = constant
d = nozzle diameter in inches
NP = nozzle pressure in psi

Appendix

G

HYDRAULICS CALCULATION
INFORMATION SHEET

Engine Pressure Formula

$PDP = NP + FL + AFL +/- EL$
where
PDP = pump discharge pressure (psi)
NP = nozzle pressure (psi)
FL = friction loss for hose (psi)
AFL = appliance friction loss (psi)
EL = elevation gain/loss (psi)

(NP) Nozzle Pressure

Type of Nozzle	Operating Pressure
Smooth-bore, handline	50 psi
Combination (fog), low pressure	75 psi
Smooth-bore, master stream	80 psi
Combination (fog), automatic	100 psi

(FL) Friction Loss Formula

$FL = c \times g^2 \times L$
where
FL = friction loss
c = coefficient for Specific Hose
g = flow rate in hundreds of gpm (g/100)
L = length of hose in hundreds of feet
    (L/100)

(c) Coefficient Information

$1\frac{1}{2}$"	24
$1\frac{3}{4}$"	15.5
2"	8
$2\frac{1}{2}$"	2
*3"	.8
4"	.2
5"	.08
6"	.05

*with $2\frac{1}{2}$" couplings

(AFL) Appliance Friction Loss

$2\frac{1}{2}$" to $2\frac{1}{2}$" wye	5 psi
$2\frac{1}{2}$" to $1\frac{1}{2}$" wye	10 psi
$1\frac{1}{2}$" to $1\frac{1}{2}$" wye	15 psi
$2\frac{1}{2}$" to $2\frac{1}{2}$" siamese	10 psi
$1\frac{1}{2}$" to $1\frac{1}{2}$" siamese	15 psi
reducer	5 psi
increaser	5 psi
monitor	15 psi
four-way valve	15 psi
standpipe	25 psi

(EL) Elevation

Floor Method
    add/subtract 5 psi for
    each floor level above/
    below the first
Elevation Method
    .5 psi per foot

Miscellaneous Information

one gallon of water = 8.35 lbs
(8.5)
one cubic foot of water = 7.48
gallons
density of water = 62.4 lb/ft³
1 psi will raise water 2.31 feet

Nozzle Reaction Formula

Smooth-bore: $NR = 1.57 \times d^2 \times NP$
Combination: $NR = .0505 \times g \times \sqrt{NP}$
        or *$NR - $ gpm $\times$ .505
NR = nozzle reaction (pounds)
d = nozzle diameter (inches)
g = flow (gpm)
NP = nozzle pressure (psi)
1.57 and .0505 = constants
*condensed formula for combination
 nozzles operated at 100 psi NP

Rated Capacity Pumps

100% @ 150 psi
70% @ 200 psi
50% @ 250 psi

Estimating Remaining Hydrant Flow

0–10% drop	3 more like lines
11–15% drop	2 more like lines
16–25% drop	1 more like line

% drop = $(SP - RP)/SP$
where
SP = static pressure (psi)
RP = residual pressure (psi)

Flow, Smooth-Bore Nozzles

gpm = $29.7 \times d^2 \times \sqrt{NP}$
where
gpm = gallons per minute
29.7 = constant
d = nozzle diameter (inches)
NP = nozzle pressure (psi)

Distance between Pumpers

$(PDP - 20) \times 100/FL$
where
PDP = pump discharge pressure
20 = residual pressure (psi) @ next pump
100 = 100'-section of hose
FL = friction loss (psi) per 100' of hose

Appendix

H

CONVERSION CHARTS

METRIC CONVERSIONS

Volume

To Convert	Into	Multiply by
Ounces (oz)	Milliliters (ml)	29.57000
Pints (pt)	Milliliters (ml)	473.20000
Pints (pt)	Liters (l)	0.47320
Quarts (qt)	Milliliters (ml)	946.40000
Quarts (qt)	Liters (l)	0.94640
Gallons (gal)	Liters (l)	3.78500
Milliliters (ml)	Ounces (oz)	0.03380
Liters (l)	Pints (pt)	2.11300
Liters (l)	Quarts (qt)	1.05700
Liters (l)	Gallons (gal)	0.26420

Mass

To Convert	Into	Multiply by
Ounces (oz)	Grams (g)	28.35000
Pounds (lb)	Grams (g)	453.60000
Pounds (lb)	Kilograms (kg)	0.45360
Tons (t)	Kilograms (kg)	1016.00000
Grams (g)	Ounces (oz)	0.03527
Grams (g)	Pounds (lb)	0.00221
Kilograms (kg)	Ounces (oz)	35.27000
Kilograms (kg)	Pounds (lb)	2.20500

Length

To Convert	Into	Multiply by
Inches (in.)	Millimeters (mm)	25.40000
Inches (in.)	Centimeters (cm)	2.54000
Feet (ft)	Centimeters (cm)	30.48000
Feet (ft)	Millimeters (mm)	3.04800
Yards (yd)	Meters (m)	0.91400
Miles (mi)	Kilometers (km)	1.60900
Millimeters (mm)	Inches (in.)	0.03900
Centimeters (cm)	Inches (in.)	0.39400
Meters (m)	Feet (ft)	3.28100
Meters (m)	Yards (yd)	1.09400
Kilometers (km)	Miles (mi)	0.62140

Area

To Convert	Into	Multiply by
Square inches (in.^{2})	Square centimeters (cm^2)	6.45200
Square inches (in.^{2})	Square millimeters (mm^2)	645.20000
Square feet (ft^2)	Square millimeters (mm^2)	92900.00000
Square feet (ft^2)	Square meters (m^2)	0.09300
Square yards (yd^2)	Square meters (m^2)	0.83600
Square miles (mi^2)	Square kilometers (km^2)	2.59000
Square centimeters (cm^2)	Square inches (in.^{2})	0.15500
Square millimeters (mm^2)	Square inches (in.^{2})	0.00155
Square meters (m^2)	Square feet (ft^2)	10.76000
Square meters (m^2)	Square yards (yd^2)	1.19600
Square kilometers (km^2)	Square miles (mi^2)	0.38600

Pressure

To Convert	Into	Multiply by
Pounds per square inch (psi)	Kilopascals (kPa)	6.89500
Pounds per square inch (psi)	Atmosphere	0.06895
Kilopascals (kPa)	Pounds per square inch (psi)	0.14500
Kilopascals (kPa)	Atmosphere	0.01000
Atmosphere	Pounds per square inch (psi)	14.70000
Atmosphere	Kilopascals (kPa)	101.30000

Flow

To Convert	Into	Multiply by
Gallons per minute (gpm)	Liters per second (lps)	0.06308
Gallons per minute (gpm)	Liters per minute (lpm)	3.78500
Liters per second (lps)	Gallons per minute (gpm)	15.85000

Hose Sizes (metric equivalent)

Inches	Millimeters
1	25.4
1 1/2	38.1
1 3/4	44.5
2	50.8
2 1/2	63.5
3	76.2
3 1/2	88.9
4	101.6
5	127
6	152.4

Pressures (metric equivalent)

psi	Kilopascals (kPa)
20	138
40	276
60	414
80	552
100	690
120	827
140	965
160	1103
180	1241
200	1379
220	1517
240	1655
260	1793
280	1931
300	2069

Tip Sizes (metric equivalent)

Inches	Millimeters
3/4	19.1
7/8	22.2
1	25.4
1 1/8	28.6
1 1/4	31.8
1 3/8	34.9
1 1/2	38.1
1 3/4	44.5
2	50.8
2 1/4	57.2
2 1/2	63.5
3	76.2

Flow Rate (metric equivalent)

Gallons per Minute (gpm)	Liters per Minute (lpm)
20	76
40	151
60	227
80	303
100	379
120	454
140	530
160	606
180	681
200	757
220	833
240	908
260	984
280	1060
300	1136

ACRONYMS

AFL	appliance friction loss		NP	nozzle pressure
API	American Petroleum Institute		NR	nozzle reaction
ARFF	airport rescue and firefighting		NST	national standard thread
CAFS	compressed air foam system		OS&Y	outside screw and yoke
DOT	Department of Transportation		PDP	pump discharge pressure
DPO	driver pump operator		PIV	post-indicating valve
FL	friction loss		PM	preventive maintenance
gpm	gallons per minute		PPE	personal protective equipment
ISO	Insurance Service Office		psi	pounds per square inch
jpr	job performance requirements		psia	pounds per square inch absolute
LDH	large diameter hose		psig	pounds per square inch gauge
MDH	medium diameter hose		PTO	power take off
MSDS	material safety data sheets		rpm	revolutions per minute
MST	midship transfer		SAE	Society of Automotive Engineers
mph	miles per hour		SCBA	self-contained breathing apparatus
NFA	National Fire Academy		SDH	small diameter hose
NFPA	National Fire Protection Association		SOP	standard operating procedure
NH	American National Fire Hose Connection Screw Threads		UL	Underwriters' Laboratory

GLOSSARY

Absolute pressure Measurement of pressure that includes atmospheric pressure, typically expressed as psia.

Acceptance test Test conducted at time of delivery to verify and document stated performance levels of the apparatus, pump, and related components.

Adapter Appliance used to connect mismatched couplings.

Annual pump service test Test conducted on an annual basis to ensure the pump and related components maintain appropriate performance levels.

Appliance friction loss The reduction in pressure resulting from increased turbulence caused by the appliance.

Appliances Accessories and components used to support varying hose configurations.

Atmospheric pressure The pressure exerted by the atmosphere (body of air) on the Earth.

Attack hose 1½-inch to 3-inch hose used to combat fires beyond the incipient stage.

Authorized emergency vehicles Legal terminology for vehicles used for emergency response, such as fire department apparatus, ambulances, rescue vehicles, and police vehicles equipped with appropriate identification and warning devices.

Auxiliary cooling system A system used to maintain the engine temperature within operating limits during pumping operations.

Auxiliary pump Pumps other than the main pump or priming pump that are either permanently mounted on or carried on an apparatus.

Available flow The amount of water that can be moved from the supply to the fire scene.

Boiling point The temperature at which the vapor pressure of a liquid equals the surrounding pressure.

Bourdon tube gauge The most common pressure gauge found on an apparatus, consisting of a small curved tube linked to an indicating needle.

Braking distance The distance of travel from the time the brake is depressed until the vehicle comes to a complete stop.

Cavitation The process that explains the formation and collapse of vapor pockets when certain conditions exist during pumping operations.

Centrifugal force Tendency of a body to move away from the center when rotating in a circular motion.

Certification test Pump test certified by an independent testing organization, typically Underwriters' Laboratory (UL).

Closed relay Relay operation in which water is contained within the hose and pump from the time it enters the relay until it leaves the relay at the discharge point; excessive pressure and flow is controlled at each pump within the system.

Combination nozzle A nozzle designed to provide both a straight stream and a wide fog pattern; most widely type used in the fire service

Compound gauge A pressure gauge that reads both positive pressure (psi) above atmospheric pressure and negative pressure (in. Hg).

Control valves Devices used by a pump operator to open, close, and direct water flow.

Density The weight of a substance expressed in units of mass per volume.

Discharge The point at which water leaves the pump.

Discharge flow   The amount of water flowing from the discharge side of a pump through the hose, appliances, and nozzles to the scene.

Discharge maintenance   Process of ensuring that pressures and flows on the discharge side of the pump are properly initiated and maintained.

Drafting   Process of moving or drawing water away from a static source by a pump.

Dry barrel hydrant   A hydrant operated by a single control valve in which the barrel does not normally contain water; typically used in areas where freezing is a concern.

Duel pumping   A pumping operation in which two pumps are connected to and supplied by a strong hydrant. The connection is typically from the hydrant to the intake of the first pump and then from an unused intake of the first pump to the intake of the second pump.

Dump site   Location where tankers operating in a shuttle unload their water.

Eductor   A specialized device used in foam operations that utilizes the venturi principle to draw the foam into the water stream.

Evaporation   The physical change of state from a liquid to a vapor.

Feathering   The process of partially opening or closing control valves to regulate pressure and flow for individual lines.

Fill site   Location where tankers operating in a shuttle receive their water.

Fire pump operations   The systematic movement of water from a supply through a pump to a discharge point.

Flow   The rate and quantity of water delivered by a pump, typically expressed in gallons per minute (gpm).

Flow meter   A device used to measure the quantity and rate of water flow in gallons per minute (gpm)

Force   A pushing or pulling action on an object.

Forward lay   Supply hose line configuration when the apparatus stops at the hydrant and a supply line is laid to the fire.

Four-way hydrant valve   Appliance used to increase hydrant pressure without interrupting the flow.

Friction loss   The reduction in energy (Pressure) resulting from the rubbing of one body against another, and the resistance of relative motion between the two bodies in contact; typically expressed in pounds per square inch (psi); measures the reduction of pressure between two points in a system.

Front crankshaft method (pump engagement)   Method of driving a pump in which power is transferred directly from the crankshaft located at the front of an engine to the pump. This method of power transfer is used when the pump is mounted on the front of the apparatus and allows for either stationary or mobile operation.

Gauge pressure   Measurement of pressure that does not include atmospheric pressure, typically expressed as psig.

Head pressure   The pressure exerted by the vertical height of a column of liquid expressed in feet.

Hose bridge   Device used to allow vehicles to move across a hose without damaging the hose.

Hose clamp   Device used to control the flow of water in a hose.

Hose jacket   Device used to temporarily minimize flow loss from a leaking hose or coupling.

Hydraulics   The branch of science dealing with the principles and laws of fluids at rest or in motion.

Hydrodynamics   The branch of hydraulics that deals with the principles and laws of fluids in motion.

Hydrostatics   The branch of hydraulics that deals with the principles and laws of fluids at rest and the pressures they exert or transmit.

Impeller   A disk mounted on a shaft that spins within the pump casing.

Indicators   Devices other than pressure gauges and flow meters (such as tachometer, oil pressure, pressure regulator, and onboard water level) used to monitor and evaluate a pump and related components.

Intake   The point at which water enters the pump.

Instrumentation   Devices such as pressure gauges, flow meters, and indicators used to monitor and evaluate the pump and related components.

Intake pressure relief valve   Pressure regulating system that protects against excessive pressure build-up on the intake side of the pump.

Jet siphon   Device that helps move water quickly without generating a lot of pressure that is used to move water from one portable tank to another or to assist with the quick off-loading of tanker water.

Laminar flow   Flow of water in which thin parallel layers of water develop and move in the same direction. During laminar flow, friction loss is typically limited because the outer layer moves along the interior lining of the hose while other layers move alongside each other.

Latent heat of fusion   The amount of heat that is absorbed by a substance when changing from a solid to a liquid state.

Latent heat of vaporization   The amount of heat that is absorbed when changing from a liquid to a vapor state.

Laws   Rules that are legally binding and enforceable.

Large diameter hose   Usually hoses with diameters ranging from 4 inches to 6 inches.

LDH   See Large diameter hose.

Main pump   Primary working pump permanently mounted on an apparatus.

Manifolds   Devices that provide the ability to connect numerous smaller lines from a large supply line.

Manufacturers' inspection recommendations   Those items recommended by the manufacturer to be included in apparatus inspections.

Manufacturers' tests   Five specific tests, required by NFPA 1901, that a manufacturer must conduct prior to delivery of new apparatus.

MDH   See Medium diameter hose

Medium diameter hose   Usually hoses with diameters ranging from 2 inches to 3½ inches.

Municipal supply   A water supply distribution system provided by a local government consisting of mains and hydrants.

National standard thread   A common thread used in the fire service to attach hose couplings and appliances.

Needed flow   The estimated flow required to extinguish a fire.

NH   A common thread used in the fire service to attach hose couplings and appliances.

Normal pressure   The water flow pressure found in a system during normal consumption demands.

Nozzle flow   The amount of water flowing from a nozzle; also used to indicate the rated flow or flows of a nozzle.

Nozzle pressure   The designed operating pressure for a particular nozzle.

Nozzle reach   The distance water travels after leaving a nozzle.

Nozzle reaction   The tendency of a nozzle to move in the direction opposite of water flow.

Onboard supply   The water carried in a tank on the apparatus.

Open relay   Relay operation in which water is not contained within the entire relay system; excessive pressure is controlled by intake relief valves, pressure regulators, and dedicated discharge lines that allow water to exit the relay at various points in the system.

Perception distance   The distance the apparatus travels from the time a hazard is seen until the brain recognizes it as a hazard.

Portable dump tank   A temporary reservoir used in tanker shuttle operations that provides the means to unload water from a tanker for use by a pump.

Pressure   The force exerted by a substance in units of weight per area; the amount of force generated by a pump or the resistance encountered on the discharge side of a pump; typically expressed in pounds per square inch (psi).

Pressure drop   The difference between the static pressure and the residual pressure when measured at the same location.

Pressure gain and loss   The increase or decrease in pressure as a result of an increase or decrease in elevation.

Pressure gauge   Device used to measure positive pressure in pounds per square inch (psi) or negative pressure in inches of mercury (in. Hg).

Pressure governor   A pressure regulating system that protects against excessive pressure buildup by controlling the speed of the pump engine to maintain a steady pump pressure.

Pressure regulating systems   Devices used to control sudden and excessive pressure buildup during pumping operations.

Pressure relief device   A pressure regulating system that protects against excessive pressure buildup by diverting excess water flow from the discharge side of the pump back to the intake side of the pump or to the atmosphere.

Preventive maintenance   Proactive steps taken to ensure the operating status of the apparatus, pump, and related components.

Priming   The process of replacing air in a pump with water.

Priming pump   Positive displacement pump permanently mounted on an apparatus and used to prime the main pump.

PTO method (pump engagement)   Method of driving a pump in which power is transferred from just before the transmission to the pump through a PTO. This method of power transfer allows either stationary or mobile operation of the pump.

Pump   Mechanical device that raises and transfers liquids from one point to another. See also Auxiliary pump; Main pump; Priming pump

Pump-and-Roll   An operation in which water is discharged while the apparatus is in motion.

Pump engagement   The process or method of providing power to the pump. See also Front crankshaft method; PTO method; Split shaft method

Pump operator   The individual responsible for operating the fire pump, driving the apparatus, and conducting preventive maintenance.

Pump panel   The central location for controlling and monitoring the pump and related components.

Pump peripherals   Those components directly or indirectly attached to the pump that are used to control and monitor the pump and related components.

Pump speed   The rate at which a pump is operating typically expressed in revolutions per minute (rpm).

Pump tests   Tests conducted to determine the performance of a pump and related components.

Rated capacity   The flow of water at specific pressures a pump is expected to provide.

Reaction distance   The distance of travel from the time the brain sends the message to depress the brakes until the brakes are actually depressed.

Relay operation   Water supply operations in which two or more pumpers are connected in line to move water from a source to a discharge point.

Required flow   The estimated flow of water needed for a specific incident.

Residual pressure   The pressure remaining in the system after water has been flowing through it.

Reverse lay   Supply hose line configuration when the apparatus stops at the scene; drops attack lines, equipment, and personnel; and then advances to the hydrant laying a supply line.

Safety-related components   Those items that affect the safe operation of the apparatus and pump, and that should be included in apparatus inspections.

Shuttle cycle time   The total time it takes for a tanker in a shuttle operation to dump water and return with another load; including the time it takes to fill the tanker, to dump the water, and the travel distance between the fill and dump stations.

Shuttle flow capacity   The volume of water a tanker shuttle operation can provide without running out of water.

Siamese   Appliance used to combine two or more lines into a single line.

Slippage   Term used to describe the leaking of water between the surfaces of the internal moving parts of a pump.

Smooth-bore nozzle   Nozzles designed to produce a compact solid stream of water with extended reach.

Soft sleeve   Shorter section of hose used when the pump is close to a pressurized water source such as a hydrant.

Spanner wrench   Tool used to connect and disconnect hose and appliance couplings.

Specific heat   The amount of heat required to raise the temperature of a substance by 1°F. The specific heat of water is 1 btu/16°F.

Speed   The rate at which a pump is operating, typically expressed in revolutions per minute.

Split-shaft method (pump engagement)   Method of driving a pump in which a sliding clutch gear transfers power to either the road transmission or to the pump transmission. This method of power transfer is used for stationary pumping only.

Spotter   An individual used to assist in backing up an apparatus.

Standards   Guidelines that are not legally binding or enforceable by law unless they are adopted as such by a governing body.

Static pressure   The pressure in a system when no water is flowing.

Static source   Water supply that generally requires drafting operations, such as ponds, lakes, and rivers.

Static source hydrants   Prepiped lines that extend into a static source.

Stream shape   The configuration of water droplets (shape of the stream) after leaving a nozzle.

Suction hose   Special noncollapsible hose used for drafting operations.

Supply hose   Hose used with pressurized water sources and operated at a maximum pressure of 185 psi.

Supply layout   The required supply hose configuration necessary to efficiently and effectively secure the water supply.

Supply reliability   The extent to which the supply will consistently provide water.

Systems   Those components that directly or indirectly assist in the operation of the pump (e.g., priming systems, pressure regulating systems, and cooling systems).

Tandem pumping   A pumping operation in which one pumper pumps all excess water from a strong hydrant to the second pumper. The connection is typically from the hydrant to the intake of the first pump which is then discharged to the intake of the second pumper as in a relay operation.

Tanker shuttle   Water supply operations in which the apparatus is equipped with large tanks to transport water from a source to the scene.

Throttle control   Device used to control the engine speed, which in turn controls the speed of the pump, when engaged, from the pump panel.

Total stopping distance   The distance of travel measured from the time a hazard is detected until the vehicle comes to a complete stop.

Traction   Friction between the tires and road surface.

Transfer valve   Control valve used to switch between the pressure and volume mode on two-stage centrifugal pumps.

Turbulent flow   Flow of water in an erratic and unpredictable pattern creating a uniform velocity within the hose that increases pressure loss because more water is subjected to the interior lining of the hose.

Vacuum   Measurement of pressure that is less than atmospheric pressure, typically expressed in inches of mercury (in. Hg).

Vapor pressure   The pressure exerted on the atmosphere by molecules as they evaporate from the surface of the liquid.

Velocity pressure   The forward pressure of water as it leaves an opening.

Venturi principle   Process that creates a low-pressure area in the induction chamber of an eductor to allow foam to be drawn into and mixed with the water stream.

Volume   A three-dimensional space occupied by an object.

Volute   An increasing void space in a pump that converts velocity into pressure and directs water from the impeller to the discharge.

Water availability   The quantity, flow, pressure, and accessibility of a water supply.

Water hammer   A surge in pressure created by the sudden increase or decrease of water during a pumping operation.

Water thieves   Similar to gated wyes, water thieves are used to connect additional smaller lines from an existing larger line.

Weight   The downward force exerted on an object by the Earth's gravity, typically expressed in pounds (lb).

Wet barrel hydrant   A hydrant operated by individual control valves that contains water within the barrel at all times; typically used where freezing is not a concern.

Wheel chock   Device placed next to wheels to guard against inadvertent movement of the apparatus.

Wye   Appliance used to divide one hose line into two or more lines.

INDEX